SECOND SCI–RSC MEDICINAL CHEMISTRY SYMPOSIUM

Special Publication No 50

Second SCI–RSC Medicinal Chemistry Symposium

The Proceedings of a Symposium organised by the Fine Chemicals and Medicinals Group of the Industrial Division of the Royal Society of Chemistry and the Fine Chemicals Group of the Society of Chemical Industry

Cambridge, England, 12th–14th September 1983

Edited by
J. C. Emmett
Smith Kline & French Research Limited, Welwyn

The Royal Society of Chemistry
Burlington House, London W1V 0BN

British Library Cataloguing in Publication Data

Medicinal Chemistry Symposium (2nd : 1983 : Cambridge)
Second SCI-RSC Medicinal Chemistry Symposium.—
(Special Publication/Royal Society of Chemistry, ISSN 0260-6291; no. 50)
1. Biological chemistry 2. Chemistry, Clinical
I. Title. II. Emmett, J. C. III. Royal Society of Chemistry. *Fine Chemicals and Medicinals Group.* IV. Society of Chemical Industry. Fine Chemicals Group. V. Series 612'.015 QP514.2

ISBN 0-85186-935-1

Printed in Great Britain by Henry Ling Ltd., at the Dorset Press, Dorchester, Dorset

Introduction

In 1981 the need for a regular UK forum for the presentation and discussion of topics of interest to the international community of Medicinal chemists was established with the 1st SCI-RSC Medicinal Chemistry Symposium, held at Churchill College, Cambridge. The theme for this first meeting, entitled "The Chemical Regulation of Biological Mechanisms", underlined the dual role of Medicinal Chemistry, namely to investigate the nature of the biological processes which drugs regulate and to discover novel therapeutic agents. The two organising bodies, the Fine Chemicals Group of the Society of Chemical Industry and the Fine Chemicals and Medicinals Group of the Royal Society of Chemistry were encouraged by the success of this first symposium and agreed to continue their collaboration in organising regular biennial meetings in the UK for medicinal chemists.

It was generally agreed that the second meeting in 1983 should attempt to cater for some of the wider interests of medicinal chemists; accordingly, four topics, covering (1) Mediator Regulation in Cardiovascular Drug Therapy, (2) The Use of Computer Graphics in Drug Design, (3) Biological Challenges to Organic Synthesis and (4) Control of the Arachidonic Acid Cascade, were chosen for the 2nd SCI-RSC Medicinal Chemistry Symposium, at Churchill College on 12-14 September 1983. The attendance of over 260 participants from thirteen countries, coupled with the quality of the papers in this volume, attest to the success of this meeting and support the 'broad church' concept of medicinal chemistry. The 3rd meeting, to be held at Churchill College on 15-18th September 1985, promises an equally interesting and stimulating programme covering the three main topics: The Modulation of Central and Peripheral Dopamine and Serotonin Receptors, Enzyme Inhibitors in the Control of Disease, and Drugs for the Control of the Endocrine System.

The excellent standard of papers and their presentation was an essential ingredient in the success of the 2nd SCI-RSC Medicinal Chemistry Symposium. In addition, however, the unstinting secretarial assistance of Mrs. C.A. Hallgren and Mr. H.L. Bennister, together with the superb efforts of the staff of Churchill College and the work of the two respective societies, in ensuring this success, are gratefully acknowledged by the joint organising committee. Warm thanks are also extended to Glaxo Group Research Ware Ltd., Hoechst UK Ltd., and ICI PLC Pharmaceuticals Division, for their generous support.

Organising Committee

Dr. G.J. Durant (Smith Kline & French Research Limited)

Dr. J.C. Emmett (Smith Kline & French Research Limited)

Dr. J.R. Housley (The Boots Company Limited)

Professor A. McKillop (University of East Anglia)

Dr. B.C. Ross (Hoechst U.K. Limited)

Dr. R. Westwood (Roussel Laboratories Limited)

Contents

Chemical Pharmacology of Calcium Antagonists

By D. J. Triggle

SCHOOL OF PHARMACY, STATE UNIVERSITY OF NEW YORK, BUFFALO, N.Y. 14260, U.S.A.

Introduction

The fundamental role of Ca^{2+} as an intermediary in stimulus-response coupling processes is well established. This role for Ca^{2+} is made possible by the existence of large inwardly directed Ca^{2+} concentration gradients, intracellular high affinity Ca^{2+} binding proteins and Ca^{2+} selective entry and mobilization, efflux and sequestration processes[1-3].

Ca^{2+} mobilization in response to membrane stimuli which initiate excitation-contraction or stimulus-secretion coupling may be from both intracellular and extracellular sources, the relative importance of which are both stimulus- and system-dependent. Stimulated Ca^{2+} entry is currently viewed as occurring through both receptor-operated (ROC) and potential-dependent (PDC) Ca^{2+} channels which are activated by receptor occupation and membrane depolarization [4-7] (Figure 1). Much remains to be learned concerning these Ca^{2+} channels. It is not known whether the channels are different structures or whether a ligand-receptor association with PDC changes their voltage sensitivity and converts them to ROC. The linkages and extent of formal association between different ligand receptors and the Ca^{2+} channels remain to be established and the question of multiple classes or states of Ca^{2+} channels has yet to be resolved.

The Ca^{2+} Channel Blockers

The structurally diverse group of compounds referred to as the Ca^{2+}

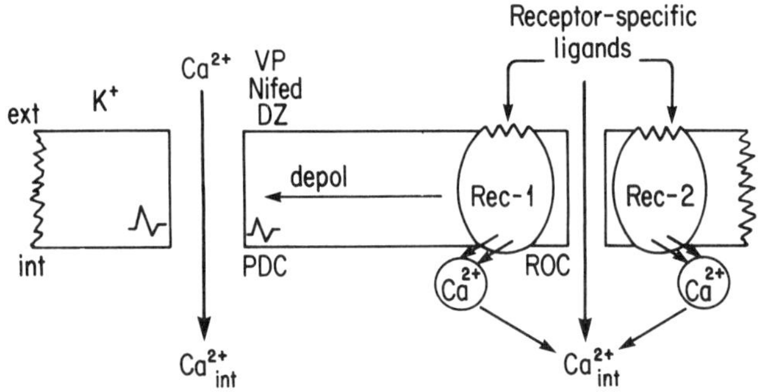

Fig.1. Ca^{2+} mobilization in response to membrane stimuli. Ca^{2+} channels of two types are depicted, potential-dependent (PDC) and receptor-operated (ROC). Specific agonist interactions (REC-1, REC-2) may mobilize Ca^{2+} from intracellular sources or through ROC and additionally may depolarize the membrane and activate PDC. K^{+}-Depolarizing stimuli depolarize the membrane and are believed to activate dominately PDC. VP, Nif and DZ represent verapamil, nifedipine and diltiazem which are depicted as inhibiting specifically the PDC. (Reproduced with permission from R.A. Janis and D.J. Triggle, J. Med. Chem. 1983, 26, 775. Copyright 1983 American Chemical Society).

channel blockers, slow channel blockers or Ca^{2+} antagonists and including verapamil, nifedipine and diltiazem (Figure 2) has played an important role in defining Ca^{2+} entry mechanisms in excitable cells [2,7,8-13]. These agents were originally demonstrated to mimic the effects of Ca^{2+} withdrawal in cardiac tissue, to act in apparently competitive fashion against Ca^{2+} in cardiac and smooth muscle and to depress, with considerable selectivity, the Ca^{2+} component of the cardiac action potential. These properties underlie their clinical and potential clinical applications[13-15] (Table 1).

Table 1. Clinical Indications for Ca^{2+} Channel Blockers

Current Indications:	Other Indications:
angina: vasospastic,	asthma
rest unstable and chronic stable	cardiac preservation
tachycardia, supraventricular	cerebral insufficiency and vasospasm
tachyarrhythmia, ventricular	esophageal motor disorders
hypertension	intestinal hyperactivity
hypertrophic cardiomyopathy	myometrial hyperactivity, premature labor
	myocardial ischemia
	peripheral vascular disorders

The structural heterogeneity of these (Figure 2) and subsequently introduced structures (Figure 3) suggests that these agents likely act at different sites and by different mechanisms to regulate potential-dependent Ca^{2+} channel function. Consistent with this verapamil/D600 and diltiazem are approximately equipotent in cardiac and smooth muscle whereas nifedipine (and other 1,4-dihydropyridines), flunarizine and cinnarjzine are markedly more effective in smooth muscle[8-10,13,15,16]. Additionally, verapamil/D600 and diltiazem, in contrast to nifedipine, exhibit the phenomena of frequency and voltage dependence whereby the activity of these agents increases with

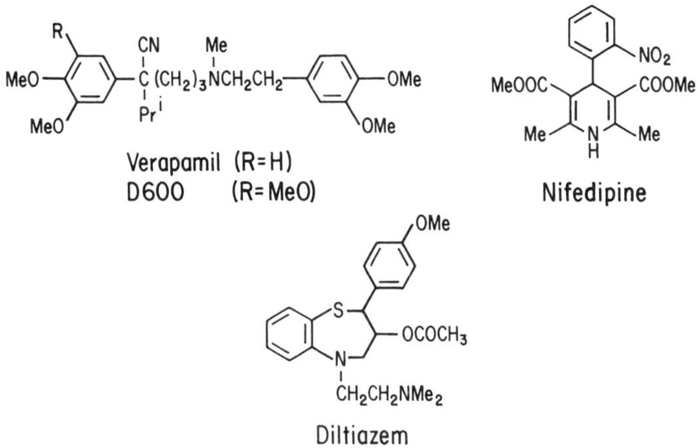

Fig. 2. Structural formulae of verapamil/D600, nifedipine and diltiazem.

increasing frequency of stimulation and with increasing membrane depolarization[17,18]. Thus, verapamil/D600 and diltiazem appear to block preferentially Ca^{2+} channels after they have been activated, whereas nifedipine and other 1,4-dihydropyridines do not show such marked discrimination.

Despite these (and other) differences the major Ca^{2+} channel blockers, verapamil/D600, nifedipine and diltiazem, do share the common property of blocking Ca^{2+} currents and stimulated Ca^{2+} entry in excitable systems. Studies in cardiac cells show that they block both inward and outward currents, but do not change the kinetics of channel activation and that the blockade of channels is reduced with elevation of extracellular Ca^{2+}. The classification term, Ca^{2+} channel blocker, thus appears to be justified[18].

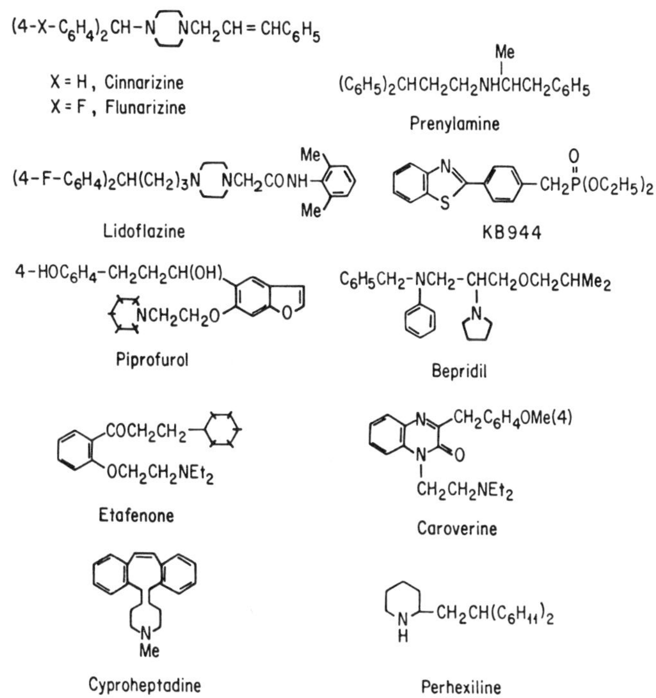

Fig.3. Structural formulae of agents with reported Ca^{2+} channel blocking properties.

Ca^{2+} Channel Blockers: Structure-Function Relationships

With the mechanistic designation of a group of agents as Ca^{2+} channel blockers, structure-function relationships assume added importance since their analysis likely probes binding sites at or in close proximity to the Ca^{2+} channel.

The most extensive structure-function analyses are available for the 1,4-dihydropyridines of the nifedipine class. Although detailed pharmacologic activities are not available for the several thousand analogs that have been synthesised, it is nonetheless clear that certain structural features are critical regardless of whether activity is determined *in vivo* (blood pressure), *in vitro* (isolated organs and tissues) or in the recently

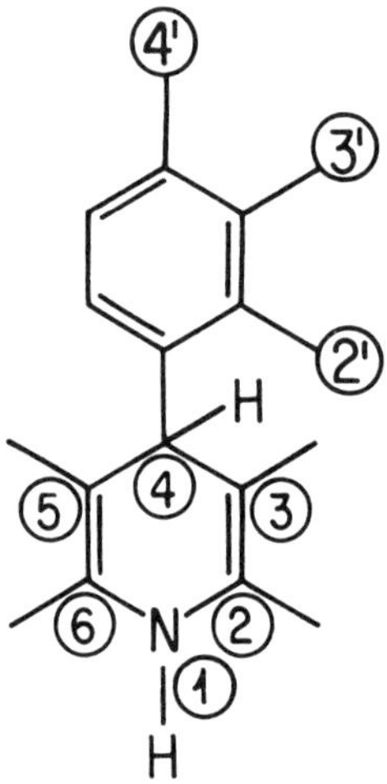

Fig. 4. Schematic representation of structural requirements for activity in the 1,4-dihydropyridines.

introduced radioligand binding assay[2,12,19-23]. These structural features may be broadly summarized as follows (Figure 4):

1. The 1,4-dihydropyridine ring is very important. Oxidation to the pyridine greatly reduces activity.

2. N_1 of the 1,4-dihydropyridine ring should be unsubstituted.

3. An aryl substituent, usually substituted phenyl, is necessary in the 4-position of the 1,4-dihydropyridine ring. The position of the substituent in the phenyl ring is critical: 2'- and 3'-substituents maintain or increase activity, but 4'-substituents are invariably detrimental regardless of substituent.

4. 2,6-Substituents, normally - Me, appear necessary for optimum activity. Limited evidence suggests that other substituents, including $-NH_2$, may also be tolerated.

5. Ester groups in the C_3 and C_5 positions of the 1,4-dihydropyridine ring appear to be optimum. Other groups including -CN and -COMe decrease activity. However, a wide range of ester substituents maintain or increase activity relative to the 3,5-dicarbomethoxy substituents of nifedipine, suggesting that this may represent a region of bulk tolerance.

Limited quantitative structure function data are available. Thus, in the series 2,6-dimethyl-3,5-dicarbomethoxy-4-substituted phenyl-1,4-dihydropyridine, the effects of the phenyl substituent on negative inotropic potency increased with increasing values of the Verloop steric constant B_1[20,22]. One function of the phenyl substituent, in addition to providing a direct binding interaction, may be to provide an orientation of the phenyl ring approximately perpendicular to the 1,4-dihydropyridine ring [19,24,25]. X-ray studies suggest that, in addition, increasing planarity of the 1,4-dihydropyridine ring may also contribute to activity[24,25] (Figure 5).

With different ester substituents at C_3 and C_5 of the 1,4-dihydropyridine ring a chiral center is created at C_4. Consistent with the thesis that 1,4-dihydropyridines bind to specific sites stereoselectivity of pharmacologic activity is observed (Table 2). Additionally, there are indications that 1,4-dihydropyridines with different ester substituents exhibit tissue selectivity and differing cardiac:vascular smooth muscle activities[12,16,32-34].

Similar, but less extensive, structure-activity correlations have been examined for agents in the verapamil/D600 and diltiazem series. These correlations also reveal a dependence of activity upon structure, including stereoselectivity[12,22,34,35]. However, it is evident that an all-encompassing structure-activity relationship is not likely for the three series of compounds.

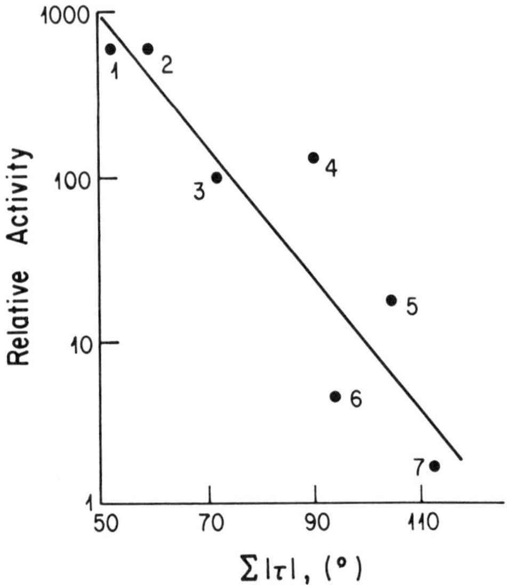

Fig. 5. Correlation between pharmacologic activity (inhibition of smooth muscle mechanical response) and distortion from planarity of 1,4-dihydropyridine ring, $\Sigma|\tau|$, (°). The compounds designated are derivatives of 2,6-dimethyl-3,5-dicarbomethoxy-4-(substituted) phenyl-1,4-dihydropyridine with substituents: 1,2 = F_5 (two crystalline forms); 3 = 2-NO_2; 4 = 3-NO_2; 5 = H (unsubstituted); 6 = 3-Me; 7 = 3-CN. (Reproduced with permission from R. Fossheim, K. Svarteng, A. Mostad, C. Rømming, E. Shefter and D.J. Triggle, J. Med. Chem. 1982, 25, 126. Copyright 1982 American Chemical Society).

Ca^{2+} Channel Blockers: Radioligand Binding to Excitable Tissues

The availability of [^{3}H]nitrendipine and [^{3}H]nimodipine has made possible direct experimental approaches to the definition of the sites of action and mechanisms of interaction between the several structural classes

Table 2. Stereoselectivity of 1,4-Dihydropyridine Activity

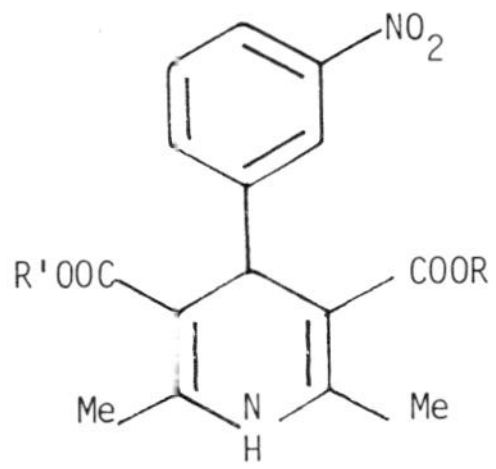

R'	R		ID_{50},M: Isomer ratio, K^+-depolarized smooth muscle: G.P.ileum		G.P.bladder		Rabbit aorta		IC_{50},M: Isomer ratio, $[^3H]$1,4-dihydropyridine binding: G.P. ileum		Rat ventricle		Rabbit brain	
Me	Et	(+)	--------		--------		$3.0x10^{-8}$	10					$8.8x10^{-9}$	20
(Nitrencipine)		(-)	--------		--------		$3.0x10^{-9}$						$4.3x10^{-10}$	
Me	$CHMe_2$	(+)	$4.1x10^{-9}$	10	$4.0x10^{-9}$	8	$6.0x10^{-10}$	100	$1.1x10^{-9}$	100	$4.9x10^{-9}$	500	$27.5x10^{-9}$	300
		(-)	$4.2x10^{-10}$		$5.6x10^{-10}$		$8.0x10^{-8}$		$1.0x10^{-11}$		$1.0x10^{-11}$		$9.0x10^{-11}$	
Et	$CHMe_2$	(+)	$4.9x10^{-9}$	2.5	--------		$1.4x10^{-8}$	10						
		(-)	$1.9x10^{-9}$		--------		$1.5x10^{-9}$							
$CHMe_2$	CH_2CH_2OMe	(+)	--------		--------		$1.6x10^{-8}$	5					$2.4x10^{-9}$	2
(Nimodipine)		(-)	--------		--------		$3.0x10^{-9}$						$1.0x10^{-9}$	
Me	$CH_2CH_2N(Me)CH_2Ph$	(+)	$3.1x10^{-10}$	4	--------		(+) > (-)							
		(-)	$1.3x10^{-9}$		--------									

of Ca^{2+} channel blockers[12,23,28,31,34,36-38]. Specific, saturable, reversible, high affinity binding has been found in tissues including smooth, cardiac and skeletal muscle and brain (Table 3). A survey of existing literature is available[39].

Table 3. Binding Characteristics of [^{3}H]1,4-Dihydropyridines

Tissue	[^{3}H]Nitrendipine			[^{3}H]Nimodipine		
	K_D nM	B_{MAX} fmoles/ mg prot.	n_H	K_D nM	B_{MAX} fmoles/ mg prot.	n_H
Guinea pig ileum (25°)	0.16	1100	1.06	0.12	750	0.95
Rat ventricle (25°)	0.25	390	0.9	0.24	400	0.9
Guinea pig brain (cortex) (25°)	0.09	505	0.94	----	---	---
Rat skeletal muscle (10°)	3.6	9200	0.96	----	---	---

The available data[39] are generally consistent with the existence (within the limitations of the filtration binding assay employed) of a single set of high affinity binding sites for [^{3}H]nitrendipine and [^{3}H]nimodipine for smooth and cardiac muscle and brain. However, binding affinity is significantly lower, and binding capacity significantly higher, in skeletal muscle.

Extensive binding-activity correlations for 1,4-dihydropyridines are available for guinea pig intestinal smooth muscle[12,23,39], where an essentially 1:1 correlation exists (Figure 6), including stereoselectivity between binding activity (displacement of [^{3}H]nitrendipine) and pharmacologic activity (inhibition of depolarization-induced mechanical responses). Very similar binding sequences are obtained in other smooth muscles, cardiac muscle and brain indicating a substantial similarity in the ligand recognition site present in these several tissues[12,39]. However, correlations between binding and pharmacologic activity differ in other

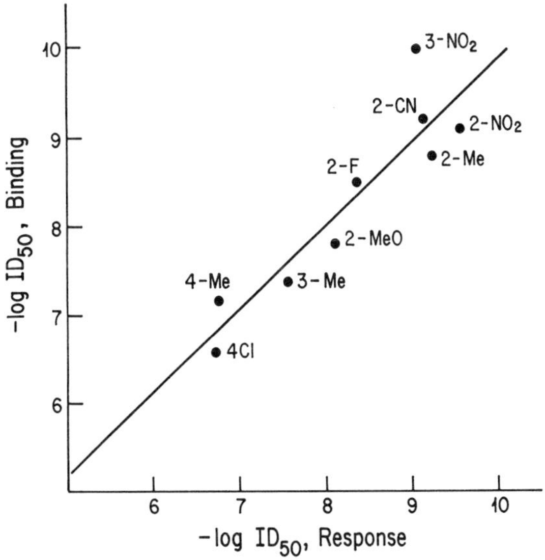

Fig. 6. Correlation between binding (inhibition of specific [^{3}H]nitrendipine binding) and pharmacologic activity (inhibition of depolarization-induced mechanical response in guinea pig ileal longitudinal smooth muscle for a series of substituted 1,4-dihydropyridines. Data from reference 23.

tissues. Thus, in cardiac muscle high affinity binding is accompanied by low affinity pharmacology and the Ca^{2+} channel blockers are clearly relatively ineffective in mammalian neuronal preparations[12,39]. The origins of these discrepancies are, as yet, unclear but may relate to changes in binding site properties during cellular disruption, loss of regulatory factors or differences in the coupling between dihydropyridine site occupancy and channel activation.

Despite these important discrepancies there is general agreement that the site determined by [^{3}H]1,4-dihydropyridine binding does represent a site of pharmacologic relevance. Binding is insensitive to a wide variety of receptor-directed ligands, but is sensitive to dihydropyridines and members of other structural categories of Ca^{2+} channel blockers[12,39]. However, it is clear that, consistent with the chemical heterogeneity of the Ca^{2+} channel blockers, these agents do not act at a single site to regulate [^{3}H]1,4-dihydropyridine binding. Thus, the dihydropyridines behave as simple competitive antagonists of radioligand binding, but verapamil/D600 behaves as a partial competitive inhibitor whilst diltiazem actually potentiates [^{3}H]nitrendipine and [^{3}H]nimodipine binding[12,23,36,37]. This suggests the existence of at least one binding site, additional to that occupied by 1,4-dihydropyridines, at which verapamil/D600 and diltiazem can exert their negative and positive heterotropic allosteric effects respectively.

Two additional features of the schematic representation of calcium channel antagonist binding shown in Figure 7 are deserving of attention. [^{3}H]1,4-dihydropyridine binding is regulated by cations, being dependent or partially dependent upon the presence of divalent cations (Ca^{2+}, Mg^{2+}, Mn^{2+}, Co^{2+} etc.) but blocked by tri-valent cations (La^{3+}, Tm^{3+} etc.)[12,23,36,37,39,40]. The relationship of this site of cation interaction to the cation permeation machinery of the Ca^{2+} channel is not known, but it is of interest that permeant cations can stimulate, to varying degrees, [^{3}H]1,4-dihydropyridine binding whilst nonpermeant cations fail to stimulate significantly binding. Thus, permeant and nonpermeant cation binding appears to control the availability of [^{3}H]1,4-dihydropyridine high affinity binding. Additionally, and of considerable potential importance to an understanding of how 1,4-dihydropyridine binding controls Ca^{2+} channel function, is the recent discovery that some dihydropyridines actually have the ability to activate Ca^{2+} channels (Figure 8) and thus serve as positive inotropic and vasoconstrictor agents[41]. This suggests that the dihydropyridines serve as allosteric regulators of Ca^{2+} channel activation

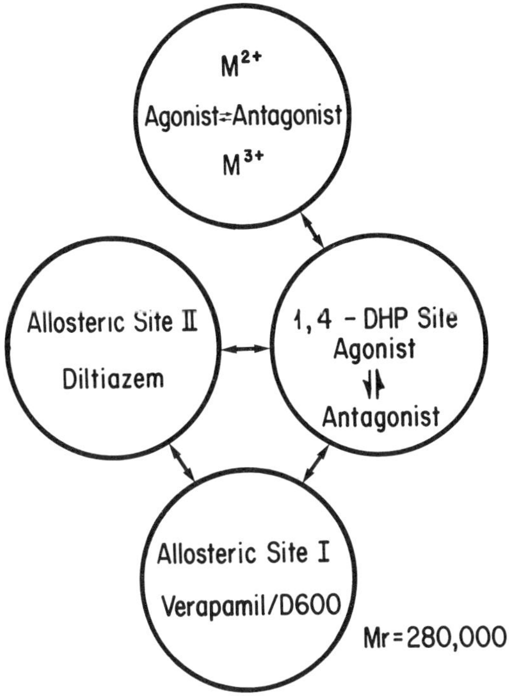

Fig. 7. Schematic representation of binding site organization in Ca^{2+} channel. Depicted are the 1,4-dihydropyridine binding site, for which both inhibitor and activator ligands exist, two allosteric sites at which organic Ca^{2+} channel blockers (VP/D600 and diltiazem) bind and a cation binding site at which cations stimulate (M^{2+}) or inhibit (M^{3+}) 1,4-dihydropyridine binding.

by binding preferentially to the activated state (agonist) or to the resting state (antagonist).

Conclusions

The chemically heterogeneous group of compounds known as the Ca^{2+} channel blockers are agents with substantial clinical and potential clinical

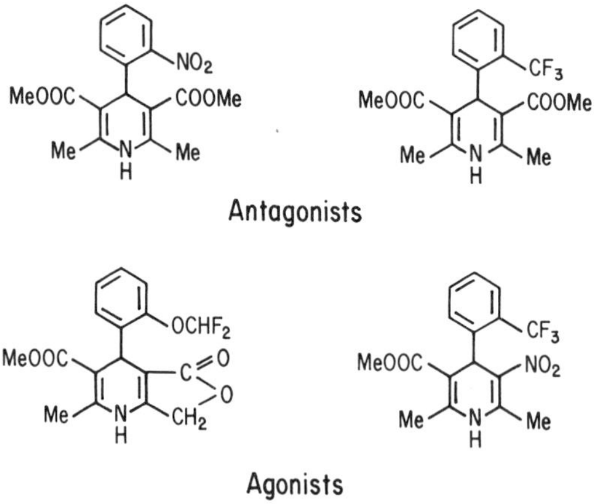

Fig. 8. Structure of 1,4-dihydropyridines with inhibitory and activating activities at the Ca^{2+} channel.

utility. Much remains to be learned about their sites and mechanisms of action. However, the delineation of structure-activity relationships is indicative of specific sites of action concerned with Ca^{2+} channel regulation. Consistent with this indication recent radioligand binding studies indicate the existence of specific binding sites in excitable tissues which possess pharmacologically appropriate properties. The future development of combined pharmacologic, chemical and biochemical approaches seems likely to be of considerable utility towards defining the sites and mechanisms of action of these agents and to providing tissue-selective second- and third-generation drugs.

Acknowledgments

This work was supported by a grant from the National Institutes of Health (HL 16003). Additional support from the Miles Institute for Preclinical Pharmacology is gratefully acknowledged.

References

1. R.H. Kretsinger, Int. Rev. Cytol., 1971, 46, 323.

2. L.B. Rosenberger and D.J. Triggle, Calcium and Drug Action, Ed. G.B. Weiss, Plenum Press, New York, 1978, Chapter 1, p. 3.

3. W.Y. Cheung (Editor), "Calcium and Cellular Function", Academic Press, New York, 1980, Vol. 1.

4. T.B. Bolton, Physiol. Rev., 1979, 3, 306.

5. K.D. Meisheri, O. Hwang and C. van Breemen, J. Mem. Biol., 1981, 59, 19.

6. R.W. Tsien, Ann. Rev. Physiol., 1983, 45, 341.

7. C. Cauvin, R. Loutzenhiser and C. van Breemen, Ann. Rev. Pharmacol. Toxicol., 1983, 23, 373.

8. A. Fleckenstein, Ann. Rev. Pharmacol. Toxicol., 1977, 17, 149.

9. G.B. Weiss (Editor), "New Perspectives on Calcium Antagonists", American Physiology Society, Bethesda, Maryland 1981.

10. P.A. van Zwieten, Prog. Pharmacol., 1982, 5, 1.

11. R.G. Rahwan and D.T. Witiak (Editors), "Calcium Regulation by Calcium Antagonists", American Chemical Society, Washington, D.C. 1982.

12. R.A. Janis and D.J. Triggle, J. Med. Chem., 1983, 26, 775.

13. A. Schwartz and D.J. Triggle, Ann. Rev. Med., in press, 1984.

14. P.H. Stone, E.M. Antman, J.E. Muller and E. Braunwald, Ann. Int. Med., 1980, 93, 886.

15. S.F. Flaim and R. Zelis (Editors), "Calcium Blockers - Mechanisms of Action and Clinical Implications". Urban and Schwarzenberg, Baltimore, Maryland, 1982.

16. K. Hashimoto, K. Takeda, Y. Katano, Y. Nakagawa, T. Tsukada, T. Hashimoto, N. Shimamoto, K. Sakai, T. Otorri and S. Imai, Arzneim-Forsch., 1979, 29, 1368.

17. D. Pelzer, W. Trautwein and T.F. McDonald, Pflüg. Arch. ges. Physiol., 1982, 394, 97.

18. K.S. Lee and R.W. Tsien, Nature (Lond.), 1983, 302, 790.
19. B. Loev, M.M. Goodman, K.M. Snador, R. Tedeschi and E. Macko, J. Med. Chem., 1974, 17, 956.
20. R. Rodenkirchen, R. Bayer, D. Steiner, F. Bossert, H. Meyer and E. Moller, Naunyn-Schmied. Arch. Pharmacol., 1979, 310, 69.
21. F. Bossert, H. Meyer and F. Wehinger, Angew. Chem., Int. Ed. Engl., 1981, 20, 762.
22. R. Rodenkirchen, R. Bayer and R. Mannhold, Prog. Pharmacol., 1982, 5, 9.
23. G.T. Bolger, P. Gengo, R. Klockowski, E. Luchowski, H. Siegel, R.A. Janis, A.M. Triggle and D.J. Triggle, J. Pharmacol. Exptl. Therap., 1983, 225, 291.
24. A.M. Triggle, E. Shefter and D.J. Triggle, J. Med. Chem., 1980, 23, 1442.
25. R. Fossheim, R. Svarteng, A. Mostad, C. Rømming, E. Shefter and D.J. Triggle, J. Med. Chem., 1982, 25, 126.
26. R. Towart, E. Wehinger and H. Meyer, Naunyn-Schmied. Arch. Pharmacol., 1981, 317, 183.
27. R. Towart, E. Wehinger, H. Meyer and S. Kazda, Arz. Forsch., 1982, 32, 338.
28. P. Bellemann, A. Schode and R. Towart, Proc. Nat. Acad. Sci. U.S.A., 1983, 80, 2356.
29. T. Shibanuma, M. Iwanami, K. Okuda, T. Takenaka and M. Murakami, Chem. Pharm. Bull., 1980, 28, 2809.
30. D.J. Triggle, unpublished observations.
31. R.A. Janis, S.C. Maurer, J.G. Sarmiento, G. Bolger and D.J. Triggle, Eur. J. Pharmacol., 1982, 82, 191.
32. S. Kazda, B. Garthoff, H. Meyer, K. Schlossmann, K. Stoepel, R. Towart, W. Vater and E. Wehinger, Arz. Forsch., 1980, 30, 2144.
33. S. Kazda, B. Garthoff, H.P. Krause and K. Schlossmann, Arz. Forsch., 1982, 32, 331.
34. R.A. Janis and A. Scriabine, Biochem. Pharmacol., in press, 1983.

35. D.J. Triggle and V.C. Swamy, Circ. Res., 1983, 52 (Suppl. I), 17.

36. H. Glossmann, D.R. Ferry, F. Lübbecke, R. Mewes and F. Hofmann, Trends. Pharmacol. Sci., 1982, 3, 431.

37. K.M.M. Murphy, R.J. Gould, B.L. Largent and S.H. Snyder, Proc. Nat. Acad. Sci. U.S.A., 1983, 80, 860.

38. F.J. Ehlert, E. Itoga, W.R. Roeske and H.I. Yamamura, Biochem. Biophys. Res. Comm., 1982, 104, 937.

39. D.J. Triggle and R.A. Janis, Modern Methods in Pharmacology (Editors N. Back and S. Spector), A.R. Liss Inc., New York, N.Y., 1983, Vol. II, in press.

40. R.J. Gould, K.M.M. Murphy and S.H. Snyder, Proc. Nat. Acad. Sci. U.S.A., 1982, 79, 3656.

41. M. Schramm, G. Thomas, R. Towart and G. Franckowiak, Nature (London), 1983, 303, 535.

α-Adrenoceptor Modulation

By S. F. Campbell

PFIZER CENTRAL RESEARCH, SANDWICH, KENT CT13 9NJ, U.K.

1. Introduction

Interest in adrenoceptor modulation can be traced to the turn of the century when Dale demonstrated that ergot alkaloids blocked most of the excitatory, but not the inhibitory, responses to adrenaline.[1] Despite intense efforts, subsequent progress was mostly qualitative until the 1940's when von Euler confirmed that noradrenaline served as the final neurotransmitter for the sympathetic nervous system and Ahlquist established the concept of α- and β-adrenoceptors. Some twenty years later, Lands subdivided β-receptors into β_1- and β_2-types, with specific locations in particular organs.[2] However, it is now known that β_1- and β_2-sites coexist in tissues, and receptor distribution is rarely homogeneous.[1,3] Drugs which selectively modulate β_1- and β_2-adrenoceptor function are important therapeutic agents; for example, agonists are used as cardiac stimulants (prenalterol, β_1) or bronchodilators (salbutamol, β_2). In addition, antagonists (propranolol, β_1/β_2; atenolol, β_1) are widely used for the treatment of angina and hypertension. All of these agents are derived from phenylethanolamine or phenoxypropanolamine skeletons, consistent with identical activation processes and strictly defined recognition sites for β_1- and β_2-receptors.

In contrast to the rapid exploitation of β-receptors, agents which selectively modulate α-adrenoceptor function emerged slowly, possibly due to difficulties in identifying different α-subtypes and disappointment over the clinical performance of early drugs. However, during the 1970's, α-receptors located on sympathetic neurones were shown to differ from the well-documented population on vascular smooth muscle and the concept of pre- and post-junctional subtypes was introduced.[4] Later, α-adrenoceptors similar to the prejunctional type were found to be widely distributed and classification by anatomical location proved inadequate. A more general terminology, α_1- and α_2-adrenoceptors, then evolved, based on pharmacological criteria employing selective agonists and antagonists.[5,6] α_1- and α_2-receptors have now been identified in many tissues but only those potentially involved in cardiovascular regulation will be considered[1,7] (Table 1).

Tissue	Adrenoceptor Type	Response
blood vessels	α_1 α_2	vasoconstriction vasoconstriction
sympathetic neurones	α_2	inhibition of noradrenaline release
brain	α_1 α_2	? hypotension, sedation
heart	α_1	increased force of contraction
platelets	α_2	aggregation
kidney	α_2	Na^+ reabsorption vasoconstriction

TABLE I: Distribution of α_1- and α_2-adrenoceptors which may contribute to cardiovascular regulation, and response to stimulation.

Most attention has focussed on α-adrenoceptor function at sympathetic neuroeffector junctions where α_2-receptors situated on nerve endings are thought to mediate inhibition of noradrenaline release.[8] By contrast, both α_1- and α_2-adrenoceptors are located in blood vessel walls and stimulation of either subtype leads to vasoconstriction[9] (Fig. I).

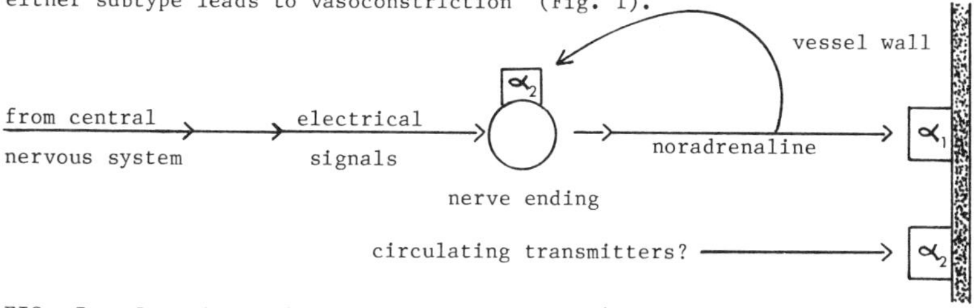

FIG. I. Location and role of pre- and post-junctional α-adrenoceptors at a sympathetic nerve synapse.

Thus, enhanced activity of the sympathetic nervous system results in elevated vascular tone, an important factor in the development and maintenance of hypertension.[10,11] Attempts to develop drugs which block the vasoconstrictor action of noradrenaline commenced over fifty years ago but clinical utility was limited[12] since most early agents did not adequately differentiate between α-receptor subtypes.[9] Blockade of postjunctional α-adrenoceptors lowers vascular tone but this can be offset by enhanced noradrenaline release as a consequence of concomitant activity at prejunctional α_2-sites. In contrast, prazosin (Fig. 2) differs from classical agents by possessing exceptional selectivity for postjunctional-(α_1) versus prejunctional-(α_2) receptors and was the first α-adrenoceptor antagonist widely effective for the treatment of hypertension.[13,14] The physiological role of vascular α_2-receptors is less well defined[7] but response to circulating, rather than neuronally released, catecholamines has been proposed.[15] Evidence is accumulating for the influence of postjunctional α_2-receptors on vascular tone in man,[16,17] and a selective α_2-antagonist, SK&F 86,466, appears to lower blood pressure in animals.[18]

α_2-adrenoceptors have been identified in the brain where stimulation by agonist drugs such as clonidine results in clinically useful antihypertensive effects, and also in sedation. Aldomet exhibits a similar profile in man presumably via metabolism to α-methylnoradrenaline, a moderately selective α_2-agonist.[7] Blockade of central α_2-receptors is now being explored, since by increasing noradrenaline turnover, prejunctional antagonists such as idazoxan (Fig. 2) may be effective for treatment of depression.[19] α_1-adrenoceptors are also found in the brain but specific functions have not been clarified so far.

α_2-receptor populations have been shown to be present on platelets and in the kidney where activation by exogenous agonists causes aggregation and sodium reabsorption, respectively.[20,21] Furthermore, α_1-receptors are located in the heart although it is generally accepted that β-receptors control cardiac function.[22] However, receptor sites can often be identified by pharmacological intervention or ligand binding techniques but for the above cases correlation with definite physiological roles has not yet been established. Possibly, certain α-receptors are dormant under normal conditions but play important roles in pathophysiological processes and cardiovascular disease when sympathetic activity is enhanced. For example, the antiarrythmic effects of prazosin in animals may result from blockade of cardiac α_1-receptors and subsequent protection from excessive catecholamines.[23,24]

Finally, some progress has been achieved in clarifying the mechanisms by which agonist-receptor interactions are translated into cellular responses. Biochemical evidence suggests that activation of α_1-adrenoceptors increases phosphatidylinositol turnover with a subsequent influx of extracellular calcium. By contrast, stimulation of α_2-adrenoceptors inhibits adenyl cyclase and a reduction in cAMP initiates the physiological response.[25] However, recent studies in rats indicate that vasoconstrictor responses to α_2-agonists require extracellular calcium whereas corresponding α_1-effects do not.[26] These results cast some doubt over the exact mechanisms involved, but the activation processes for α_1- and α_2-adrenoceptors do appear to be quite different. Moreover, calcium antagonists block postjunctional, rather than prejunctional, α_2-mediated events[27] suggesting subtle differences between even such closely related receptor subtypes. Further progress now depends on the synthesis of more selective agents in order to realise the full clinical potential offered by specific modulation of α-adrenoceptor function.

2. Structure Activity Relationships (SARs) for Quinazoline α_1-Adrenoceptor Antagonists

In contrast to the well defined SARs for β-receptor ligands, α-adrenoceptor antagonists comprise a particularly heterogeneous collection (Fig. 2) and attempts to rationalise data from such diverse structural series would not be realistic. Therefore, this paper will concentrate on our own work with 2,4-diamino-6,7-dimethoxyquinazoline derivatives. These compounds are potent, highly selective, postjunctional α_1-adrenoceptor antagonists which have proved clinically useful for the treatment of hypertension.[28,29]

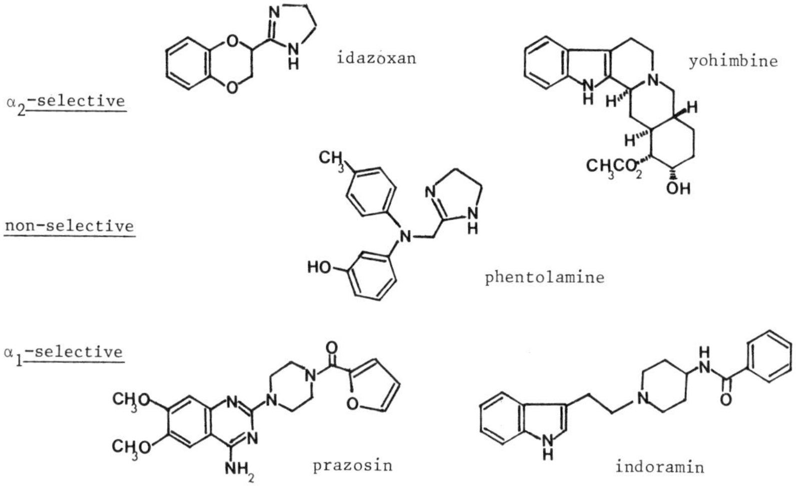

FIG. 2. Representative α-adrenoceptor antagonists.

Prazosin (I), doxazosin and related analogues (Table 2) are potent, competitive antagonists of α_1-mediated responses to noradrenaline (II). Moreover, these 2,4-diaminoquinazoline derivatives display little affinity for other receptors and therefore provide an attractive starting point for understanding structural requirements for α_1-adrenoceptor recognition. It will be assumed that the agonist (II) and antagonist (I) compete for common receptor sites and the medicinal chemical challenge is to identify structural features which could perform similar functions (Fig. 3). In the case at hand, the aromatic rings of prazosin and noradrenaline serve as initial reference points with the respective 6,7-dimethoxy- and 3,4-dihydroxy-functions being similarly accepted by the receptor, although only in the latter case would activation occur.[30] An obvious structural difference is that only noradrenaline contains a benzylic hydroxyl function but, although important for potency, it is not essential for agonist activity.[31] Moreover, this hydroxyl moiety is absent from several other α-antagonist series and does not appear to be a fundamental requirement.

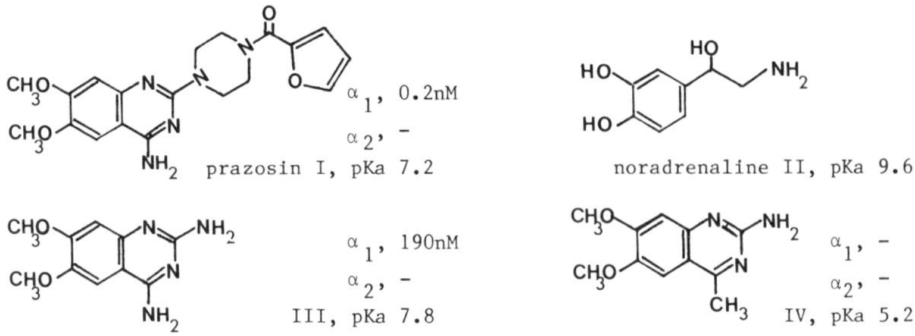

FIG. 3. Structural/basicity requirements for quinazoline α_1-adrenoceptor binding affinity.[32,33]

Usually, only basic substances interact with α_1-adrenoceptors and the active site may contain an anionic or nucleophilic centre which binds to positively charged agonists or antagonists. Noradrenaline (pKa 9.6)[36] and prazosin (pKa 7.2) can be protonated at physiological pH (7.4) and, at first sight, the primary amino function and the quinazoline 2-nitrogen atom appear to be similarly located basic centres. As expected, simplified derivatives such as 2,4-diamino-6,7-dimethoxyquinazoline (III, pKa 7.8) are also recognised at the α_1-receptor and although less potent than prazosin, the selectivity characteristic of the whole series is still evident. On the other hand, the 2-amino-4-methyl analogue (IV, pKa 5.2) is essentially inactive since, despite structural similarity, this compound would not be protonated under physiological conditions. These results indicate that a protonated quinazoline nucleus is required for α_1-adrenoceptor affinity and that potency can be enhanced by modification of the 2-substituent.

3. A Role for the Quinazoline Nucleus at α_1-Adrenoceptors

Noradrenaline contains a single basic nitrogen centre and the protonated species (V) predominates (*ca* 95%) at physiological pH.[37] By contrast, the 2,4-diaminoquinazoline system contains four potential protonation sites but the exocyclic nitrogen atoms need not be considered since these electron pairs are extensively conjugated with the aromatic Π-system. For the ring nitrogen atoms, molecular orbital calculations indicate that protonation at N-1 (VI) is preferred over N-3[38] and similar conclusions have been reached for quinazoline dihydrofolate reductase inhibitors.[39] Protonation[40] and quaternisation[41] of related 2,4-diaminopyrimidine derivatives also predominate at N-1.

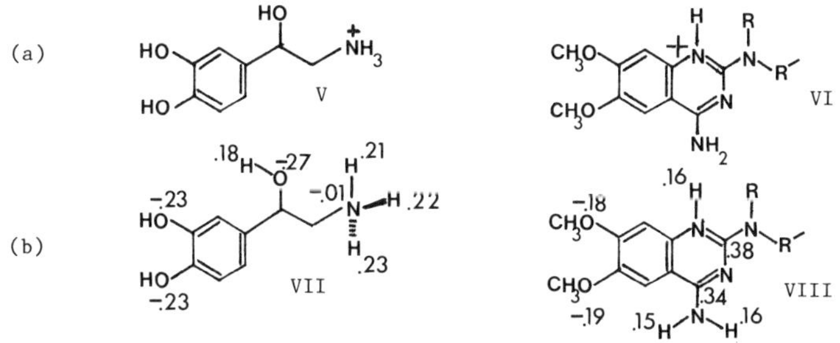

FIG. 4. Protonated forms of noradrenaline and a quinazoline derivative, (a) localised; (b) delocalised; CNDO/2 charge distribution by Mulliken population analysis (only selected centres shown).

These results suggest major electronic differences between the amino function of noradrenaline and the quinazoline 2-nitrogen atom, and their ability to participate in similar receptor interactions must be doubted. However, structures (V) and (VI) suggest positive charge location on a particular nitrogen centre whereas calculations indicate considerable dispersion over neighbouring atoms.[42,43] The importance of charge delocalisation in protonated drug molecules has been pointed out on several occasions,[44] but with little apparent impact on SAR interpretation.

For noradrenaline (VII), the three hydrogen atoms bear most of the positive charge and there seems little reason to emphasise either the location or Coulombic interaction of an essentially neutral nitrogen centre in agonist-receptor recognition. Charge distribution is more extensive for the protonated quinazoline (VIII) and, although both the N_1-H and 4-NH_2 functions appear equally suited as focal points for receptor interaction, only the former will be considered at present. Thus, agonist and antagonist molecules probably participate in charge-reinforced hydrogen bonding with an anionic/nucleophilic moiety on the receptor, but this site must be accessible to both noradrenaline and the quinazoline series. As positive charge is symmetrically distributed around hydrogen atoms, directional flexibility for electrostatic interaction with a negative centre is allowed. On the other hand, identification of the nucleophilic site is purely speculative but as the recognition area is part of a complex protein structure, a carbonyl moiety in the polypeptide backbone could serve as an effective hydrogen bond acceptor.[45,46] Moreover, considerable directional latitude for these carbonyl-hydrogen bonds would be possible, with little detrimental effect on bonding energy.[47,48]

Thus, the formally charged nitrogen atoms need not play direct roles in drug-receptor interactions but attention should concentrate instead on charge-reinforced hydrogen bonding. The α_1-adrenoceptor could comprise a carbonyl function which accepts an in-plane, partially charged hydrogen atom from noradrenaline or the quinazoline series,[49] and an hydrophobic area which accommodates the aromatic rings from either structure[50] (Fig. 5). For each case, hydrogen bonds are approximately 2Å, and since deviation from linearity is only thirty degrees significant stabilising interactions should occur. Agonists and antagonists can therefore bind at common receptor sites and the high α_1-adrenoceptor affinity of the quinazoline derivatives may reflect the fact that N-1 protonation is particularly appropriate for the nucleophilic centre on the receptor protein.

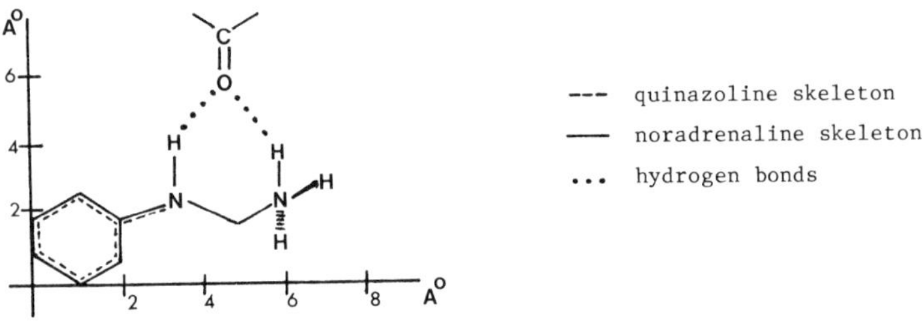

FIG. 5. Composite diagram illustrating interaction of noradrenaline and the quinazoline nucleus with common hydrophobic and hydrogen bond acceptor sites (partial structures only).

This simple receptor model focusses attention on charge-reinforced hydrogen bonding but the initial choice of a carbonyl function was somewhat arbitrary. Thus, it did not seem appropriate to carry out detailed refinement at this stage and instead, alternative anionic centres were considered. Therefore, molecular mechanics techniques were employed to assess the interaction of the protonated form of noradrenaline (VII) with selected counterions but first, two provisos were made:-

a) the aromatic ring and the anion should be essentially coplanar, consistent with the previous receptor concept.
b) the anion should be capable of simultaneous hydrogen bonding with both the benzylic hydroxyl group and the ammonium head.

The latter condition avoids separate binding sites for each functionality as, intuitively at least, a single acceptor moiety should be preferred. The interaction of various counterions with noradrenaline was studied using molecular mechanics simulation of non-bonded forces (Van der Waals and Coulombic) to identify favourable binding positions, then full relaxation energy minimisation was initiated to determine the preferred geometries of interaction. Final binding energies (enthalpies of interaction) were calculated by standard INDO methods. A carbonyl ligand was first evaluated but concomitant interaction with both hydrogen bonding moieties could not be accommodated. However, preliminary calculations employing N-acetylglycine as the receptor fragment have been more successful and will be reported in due course.

Chloride, phosphate and carboxylate anions satisfied both criteria and cyclic hydrogen bonded structures have been observed in the crystal for the first two cases.[51,52] N.M.R. studies support similar cyclic arrangements for hydrochloride salts of noradrenaline-like molecules in chloroform but not aqueous solution.[53] These results may mirror conformational preferences at the receptor and in physiological fluids respectively, since the target site is most likely a lipoid environment but agonists and antagonists must be transported through an aqueous milieu. Moreover, drug-receptor interactions may be initiated by transfer of charged counterions between aqueous and hydrophobic phases and de-solvated species could well be involved.[54] These theoretical and experimental results may therefore be very helpful in understanding drug action, although gas phase binding enthalpies must be treated qualitatively rather than quantitatively.

The three counterions mentioned previously appear equally suited for detailed evaluation[55] but the carboxylate anion was selected mainly because salt bridges involving aspartate and glutamate residues with protonated heterocycles have been detected in other enzyme systems.[56] Thus, for noradrenaline a coplanar cyclic hydrogen bonding arrangement (Fig. 6, IX) is preferred (binding energy, -155.93 kcal/mol) and as this conformation lies close (<2 kcal/mol) to the global minimum (phenyl ring rotated through 60°) it would be accessible to the natural α_1-transmitter. A similar interactive study with a protonated quinazoline derivative also demonstrated charge-reinforced hydrogen bonding (X, binding energy, -72.2 kcal/mol) but closer approach (<2.5Å) of the carboxylate anion was prevented by the piperidine ring.[57] Comparison of structures (IX) and (X) indicates that the carboxylate counterions are differently located (ca 4Å separation) relative to the parent aromatic rings and appear incompatible with the previous concept of fixed recognition sites. However, the earlier carbonyl model was static and while initial agonist and antagonist receptor recognition may be similar, the consequences of binding are quite different. Agonists activate the receptor to produce a physiological response whereas antagonists exert "squatters' rights" and need not disturb the active site. For example, in the β-receptor series, agonist binding is enthalpy-driven, consistent with strong bonding to the receptor in order to overcome an unfavourable decrease in entropy.[58] On the other hand, antagonist interaction is entropy-driven since the major contribution to binding affinity results from the entropy increase associated with released water molecules.[59] Prazosin binding to the α_1-adrenoceptor may also be entropy-driven.[60]

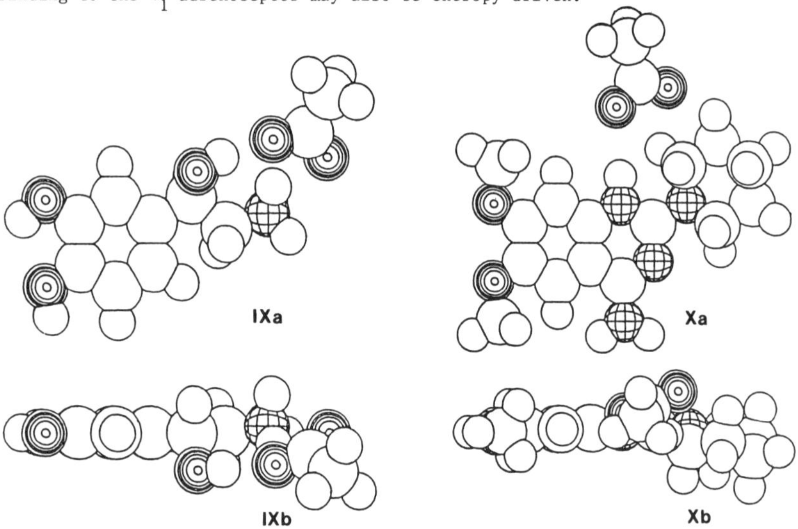

FIG. 6. Interaction of noradrenaline (IX) and a quinazoline derivative (X) with a carboxylate counterion (see text for provisos). Face-on (a) and side-views (b) illustrated.

An additional, important factor is that these quinazolines (Table 2) are highly selective α_1-adrenoceptor ligands with little affinity for other receptor systems.[61] Thus, the quinazoline nucleus appears to be exquisitely suited to the α_1-adrenoceptor, possibly simulating a conformationally restricted form of noradrenaline. Complex (X) may therefore represent the interaction of the protonated heterocycle with the ground state of the α_1-adrenoceptor and the high binding affinity reflects an hydrophobic attraction, charge-reinforced hydrogen bonding and a favourable entropy component associated with release of bound water molecules.[62] This drug-receptor complex (X) is also a minimum enthalpy arrangement, given the original constraints, and any conformational reorganisation required for receptor activation would be energetically unfavourable.

When noradrenaline approaches this receptor ground state, the carboxylate counterion forms a hydrogen bond with the benzylic hydroxyl function and initiates a medium-range electrostatic interaction with the ammonium head (Fig. 7, XI, binding energy -74.54 kcal/mol). Charge-reinforced hydrogen bonding can then be increased by a 4Å migration of the counterion (IXa, binding energy, -155.93 kcal/mol). Thus, the conformational change in the protein structure usually associated with receptor activation could be promoted by the free energy decrease accompanying transformation of the initial agonist complex (XI) into the more stable arrangement (IXa).[63,64]

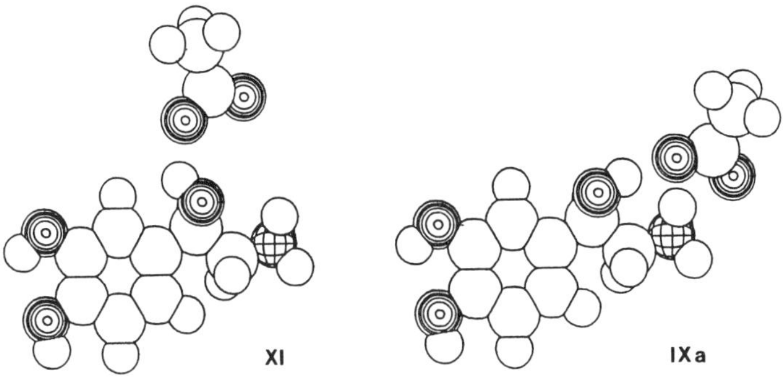

FIG. 7. Interaction of noradrenaline with a carboxylate counterion, α_1-adrenoceptor ground state (XI); activated state, (IXa).

A simple α_1-adrenoceptor model has therefore been developed which rationalises the different consequences of agonist and antagonist receptor occupancy. Importantly, the model has been extended to S(+)- noradrenaline where the counterion complex is less favoured (Fig. 8, binding energy -141.86 kcal/mol) due to repulsive interactions between the hydroxyl proton and the positively charged ammonium head. Similar binding energies are also obtained for dopamine (the deshydroxy analogue) and are consistent with pharmacological evidence that both compounds are full α_1-agonists but are considerably less potent (ca. 100 fold) than R(-)-noradrenaline.[65] Presumably, counterion migration results from medium-range Coulombic interaction and is not steered by the hydroxyl function.

Hydrophobic interactions involving aryl rings have been assumed to be the initial recognition processes for both agonists and antagonists at the α_1-adrenoceptor and charge-reinforced hydrogen bonding is viewed as a subsequent event. This seems reasonable as hydrophobic areas on a receptor surface should be accessible to drug molecules while a solvated carboxylic acid moiety would be much less available. Thus, the initial hydrophobic association not only locks the drug onto the receptor surface but it may also force the counterion into the lipoid pocket for subsequent electrostatic interaction.

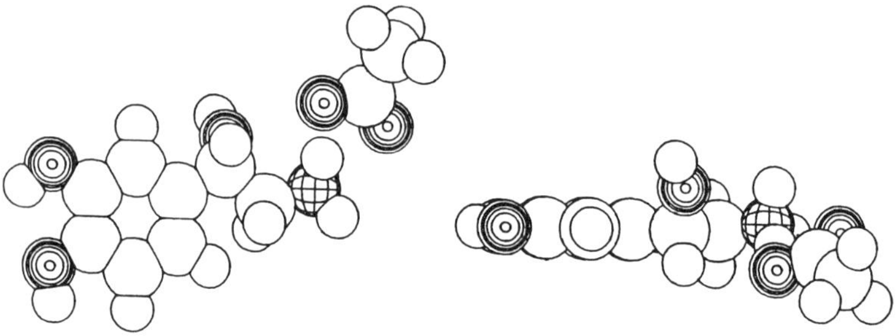

FIG. 8. Interaction of S(+) noradrenaline with a carboxylate anion on the α_1-adrenoceptor, activated state. Counterion location as for (IX).

Finally, the remarkable α_1-adrenoceptor selectivity of this quinazoline series (Table 2) is consistent with a receptor model containing the hydrophobic binding area and carboxylate counterion in parallel planes. By contrast, initial analysis of α_2-adrenoceptor SARs suggests that these key areas may be orthogonal to one another[66,67] and a flat quinazoline nucleus could not accommodate such alternative geometry. These observations form the basis of current work to develop a corresponding model for the α_2-adrenoceptor.[68]

4. The Influence of the Quinazoline 2-Substituent

The amino-function of noradrenaline (II) and the quinazoline 2-nitrogen atom (III) have quite different electronic characteristics but since they should occupy similar regions on the α_1-adrenoceptor, SARs for the catecholamine series may be transferable. Modification of the nitrogen substituent in the latter compounds has marked effects on receptor selectivities as Ahlquist observed in his original distinction between α- and β-receptors. For example, isoprenaline has only weak α-adrenoceptor activity, suggesting limited steric tolerance around the nitrogen atom, but as the substituent is further extended, affinity is regained (Fig. 9).

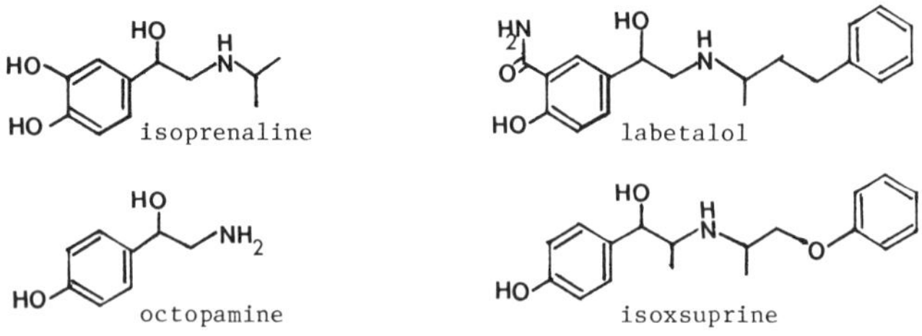

labetalol: α_1, 150nM; α_2, 1,600nM

isoxsuprine: α_1, 26nM; α_2, 1,100nM

FIG. 9. Effect of N-substitution on α-adrenoceptor affinity of "catecholamine" derivatives. (Note octopamine is an α-agonist).

Labetalol, isoxsuprine and related derivatives are antagonists and their extended nitrogen substituents could occupy the receptor area where the conformational consequences of agonist-induced counterion migration are normally expressed. Comparable SAR data are not available for many of these analogues,[69] but structural resemblance suggests they occupy the α_1-receptor in a similar manner. The second aromatic rings could then capitalise on a distal, hydrophobic region which need not be an integral part of the receptor active site but simply a lipoid area commonly found in protein structures. Even so, these α-antagonists appear to interact with a specific hydrophobic site centered some 7.5Å from the parent nitrogen atom, in addition to the key anionic and aromatic regions previously discussed.

Can these observations be applied to the quinazoline series? (Table 2). 2,4-Diamino-6,7-dimethoxyquinazoline (No. 1) displays α_1-adrenoceptor affinity similar to more elaborate antagonists such as labetalol and isoxsuprine. The rigid quinazoline nucleus therefore enjoys an effective receptor fit since the more flexible structures require an additional hydrophobic interaction to offset any entropy loss associated with conformational restriction. However, receptor recognition of this prototype quinazoline (No. 1) may be atypical as the 2-amino function could also interact with the carboxylate counterion[56] and so restrict optimum alignment of the aromatic ring. Trisubstituted derivatives avoid this problem, and the observed hundred-fold increase in affinity (Nos. 2,3) could support the importance of a single N_1-H-carboxylate interaction (Fig. 6, X) rather than a direct role for these alkyl substituents at the receptor. Potency increases a further ten-fold when a carbonyl function is linked to the piperidine 4-position (No. 5), but unexpectedly, the corresponding 4-phenyl derivative (No. 6) was less active. When the aromatic ring is fused in a bicyclic arrangement (No. 7) a major increase in activity occurred (cf. No. 3) and the exceptional binding affinity (Ki, 0.02nM) suggests the compound is capitalising on an additional receptor site. However, as the centre of the aromatic ring is only some 4Å from the 2-nitrogen atom, this may not be the hydrophobic area implicated for the labetalol types. SARs for piperazine derivatives generally correspond to the piperidine series and in particular, prazosin (No. 10) and doxazosin (No. 11) are the most potent members of this group. Apparently, incorporation of a planar π-system via a carboxamide moiety appears to be an effective alternative to an aromatic ring.[70] In summary, the simple diaminoquinazoline (No. 1) is a selective α_1-adrenoceptor ligand but, by appropriate modification of the 2-substituent, affinity can be increased some ten thousand-fold and compounds of outstanding potency have been identified. Equally important, these derivatives show

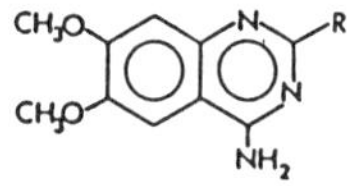

NO.	R	α_1	α_2	$EC_{(40)}$ POST	$EC_{(40)}$ PRE	% reduction in rat b.p.
1	$-NH_2$	190	—	NT	NT	5*
2	$-N(C_2H_5)_2$	2.0	—	130	1,500	50*
3	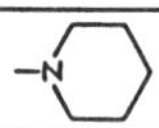	2.5	—	300	—	18*
4	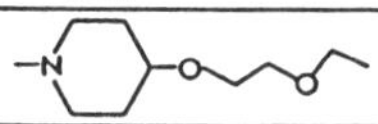	1.5	NT	38	—	83
5	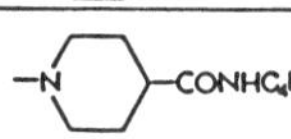	0.3	—	50	—	100
6	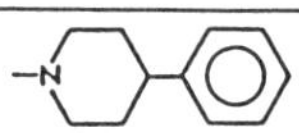	1.0	—	NT	NT	26*
7	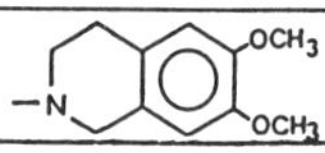	0.02	—	NT	NT	100*
8	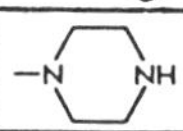	33	—	1750	—	21*
9	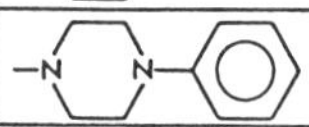	3.4	—	NT	NT	45
10	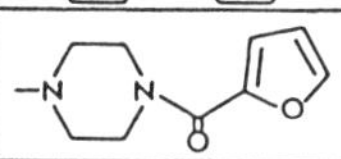	0.2	—	4.5	1,300	70*
11	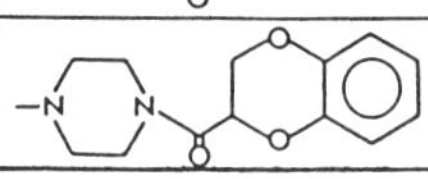	0.5	—	50	—	77

TABLE 2. Binding (Ki, nM.),[33] functional (EC_{40}, nM.)[71] and antihypertensive[72] activities for selected quinazoline derivatives. *Okamoto strain. No. 10, prazosin; No. 11, doxazosin (UK-33,274).

little, if any, propensity to displace clonidine from α_2-receptor sites and exceptional α_1-selectivity is characteristic of this entire quinazoline series.

Binding studies provide a useful measure of affinity for α_1-adrenoceptors but the compound must also be able to block the action of noradrenaline. Results from experiments determining functional antagonism[71] (Table 2) roughly agree with binding data and all of the compounds examined are selective postjunctional α_1-antagonists with minimal effects at prejunctional sites. Clearly, these quinazoline derivatives avoid the detrimental consequences of non-selective blockade. However, in vitro receptor affinity is only a guide to in vivo behaviour and antihypertensive activity in animal models must be established in order to select compounds worthy of secondary pharmacological evaluation. In the spontaneously hypertensive rat,[72] all of these quinazoline derivatives (except No. 1) lower blood pressure but efficacy and duration of action are markedly influenced by the 2-substituent (cf. Nos. 3,4,5). The tetrahydroisoquinoline derivative (No. 7) also displays excellent, long-lasting activity in the rat but more detailed profiling did not justify clinical evaluation. By contrast, the desmethoxy analogue (see No. 7, H for OCH_3) was essentially inactive in vivo despite outstanding binding affinity (Ki, 0.05nM). Thus, substitution may stabilise a metabolically vulnerable aromatic ring although it is not required for receptor interaction. Prazosin and doxazosin display similar efficacy in the rat but, on first sight, are not the most active members of the series. However, direct comparison of these data can be misleading since single-dose experiments may represent quite different points on the dose-response curves. In man, both compounds antagonise the α_1-mediated vasoconstrictor responses to noradrenaline and are potent, effective antihypertensive agents.[28,29] Importantly, the elimination half-lives of prazosin and doxazosin in man are approximately 2.5 and 11 hours, respectively, and the latter compound offers the convenience of once-a-day dosing.[73]

5. Relative Importance of the Quinazoline Nucleus and 2-Substituent

In analysing SARs, structural modifications cannot be viewed as independent events but the properties of the whole molecule must be considered. For example, elaboration of a quinazoline 2-dimethylamino substituent to the prazosin side chain has only a moderate effect on binding affinity, suggesting little influence for the furoylpiperazine moiety (Fig. 10). However, the former compound is a stronger base (pKa 8.1 vs 7.2) and this could be of paramount importance if a protonated quinazoline nucleus is required at the receptor. Thus, it is not possible to quantify the contribution of the extended 2-substituent since the unexpectedly high α_1-adrenoceptor affinity of (XII) may simply result from increased basicity.

The "deaza" prazosin derivative (XIII) displays α_1-affinity similar to labetalol whereas a simple alkyl analogue would be essentially inactive. At first sight, the more elaborate 2-substituent now appears to dominate receptor interactions but no conclusions can be drawn when two inactive fragments are combined in a single molecule.[74] Perhaps a more informative comparison is between (I) and (XIII); prazosin is substantially more basic and the six hundred-fold increase in α_1-affinity compared to the deaza analogue again reflects the crucial role played by the protonated quinazoline nucleus.[75] Equally important, (XIII) is weakly active in the rat whereas prazosin is particularly effective in this species and in man.

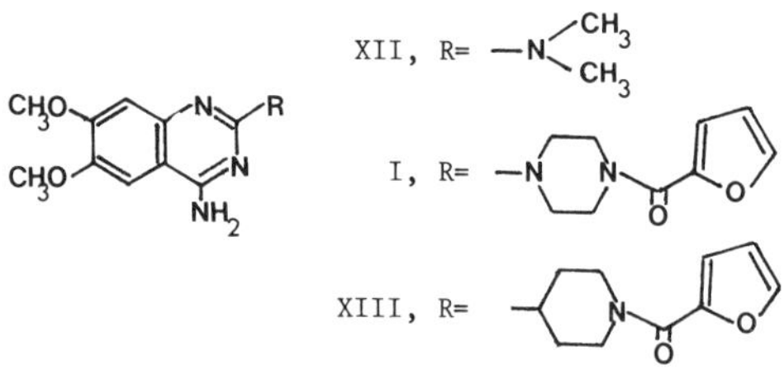

XII, R= $-N(CH_3)_2$ — α_1, 5.0nM; pKa, 8.1

I, R= — α_1, 0.2nM; pKa, 7.2

XIII, R= — α_1, 120nM; pKa, 6.4

FIG. 10. Importance of nuclear and 2-substituent interactions for quinazoline derivatives at α_1-adrenoceptors.

6. Acknowledgements

The author gratefully thanks his colleagues in the Biology (Drs. V.A. Alabaster, D.A. Cambridge, G.M.R. Samuels) and Chemistry Departments (Dr. J.D. Hardstone, Mr. M.J. Palmer, Miss R.M. Plews) without whose assistance this work could not have been completed. All of the theoretical calculations were performed by Dr. M.S. Tute who also provided invaluable help during informal discussions. Finally, the initial programme which led to the discovery of prazosin was carried out in our Groton laboratories and was directed by Dr. H-J. Hess.

References

[1] For a recent review see, K.E. Andersson, Acta Med. Scand. Suppl., 665, 9 (1982).

[2] A.M. Lands, A. Arnold, J.P. McAuliff, F.P. Luduena and T.G. Brown, Nature, 214, 597 (1967).

[3] For a recent review see, J.E.S. Wikberg, Acta Med. Scand., Suppl. 665, 19 (1982).

[4] K. Starke, H. Montel, W. Gayk and R. Merker, Naunyn-Schmiedeberg's Arch. Pharmacol., 285, 133 (1974).

[5] S. Berthelsen and W.A. Pettinger, Life Sci., 21, 595 (1977).

[6] J.E.S. Wikberg, Acta Physiol. Scand., Suppl., 468 (1979).

[7] P.B.M.W.M. Timmermans and P.A. van Zwieten, J. Med. Chem., 25, 1389 (1982).

[8] S.Z. Langer, Pharmacol. Rev., 32, 337 (1981).

[9] For a recent review see, K. Starke and J.R. Docherty, J. Cardiovasc. Pharmacol., 2, S269 (1980).

[10] D.S. Goldstein, Hypertension, 5, 86 (1983).

[11]F.R. Bühler, F.W. Amann, P. Bolli, L. Hulthén, W. Kiowski, R. Landmann and E. Burgisser, J. Cardiovasc. Pharmacol., 4, S134 (1982).
[12]For example, see, O. Schier and A. Marxer, Progress in Drug Research, 13, 101 (1969).
[13]I. Cavero and A.G. Roach, Life Sci., 27, 1525 (1980).
[14]W.S. Colucci, Ann. Int. Med., 97, 67 (1982); idem, Am. J. Cardiol., 51, 639 (1983).
[15]E.J. Ariens and A.M. Simonis, Biochem. Pharmacol., 32, 1539 (1983).
[16]H.L. Elliott and J.L. Reid, Clin. Sci., 65, 237 (1983).
[17]P. Bolli, P. Erne, W. Kiowski, B.H. Ji, F.W. Amann and F.R. Bühler, First European Meeting on Hypertension, Milan 1983, Abs. 36.
[18]J.M. Roesler, L.J. Kopaciewicz and J.P. Hieble, Pharmacologist, 25, 242 (1983).
[19]C.B. Chapleo, P.L. Myers, R.C.M. Butler, J.C. Doxey, A.G. Roach and C.F.C. Smith, J. Med. Chem., 26, 823 (1983).
[20]J.A. Grant and M.C. Scrutton, Br. J. Pharmacol., 71, 121 (1980).
[21]R.M. Graham, W.A. Pettinger, A. Sagalowsky, J. Brabson and T. Gandler, Hypertension, 4, 881 (1982).
[22]H.J. Schümann, Europ. Heart J., 4 Suppl. A, 55 (1983).
[23]D.J. Sheridan, P.A. Penkoske, B.E. Sobel and P.B. Corr, J. Clin. Invest., 65, 161 (1980).
[24]M.J. Davey, Vasodilatation, ed. by P.M. Vanhoutte and I. Lensen, p 499, Raven Press, New York (1981).
[25]J.N. Fain and J.A. Garćia-Sáinz, Life Sci., 26, 1183 (1980).
[26]J.C.A. van Meel, P.B.M.W.M. Timmermans and P.A. van Zwieten, J. Cardiovasc. Pharmacol., 5, 580 (1983).
[27]K. Starke and H.J. Schümann, Arzneim-Forsch., 23, 193 (1973).
[28]W. Singleton, C.A.P.D. Saxton, J. Hernandez and B.N.C. Prichard, J. Cardiovasc. Pharmacol., 4, S145 (1982).
[29]R.M. Graham and W.A. Pettinger, N. Eng. J. Med., 300, 232 (1979).
[30]P. Pratesi and E. Grana, Adv. Drug Res., 2, 127 (1965).
[31]P.N. Patil, D.D. Miller and U. Trendelenburg, Pharmacol. Rev., 26, 323 (1974).
[32]H. Glossmann, R. Hornung and P. Presek, J. Cardiovasc. Pharmacol., 2, S303 (1980).
[33]α_1- and α_2-adrenoceptors in a rat brain membrane preparation were labelled with tritiated prazosin and clonidine respectively[34,35] and the abilities of test compounds to displace these ligands measured. An entry, α_1, 2nM; α_2, -. indicates that the Ki for the compound at α_1-adrenoceptors is 2×10^{-9}M while for the α_2-site, the Ki is $>1 \times 10^{-6}$M.

[34]P. Greengrass and R. Bremner, Eur. J. Pharmacol., 55, 323 (1979).
[35]D.A. Greenberg, D.C. U'Prichard and S.H. Snyder, Life Sci., 19, 69 (1976).
[36]F. Mack and H. Bönisch, Naunyn-Schmiedeberg's Arch. Pharmacol., 310, 1 (1979).
[37]C.R. Ganellin, J. Med. Chem., 20, 579 (1977).
[38]M.S. Tute, unpublished INDO calculations.
[39]G.M. Crippen, J. Med. Chem., 22, 988 (1979).
[40]D.V. Griffiths and S.P. Swetnam, J.C.S. Chem. Comm., 1224 (1981).
[41]D.J. Brown and T. Teitei, J.C.S., 755 (1965).
[42]D.H. Aue, H.M. Webb and M.T. Bowers, J. Am. Chem. Soc., 98, 311 (1976).
[43]L.J. Saethre, T.A. Carlson, J.J. Kaufman and W.S. Koski, Mol. Pharmacol., 11, 492 (1975).
[44]B. Pullman, P. Courrière and H. Berthod, ibid., 11, 268 (1975).
[45]P.A. Kollman, J. Am. Chem. Soc., 94, 1837 (1972).
[46]A. Johansson, P. Kollman, S. Rothenberg and J. McKelvey, ibid, 96, 3794 (1974).
[47]N. Sakabe, K. Sakabe and K. Sasaki in Structural Studies on Molecules of Biological Interest, Eds., G. Dodson, J.P. Glusker and D. Sayre, Clarendon Press, Oxford (1981), Chapter 41.
[48]Z. Berkovitch-Yellin and L. Leiserowitz, J. Am. Chem. Soc., 102, 7677 (1980).
[49]See also, C.W. Thornber, Chem. Soc. Reviews, 8, 563 (1979).
[50]T. Dipaolo, L.H. Hall and L.B. Kier, J. Theor. Biol., 71, 295 (1978).
[51]Noradrenaline hydrochloride, D. Carlström and R. Bergin, Acta Cryst., 23, 313 (1967).
[52]Ephedrine monohydrogenphosphate, R.A. Hearn, G.R. Freeman and C.E. Bugg, J. Am. Chem. Soc., 95, 7150 (1973).
[53]J. Zaagsma, J. Med. Chem., 22, 441 (1979).
[54]M.A. Holmes and B.W. Matthews, Biochemistry, 20, 6912 (1981).
[55]Earlier workers have proposed that the α-receptor contains a carboxylate or phosphate anion. For a review see, B.M. Bloom and I.M. Goldman, Adv. Drug Res., 3, 121 (1966).
[56]J.T. Bolin, D.J. Filman, D.A. Matthews, R.C. Hamlin and J. Kraut, J. Biol. Chem., 257, 13650 (1982).
[57]Calculations for a 2-piperidino substituent should be relevant for a wide range of quinazoline derivatives (see Table 2).
[58]G.A. Weiland, K.P. Minneman and P.B. Molinoff, Mol. Pharmacol., 18, 341 (1980).
[59]W. Kauzmann, Adv. Prot. Chem., 14, 1 (1959).
[60]H. Glossmann and R. Hornung, Naunyn-Schmiedeberg's Arch. Pharmacol., 314, 101 (1980).

[61] For example, prazosin has minimal affinity for α_2, 5HT, β_1, β_2, dopamine and cholinergic receptors.

[62] For a review of drug-receptor interactions see, P.A. Kollman in Burger's Medicinal Chemistry, 4th Ed., Part 1, M.E. Wolff, Ed., Wiley, New York, 1980, p. 313.

[63] T.J. Franklin, Biochem. Pharmacol., 29, 853 (1980).

[64] A.S.V. Burgen, Federation Proc., 40, 2723 (1981).

[65] R.R. Ruffolo, E.L. Yaden and J.E. Waddell, J. Pharmacol. Exp. Ther., 222, 645 (1982).

[66] A. Carpy, J.M. Leger, G. Leclerc, N. Decker, B. Rouot and C.G. Wermuth, Mol. Pharmacol., 21, 400 (1982).

[67] S.F. Campbell, unpublished observations.

[68] The important role of aromatic substituents in influencing α_1- vs α_2-selectivity will not be discussed in this paper.

[69] M. Aggerbeck, G. Guellaën and J. Hanoune, Br. J. Pharmacol., 65, 155 (1979).

[70] G.L. Olson, H-C. Cheung, K.D. Morgan, J.F. Blount, L. Todaro, L. Berger, A.B. Davidson and E. Boff, J. Med. Chem., 24, 1026 (1981).

[71] Assessment of functional antagonism against noradrenaline at pre- and post-junctional α-adrenoceptors was determined in a rabbit pulmonary artery preparation and results expressed as EC_{40} values (nM.)[4]

[72] Antihypertensive activity was evaluated after oral administration (5mg/kg) to spontaneously hypertensive rats (New Zealand or Okamoto strain). The falls in blood pressure (mm Hg) during the test period (4.5hr) were measured using an indirect tail cuff method and the maximum value expressed as follows,

$$\%\ \text{reduction in hypertension} = \frac{\text{fall in blood pressure}}{(\text{control blood pressure} - 130)} \times 100$$

where 130 represents the systolic blood pressure of normotensive rats.

[73] H.L. Elliott, P.A. Meredith, D.J. Sumner, K. McLean and J.L. Reid, Br. J. Clin. Pharmacol., 13, 699 (1982).

[74] M.I. Page, Angew. Chem. Int. Ed. Engl., 16, 449 (1977).

[75] L. Cocco, C. Temple, J.A. Montgomery, R.E. London and R.L. Blakley, Biochem. Biophys. Res. Comm., 100, 413 (1981).

Structure–Activity Studies with Angiotensin Converting Enzyme Inhibitors Related to Enalapril and MK-521

By C. S. Sweet*, A. A. Patchett, E. H. Ulm, and D. M. Gross

MERCK INSTITUTE FOR THERAPEUTIC RESEARCH, WEST POINT, PA. 19486, U.S.A.

The discovery of captopril by Ondetti and coworkers[1,2] is one of the recent outstanding achievements in medicinal chemistry. Their work was based upon previous knowledge about the biological properties of South American snake venom peptides[3] and the encouraging clinical efficacy of teprotide[4]. Using insights from and extending the work with byproduct inhibitors of carboxypeptidase A[5], the Squibb research group synthesized compounds which had a high affinity for Zn^{++}, a key component in the catalytic site of converting enzyme. Several reports from our laboratory have detailed efforts leading to the design[6] and biological characterization[7-10] of the ACE inhibitors enalaprilat (MK-422), enalapril (MK-421) and lisinopril (MK-521)[11-12]. One extension of this work has been synthesis of 7-, 8- and 9-membered monocyclic lactams[13] which bind to ACE in a conformation close to the minimum energy conformation of enalaprilat. The biological properties of some of these conformationally rigid compounds[13,14], a series of more elaborate derivatives[15], and a limited series of analogs related to MK-521 will be described in this report.

Biological Activity of Standard ACE Inhibitors. Over the last several years, our laboratory has evaluated a considerable number of ACE inhibitors including some bench mark inhibitors as described below. The inherent enzyme inhibitory activity was quantified in our laboratory using hog plasma as a source of converting enzyme, as described previously[6], and *in vivo* inhibitory activity was estimated using anesthetized Sprague-Dawley rats in an assay in which cumulative doses of the ACE inhibitors are given intravenously after a pressor challenge of angiotensin I. The dose of each inhibitor producing a 50% blockade of angiotensin I (ID_{50}) is computed[8]. Oral activity, assessed in unanesthetized rats, is expressed as the dose of a given inhibitor

Table 1

Comparative ACE Inhibitory Activity of Standards

ACE Inhibitor	I_{50} (M)	ID_{50} µg/kg i.v. ± 95% C.I.	ID_{50} mg/kg p.o. ± 95% C.I.	AUC at ED_{50}
Enalaprilat	1.2×10^{-9}	8.2(6.6-10.6)	2.29(1.54-3.40)	266
Enalapril	2×10^{-7}	14(12.2-16.2)	0.29(0.20-0.42)	219
Lisinopril	1.2×10^{-9}	2.3(1.7-3.1)	0.19(0.13-0.27)	223
Captopril	2.1×10^{-8}	26(21-33)	0.33(0.22-0.49)	190
Pivalopril	___	___	0.60(0.37-0.97)	133
SA-446[a]	1.0×10^{-9}	___	1.18(0.78-1.75)	153

which produces a 50% inhibition of angiotensin I at its maximum[9-10], and the AUC is the area under the % inhibition vs time curve during a seven-hour period as described by Sweet[10]. The diacid inhibitor 1b, enalaprilat (MK-422), and lisinopril (2) were highly active in inhibiting hog plasma ACE *in vitro* and in blocking the acute pressor response to i.v. angiotensin I in rats (Table 1) and in anesthetized dogs[8,10,11]. Unlike 1b, which was not well absorbed in rats (ID_{50} = 2.29 mg/kg p.o.), lisinopril was more than 10 times more active orally (Table 1) and, thus, similar to enalapril (1a). In terms of ACE inhibition, the duration of action of some of the standards vary considerably (Table 1). When the duration of action, as indicated by AUC of six of these standards, was compared at equivalent ACE inhibitory levels, enalaprilat had the longest duration of action while pivalopril[16] had the shortest duration. In practical terms, AUC values approaching 200 at the ED_{50} translate into an effective duration of action of approximately 6-8 hours. Oral doses larger than those summarized on Table 1 are needed to reduce blood pressure in the spontaneously hypertensive rat[7,9], in part presumably because this is a low renin

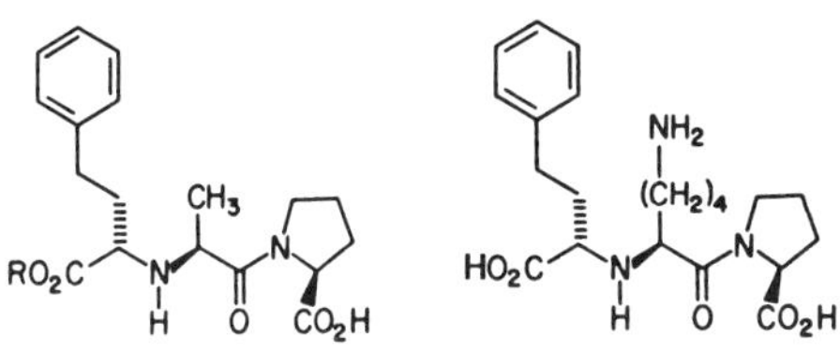

1 a) R = C_2H_5, enalapril
b) R = H, enalaprilat

lisinopril
2

---

[a]Jap. J. Pharmacol. 31: 875, 1981

model. For example, we routinely need to administer enalapril at 3.0 and captopril at 30.0 mg/kg p.o. to observe blood pressure reductions of 15–25 mm Hg in the conscious SHR.

Analogs of Lisinopril (MK-521). Introduction of monoesters at the N-carboxyalkyl group of enalaprilat increased absorption, as shown by the improved ID_{50} data in rats (Table 1). Another approach for improving oral activity has been to replace L-alanine with L-lysine. A number of substituents have been made at the L-alanine position, some of which are shown in Table 2.

Table 2

ACE Inhibitory Activity as a Function of Side Chain Length of Analogs Related to Lisinopril (MK-521)

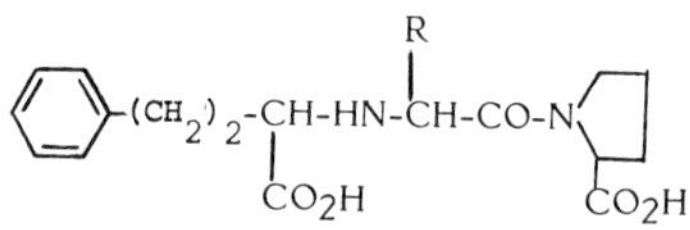

R Substituent[c]	I_{50} (M) Hog Plasma ACE	Inhibition of Ang I Pressor Response AUC_0^7 hr (3.0 mg/kg p.o.)
$-CH_2NH_2$(R,S)[a]	4.8×10^{-7}	---
$-(CH_2)_2NH_2$(R,S)[a]	3.8×10^{-8}	37
$-(CH_2)_3NH_2$(R,S)	4.5×10^{-9}	225
$-(CH_2)_4NH_2$(R,S)	6.8×10^{-9}	352
$-(CH_2)_4NH_2$(S) (Lisinopril)	1.2×10^{-9}	426
$(CH_2)_5NH_2$(RS,RS)[b]	1.1×10^{-8}	55

[a] Compounds synthesized by M.T. Wu, L. Payne and T. Ikeler.

[b] Compounds synthesized by W.V. Ruyle and J. Celebuski.

[c] Most of the compounds reported have the S,S configurations in the dipeptide portion of the molecule but were mixtures of diastereoisomers which have R and S configurations at the carbon bearing the carboxyl. This is designated by (R,S).

On an empirical basis, we have found that the lysine analog had the best oral activity of those compounds shown. It is rather slowly absorbed from the gastro-intestinal tract which contributes to a prolonged oral activity. In addition, Bull, Thornbery and Cordes (personal communication) have evaluated the inhibition of rabbit lung converting enzyme, and based on their kinetic analysis lisinopril had a K_i of 1×10^{-10}<u>M</u> and a half-life of the enzyme inhibitor complex of 105 min. These findings indicate that lisinopril tightly binds to and slowly dissociates from the enzyme. Enalaprilat had a K_i of 2×10^{-10}M but its off rate is also faster with a half-life value of 27 minutes. In SHR, the antihypertensive response to lisinopril following single doses of 0.312 to 5.0 mg/kg p.o. (Figure 1) resembles many other ACE inhibitors. There is an early fall in MAP (15-20 mm Hg) which is evident at 1 hour and a recovery at about 8-12 hours. Lisinopril at 20 mg/kg p.o. was particularly impressive with a maximum fall in blood pressure averaging about 40 mm Hg. The reason for this larger response is not known but may be related to the compound's impressive duration of action as an inhibitor of ACE.

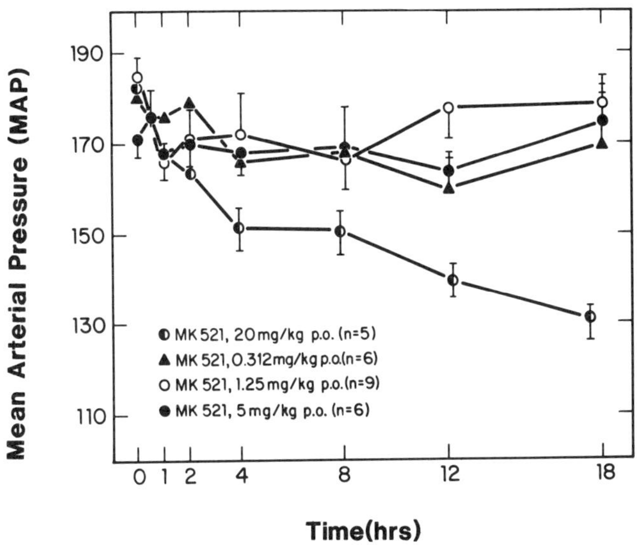

Figure 1

Other potent inhibitors in which histidine or arginine have been substituted in the Ala position or in which various substituents have been introduced on the epsilon amino group of lisinopril have been synthesized (Table 3).

Table 3

In Vitro and In Vivo ACE Inhibitory Activity of

N-(1-carboxy-3-phenylpropyl)-dipeptides

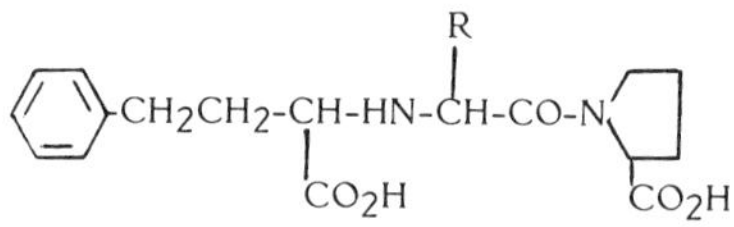

R -NHCHCO-	I_{50} (M) Hog Plasma ACE	Oral AUC ACE Inhibitory Activity at 3.0 mg/kg p.o.
Histidine	7.4×10^{-8}	15
Σ-N-Acetyllysine	47% @ 2×10^{-8}	negligible
Arginine	6.5×10^{-9}	negligible
Σ-N,N-Dimethyllysine	9.1×10^{-9}	10
Σ-N-Methyllysine	4.6×10^{-9}	N.T.

As indicated, most of these compounds possessed strong ACE inhibitory activity on the isolated enzyme, but none equated with the oral activity of lisinopril.

Lactams. The potency of enalaprilat led Thorsett and colleagues[13] to reason that the inhibitor might bind to ACE in a state close to its minimum energy conformation. They utilized the Merck Molecular Modeling System[17] which uses semi-empirical force field calculations to generate an energy-minimized conformation of enalaprilat from its X-ray structure. Using this modelling system they also compared the structures of potentially new conformationally restricted analogs. From this work it was decided to synthesize several 7-, 8- and 9-membered monocyclic lactams. Each of these compounds have slightly different conformations which are close to that of L-alanyl-L-proline in its minimal energy conformer. The most potent is the 8-membered ring structure (3). An extensive biological evaluation was undertaken to determine whether these lactam inhibitors were superior to 1b. In Table 4 the relevant ACE inhibitory profile is shown. Within experimental error, the diacid inhibitor, 3b, was similar in potency to enalaprilat (Table 1).

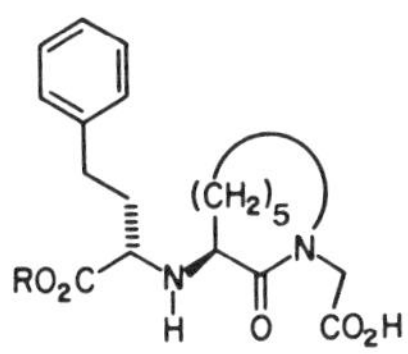

3

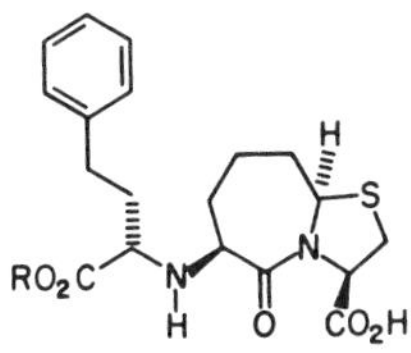

4

a) R = C_2H_5

b) R = H

Table 4

Comparative ACE Inhibitory Activity of the

8-membered Lactams and Bicyclic Decane Derivatives

ACE Inhibitor	I_{50} (M)	ID_{50} µg/kg i.v. ± 95% C.I.	ID_{50} mg/kg p.o. ± 95% C.I.	AUC at ED_{50}
3b	**1.5×10^{-9}**	**12 (9.3-15.7)**	**3.0**	**——**
3a	**6.1×10^{-7}**	**26.8 (22.6-31.7)**	**0.10 (0.07-0.16)**	**145**
4b	**6×10^{-10}**	**9.4 (7.6-11.6)**	**0.42 (0.29-0.61)**	**234**
4a	**2.1×10^{-7}**	**25.6 (19.9-32.8)**	**0.15 (0.09-0.26)**	**233**

This inhibitor blocked angiotensin I pressor responses in the anesthetized rat with an ID_{50} of about the same order of magnitude as enalapril (Table 1). As with most diacid ACE inhibitors, the oral activity of 3b was poor. In preliminary experiments we observed 28, 36, 54, 23 and 21% inhibition of angiotensin I at 1/2, 1, 2, 4 and 7 hours, respectively, following 3.0 mg/kg p.o. Esterification provided a solution to this problem as shown by the pronounced oral activity of 3a. Its oral ID_{50} is the lowest of any of the inhibitors which we are reporting in this study. However, when compared to other ACE inhibitors, its short duration (AUC = 145) became a liability. There was little ACE inhibitory activity on the 7th hour of rats treated with a dose corresponding to approximately its ID_{50}, whereas in enalapril and MK-521-treated rats, angiotensin I pressor responses were blocked by greater than 15-20%. At 10 times its ID_{50}, there was only 20% inhibition of angiotensin I. 3a was evaluated at 0.078 to 80 mg/kg p.o. for its ability to lower mean arterial pressure in spontaneously hypertensive rats. This compound produced a maximum fall in mean arterial pressure of 8, 18, 15, 20 and 25 mm Hg following doses of 0.078, 0.312, 5, 20 and 80 mg/kg p.o., respectively. The onset of action was apparent during the first 1/2 hour, and the fall in blood pressure persisted for up to 8 hours with the largest doses. Thus, as with other ACE inhibitors, the antihypertensive activity of 3a occurred at doses in excess of those required to block angiotensin I pressor responses. A partial explanation for the short duration of action of 3a was provided from physiological disposition studies which indicated that the molecule may undergo some metabolism other than bioactivation to 3b. These studies demonstrated that up to 16% of an oral dose was metabolized beyond bioactivation.

The final two inhibitors which will be described were also synthesized as mimics for the L-alanyl-L-proline unit of ACE inhibitors. As an extension of the bicyclic lactam work[13,14], Wyvratt and colleagues[15] made 4,a,b in a multi-step synthesis which involved careful control of stereochemistry. Four diastereomeric diacid inhibitors were evaluated from this series. The bicyclic inhibitor 4b with I_{50} of 6.0 x 10^{-10}M is particularly noteworthy when compared to enalaprilat (I_{50} = 1.2 x 10^{-9}M). The binding affinity of 4b for rabbit angiotensin-converting enzyme has been determined by competitive equilibrium dialysis against enalaprilat (K_i = 2.1 x 10^{-10}M). The measured K_i value of 7.6 x 10^{-11}M at pH 7.5 makes 4b the most potent inhibitor of ACE yet reported. Surprisingly, this increased *in vitro* potency did not translate over into an increased i.v. potency in anesthetized rats (Table 4). However, this diacid had very good oral activity with an ID_{50} of 0.42 (0.29-0.61) mg/kg p.o., a value within the same experimental range as captopril (Table 1), and a duration of action which rivals lisinopril. As one might predict, its antihypertensive action was also impressive. In SHR, blood pressure decreased to 134 $\pm$ 3 ($P < 0.05$) at 1 hour from a pretreatment value of 181 $\pm$ 4 (S.E.) mm Hg, and blood pressure was still significantly reduced at 12 hours (22 mm below pretreatment). The ester 4a, as expected did not display significant in vitro activity but it had excellent oral activity (ID_{50} = 0.15 mg/kg) with a duration of action in excess of 7 hours at 0.3 and 1.0 mg/kg p.o. At its ID_{50}, angiotensin I pressor responses were blocked by 17% at 7 hours. At 7 hours, the corresponding diacid (4b) was somewhat more active in that angiotensin I pressor responses were attenuated by about 35%. 4a also reduced blood pressure in the SH rat at 5-20 mg/kg p.o., but not to the same blood pressure "floor" as 4b. A duration of action of 12 hours was also expected from oral ACE inhibitory studies (Table 4).

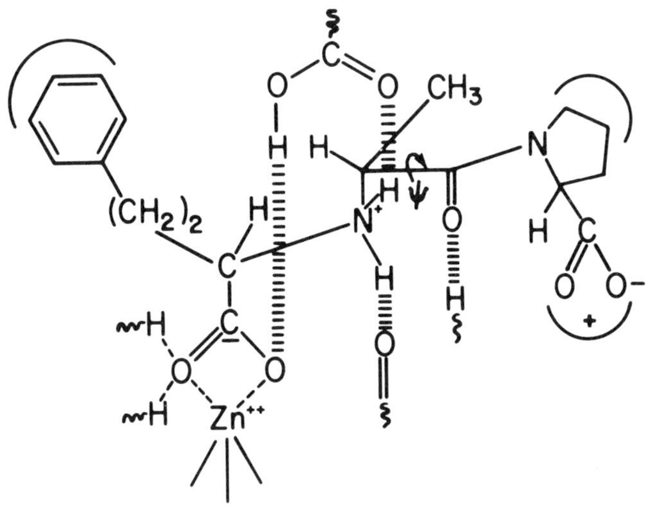

Figure 2

Summary: The Binding of Enalaprilat to ACE

The present work with highly potent ACE inhibitors has extended our understanding of the enzyme's active site. In particular, the angles and molecular volumes of these inhibitors in their lowest energy conformers permit approximate definition of the enzyme's active site topology in its S_1' and S_2' subsites.

Further insights into the binding of these inhibitors to ACE have been drawn from x-ray crystallography. Since angiotensin converting enzyme has not been crystallizied, these studies employed thermolysin, a related crystalline Zn^{++} enzyme and a thermolysin inhibitor reported by Maycock et al.[18] which is structurally identical to enalaprilat except the Ala-Pro unit was replaced by Leu-Trp to accomodate the specificity requirements of thermolysin. The unpublished structural studies of this enzyme/inhibitor complex by Monzingo and Matthews have led to several striking findings. In the first place, the Zn^{++} bound carboxyl is a bidentate ligand in which both oxygen atoms are bound to Zn^{++}. Furthermore, there is a very precise network of hydrogen bonds involving the enzyme and inhibitor. We have extrapolated these interactions to converting enzyme (Figure 2) on the reasonable assumption that similar enzymes will bind identical inhibitor designs similarly. Thus, we attribute the high potency and specificity of our converting enzyme inhibitors to these interactions and to a hydrophobic bond made to the inhibitors' phenethyl group. Together these multiple, sterically precise interactions replace the thiol-Zn^{++} ligand which is responsible for the potency of captopril.

Acknowledgements

The authors would like to thank Mrs. J. Brooke for typing the manuscript and Dr. F. Robinson for his valuable editorial advice. Mr. S.L. Gaul, Mr. T. Vassil and Mr. C.T. Ludden are acknowledged for their technical expertise in conducting many of the experiments.

References

1. M.A. Ondetti, B. Rubin and D.W. Cushman, Science, 1977, 196, 441.

2. D.W. Cushman, H.S. Cheung, E.F. Sabo and M.A. Ondetti, Biochemistry, 1977, 16, 5484.

3. S.H. Ferreira, Br. J. Pharmacol., 1965, 24, 163.

4. H. Gavras, H.R. Brunner, J.H. Laragh, J.E. Sealy, I. Gavras and R.A. Vukovitch, N. Engl. J. Med., 1974, 291, 1974.

5. L.D. Byers and R. Wolfenden, Biochemistry, 1973, 12, 2070.

6. A.A. Patchett, E. Harris, E.W. Tristram, M.J. Wyvratt, M.T. Wu, D. Taub, E.R. Peterson, T.J. Ikeler, J. ten Broeke, L.G. Payne, D.L. Ondeyka, E.D. Thorsett, W.J. Greenlee, N.S. Lohr, R.O. Hoffsommer, H. Joshua, W.V. Ruyle, J.W. Rothrock, S.D. Aster, A.L. Maycock, F.M. Robinson, R. Hirschmann, C.S. Sweet, E.H. Ulm, D.J. Gross, T.C. Vassil and C.A. Stone, Nature, 1980, 288, 280.

7. C.S. Sweet, D.M. Gross, P.T. Arbegast, S.L. Gaul, P.M. Britt, C.T. Ludden, D. Weitz and C.A. Stone, J. Pharmacol. Exp. Ther., 1981, 216, 558.

8. D.M. Gross, C.S. Sweet, E.H. Ulm, E.P. Backlund, A.A. Morris, D. Weitz, D.L. Bohn, H.C. Wenger, T.C. Vassil and C.A. Stone, J. Pharmacol. Exp. Ther., 1981, 216, 552.

9. C.S. Sweet, P.T. Arbegast, S.L. Gaul, E.H. Blaine and D.M. Gross, Eur. J. Pharmacol., 1981, 76, 1981.

10. C.S. Sweet, Fed. Proc., 1983, 42, 167.

11. C.S. Sweet, D.M. Gross, E.H. Blaine, E.H. Ulm and C.A. Stone, "New Trends in Arterial Hypertension," Biomedical Press, Elsevier/North-Holland, Amsterdam, NY, 1981, p. 263.

12. E.H. Ulm, M. Hichens, H.J. Gomez, A.E. Till, E. Hand, T.C. Vassil, J. Biollaz, H. Brunner and J.L. Schelling, Br. J. Clin. Pharmacol., 1982, 14, 357.

13. E.D. Thorsett, E.E. Harris, S. Aster, E.R. Peterson, A.A. Patchett and T.C. Vassil, 184th Natl. Am. Chem. Soc. Mtg., Kansas City, MO, 1982, MEDI 78.

14. E.D. Thorsett, E.E. Harris, E.E. Aster, E.R. Peterson, E.W. Tristram,

E.H. Ulm, J.P. Snyder, J.P. Springer and A.A. Patchett, Proc. Am. Peptide Sym. 8th Mtg., Tucson, AZ, 1983, in press.

15. M.J. Wyvratt, M.H. Tischler, T.J. Ikeler, J.P. Springer, E.W. Tristram and A.A. Patchett, Proc. Am. Peptide Sym. 8th Mtg., Tucson, AZ, 1983, in press.

16. T.A. Solomon, J.O. Knipe, F.S. Caruso, R.A. Vukovich and E.S. Neiss, Clin. Pharmacol. Ther. 1982, 31, 271.

17. P. Gund, J.D. Andose, J.B. Rhodes and G.M. Smith, Science, 1980, 208, 1428.

18. A.L. Maycock, D.M. DeSousa, L.G. Payne, J. ten Broeke, M.T. Wu and A.A. Patchett, Biochem. Biophys. Res. Commun., 1981, 102, 963.

Phosphodiesterase Inhibitors as Cardiotonic Agents

By S. F. Campbell, N. J. Cussans, J. C. Danilewicz, A. G. Evans, A. L. Ham, A. A. Jaxa-Chamiec, D. A. Roberts*, and J. K. Stubbs

PFIZER CENTRAL RESEARCH, SANDWICH, KENT CT13 9NJ, U.K.

1. Introduction

Congestive heart failure results primarily from a defect in the contractile pumping mechanism of the heart, leading to reduced cardiac output.[1] This is accompanied by compensatory vasoconstriction in order to maintain perfusion of vital organs, and fluid retention resulting from reduced renal blood flow. Current drug therapy addresses these individual haemodynamic problems with inotropes, vasodilators[2] and diuretics. For example, digitalis has been used for many years to provide inotropic support, but there is considerable doubt over chronic efficacy and the drug has a narrow therapeutic ratio. Vasodilators such as prazosin (α_1-adrenoceptor antagonist) and captopril (angiotensin converting enzyme inhibitor) relieve excessive veno- and arteriolar-constriction, unloading the heart so that cardiac output improves. Current clinical trials should establish the role of these agents in the long-term treatment of congestive heart failure. Diuretics remain the first-line therapy, reducing preload by preventing fluid retention. Although all of these agents are undoubtedly useful in improving some of the symptoms of heart failure, there is still strong clinical demand for superior drugs, particularly those which combine inotropic and vasodilator properties. Such agents would address the major haemodynamic derangements characteristic of congestive heart failure by improving myocardial performance and reducing elevated peripheral resistance.

Combination of dual pharmacological activities in a single molecule is a perennial problem for medicinal chemists, but nevertheless our objective has been to obtain safe, orally active inotrope-vasodilators for chronic use in heart failure. Therefore, we have focussed on elevation of intracellular cyclic adenosine monophosphate[3] (cAMP), since this cyclic nucleotide can indirectly control both cardiac muscle contraction and vascular tone. cAMP elicits physiological responses <u>via</u> the activation of a series of protein kinases which, by phosphorylation of key enzymes, regulate Ca^{++} movement.[4] Thus an increase in intracellular cAMP concentration in myocardial cells results in enhanced cytosolic Ca^{++} levels, thereby triggering muscle contraction. Increased cytosolic Ca^{++} levels are also required for vascular smooth muscle contraction but in this tissue cAMP appears to activate a calcium binding process, thereby producing relaxation.[5] Intracellular cAMP levels can be enhanced by stimulation of cardiac and vascular β-receptors, and agonists such as prenalterol and pirbuterol do provide haemodynamic benefit in heart failure. Alternatively, we have concentrated on inhibition of phosphodiesterase,[6] (PDE) the enzyme responsible for cAMP degradation. However, multiple forms of PDEs exist that differ in both tissue distribution and relative substrate selectivity. Thus, in attempting to design an inotrope/vasodilator, it would be preferable to inhibit only cAMP PDE in cardiac and vascular smooth muscle.

There are also several mechanisms independent of cAMP by which cytosolic Ca^{++} concentration can be modulated, for example, by a Na^{+}/Ca^{++} exchange process. Thus, intracellular Ca^{++} concentration can be increased by blocking the removal of Na^{+} by Na^{+}/K^{+} ATPase (sodium pump), and this is believed to be the mode of action of the cardiac glycosides. Cytosolic Ca^{++} can also be increased by Ca^{++} channel agonists[7] though vasopressor activity may make such agents unsuitable for use in congestive heart failure.

2. Phosphodiesterase inhibition and inotropic activity.

Examination of the literature revealed that numerous compounds from several diverse structural classes had been reported to inhibit PDE *in vitro*[8,9] but little quantitative information was available to correlate SARs for enzyme inhibition with cardiac stimulant activity. For example, high concentrations of inhibitors (10^{-4} M) were frequently required and furthermore, early experiments did not distinguish between high and low affinity forms of the enzyme or between competitive and non-competitive interactions. Moreover, *in vitro* enzyme inhibition does not automatically result in cardiac stimulant activity. For example, dipyridamole, theophylline and papaverine (Fig. 1) combine both PDE inhibition and inotropic activity, but these latter effects are now thought to be mediated by alternative mechanisms.[10,11,12] On the other hand, R07-2956 is a PDE inhibitor and produces an inotropic response in dogs.[13] However, marked tachycardia and elevated blood glucose levels were also observed and the compound did not appear to provide an attractive starting point for further synthesis. More recently it has been suggested that PDE inhibition may contribute to the favourable haemodynamic effects of amrinone[14] and MDL-17,043[15] but this information was not available at the start of our studies.

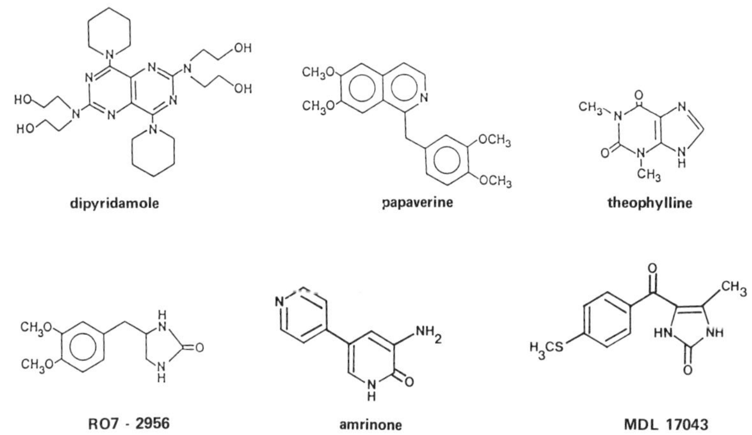

FIG. 1. Representative PDE inhibitors and inotropic agents.

Since the prior art was not particularly helpful we decided to examine several series of PDE inhibitors for in vitro cardiac stimulant activity in isolated guinea-pig double atria,[16] to determine whether any correlation between inotropic potency and enzyme inhibition could be established. In particular, attention focussed on compounds which increased the force of contraction rather than heart rate, since increases in heart rate elevate myocardial oxygen

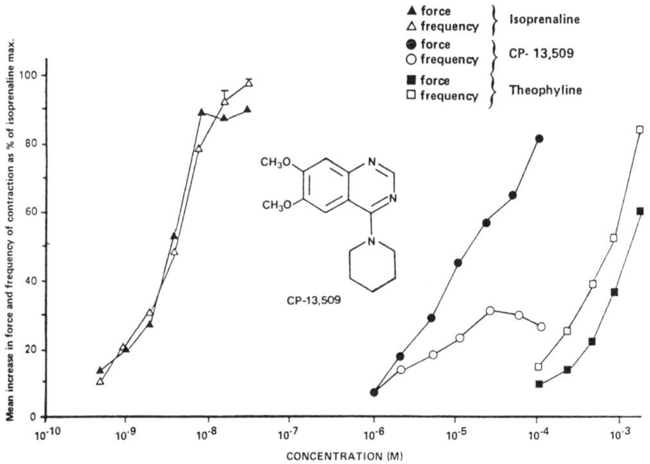

FIG. 2. Effects of isoprenaline, CP-13,509 and theophylline on the force and frequency of contraction in the guinea pig double atrium[16] at various concentrations.

consumption and this is undesirable for patients in heart failure. From these studies, 4-piperidino-6,7-dimethoxyquinazoline (CP-13,509, Fig. 2) emerged as a potential lead. The compound was a potent, competitive[17] cAMP PDE inhibitor (IC_{50} 3.3μM vs beef heart PDE commercially available crude enzyme from Sigma), and produced an 80% and 30% increase in the force of contraction and heart rate respectively, a clear improvement in selectivity over isoprenaline and theophylline. CP-13,509 was also active in the anaesthetised dog (0.25mgkg^{-1}, i.v. administration) causing a 70% increase in $LVdP/dt_{max}$[18] (the maximum rate of increase of left ventricular pressure) with little effect on heart rate, and producing a dose related reduction in blood pressure. Unexpectedly, only marginal inotropic activity was detected in the conscious animal, even at an oral dose of 40mgkg^{-1}. However, having demonstrated the potential of this approach in the laboratory, we were keen to identify agents suitable for clinical evaluation and the first objectives were to improve force/rate selectivity and oral activity. Initial modification of either the aromatic moiety or the 2-position of the quinazoline was not beneficial but substitution of the piperidine ring led to buquineran (UK-14,275), an orally effective inotrope in the dog[19] (10mg kg^{-1} p.o.) with little tendency to increase heart rate (Fig. 3). Cardiac stimulant activity was also demonstrated in normal volunteers[20,21] and in heart failure patients[22,23] where buquineran was found to be more effective during exercise than at rest.[24] This is highly desirable from a therapeutic viewpoint since increased inotropic support is available when most required. Unexpectedly, buquineran was found to prolong the QT interval in the ECG complex,[25,26] and while this need not be deleterious,[27] it may be indicative of dysrythmogenic potential and a search for alternative candidates began.

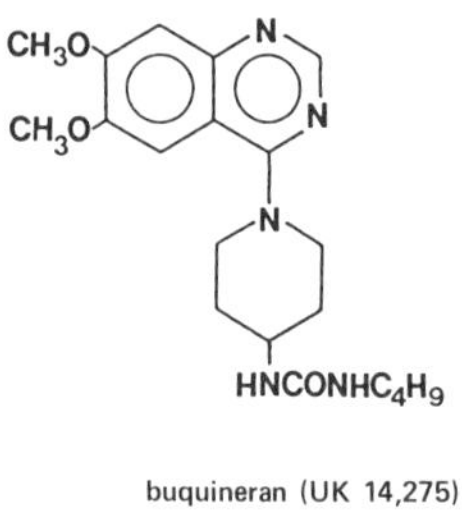

buquineran (UK 14,275)

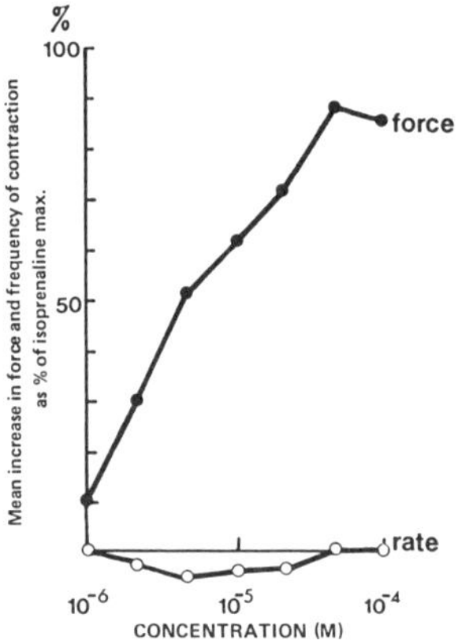

IC_{50} values
cAMP PDE 16μM
cGMP PDE 55μM

FIG 3. Effect of buquineran on the force and frequency of contraction in the guinea pig double atrium[16] at various concentrations.

3. The Interaction of Quinazolines with PDE

At this stage, an attempt was made to rationalise how these quinazoline inhibitors could compete with cAMP for specific recognition sites on PDE. The objective was to identify common structural features in both substrate and inhibitor which could play similar roles on the enzyme. Moreover, unexploited binding sites might also be identified which, by appropriate chemical modification of the inhibitor, could provide superior agents. cGMP was also included in the analysis since elevated intracellular levels of this cyclic nucleotide may mediate negative inotropic and chronotropic responses[5] and, according to the Yin Yang[28] hypothesis, may counteract cAMP mediated events. In agreement with this proposal, although buquineran is five times less potent than CP-13,509 (IC_{50} values vs cAMP PDE 3.3μM; cGMP PDE 5.4μM) as a cAMP PDE inhibitor, it is relatively weak vs cGMP PDE (IC_{50} 55μM) and therefore possesses a better selectivity ratio (IC_{50} cGMP/IC_{50} cAMP; 3.4 vs 1.6). Inotropic activity differs only by twofold and so an enhanced selectivity ratio may offset the reduction in cAMP PDE inhibitory potency. In vascular smooth muscle cGMP appears to impede Ca^{++} entry,[6] relaxing the vessel. Thus selectivity for cAMP vs cGMP PDE may be less important in this tissue since both cyclic nucleotides appear to produce the same overall effect.

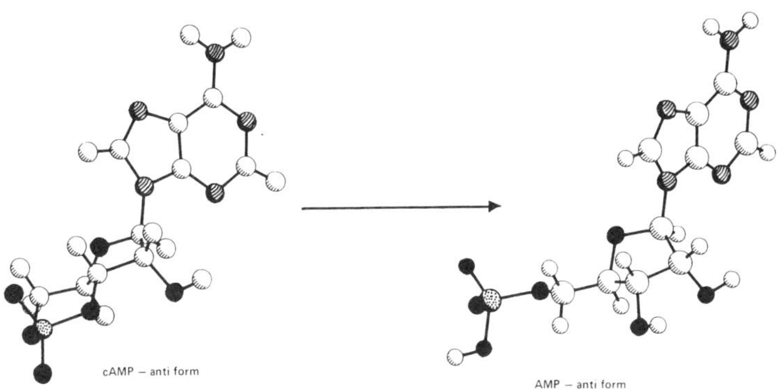

FIG. 4. Biological degradation of cAMP (anti form) by PDE to AMP.

The biological inactivation of cAMP involves hydrolysis of the cyclic phosphate moiety (Fig. 4) to produce adenosine-5′-monophosphate (AMP). The stereochemistry of hydrolysis by several PDE enzymes has now been established and it is clear that a uniform path is not followed.[29] Although no precise details of the mechanism of hydrolysis of cAMP are known, it is likely that once the nucleotide is located on the PDE active site, nucleophilic attack on phosphorus is initiated and hydrolysis proceeds via a trigonal bipyramidal intermediate with overall inversion of configuration.[30] Recently, a covalent intermediate has been detected[31] in the hydrolysis of tritium labelled cAMP by bovine intestine PDE in which the nucleotide is bound to either a serine or threonine residue through a phosphodiester bond (Fig. 5). Subsequent nucleophilic attack on this intermediate by a water molecule would then afford AMP. A different mechanism, which also proceeds with inversion of configuration, has recently been proposed in an elegant study by Cotton et al[32] for the hydrolysis of 5′-phosphodiester bonds of both RNA and DNA by the enzyme staphylococcal nuclease. A high resolution X-ray crystal analysis of the enzyme/thymidine-3′,5′-biphosphate-calcium ion complex revealed that the phosphate moiety is bound by four hydrogen bonds to two arginine residues, which may also activate the substrate to nucleophilic attack. Further support for this activation process has been shown by Springs and Haake[33] who demonstrated that the presence of guanidinium ions produced a twenty-seven-fold rate enhancement for the cleavage of phosphodiesters by fluoride ion. This type of coordination of phosphate groups by basic enzymic amino acid residues with subsequent activation towards nucleophilic attack may be generally applicable to the hydrolysis of phosphodiester bonds of other nucleotides by different phospholytic enzymes. Thus a similar mechanism has been suggested[34] involving the arginine 77 residue of another phospholytic enzyme RNase T_1 in the cleavage of RNA phosphodiester bonds.

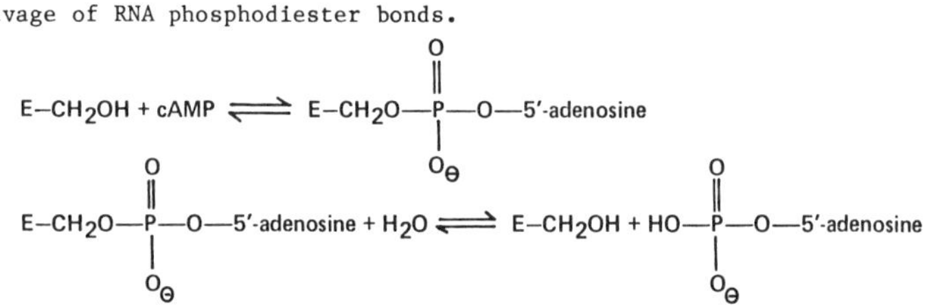

FIG. 5. Mechanism of cAMP hydrolysis proposed by Butler.[31]

In attempting to understand the interaction of cAMP with phosphodiesterase it is important to consider the physicochemical properties of the nucleotide. Firstly, the purine moiety (pKa 3.8) should not be protonated at physiological pH, and therefore the enzyme should recognise the parent heterocycle. Secondly, conformational energy calculations predict[35] that although the anti and syn conformational ranges are both accessible, the anti conformation is slightly favoured, (ca 1kcal $mole^{-1}$). NMR studies[36,37] and X-ray data[38,39] also indicate that the anti form is favoured, and we have therefore assumed that this conformation is preferred by the enzyme. Furthermore, cAMP derivatives which are either restricted to (8-bromo),[40,41] or prefer (5′-methylene)[41,42] the syn conformation are inhibitors rather than substrates for the enzyme. This suggests that the purine ring could be recognised similarly to the natural substrate but that the cyclic phosphate moiety is sterically indisposed for interaction with the key nucleophilic centre on PDE. The interaction of cGMP with its particular PDE can also be analysed in a similar fashion, but in this case physical determinations[35,36,37,43] indicate that the syn conformation is favoured (ca 2Kcal mole^{-1}) and we have therefore assumed that this spatial arrangement is preferred by cGMP PDE.

The pKa of buquineran is 5.95 and the unprotonated heterocycle, which predominates (90:10) at physiological pH, should be recognised by the enzyme by analogy with cAMP. These quinazoline derivatives are competitive PDE inhibitors and presumably interact with the recognition site for cAMP. Thus in trying to rationalise how a purine and a quinazoline nucleus could interact with a common site, we initially calculated and compared electron densities on the various ring nitrogen atoms, assuming that these could be key recognition centres. Thus if substrate (I) and inhibitor (III) are recognised in the orientation shown (Fig. 6), where the electron rich nitrogen atoms of the aminopyrimidine portion of the substrate correspond to the similarly electron rich nitrogen atoms of the inhibitor, it is evident that the ribose recognition area is not being exploited by the quinazoline series. Hydrophilic groups were therefore introduced into the 7- and 8-positions of the aromatic ring, but no significant increases in potency were achieved; indeed, inferior compounds were frequently encountered. These results indicated that this particular quinazoline orientation may not be appropriate, possibly implying that individual centres of highest charge density are not specifically involved in the interaction of cAMP with phosphodiesterase. In agreement with this proposal,[44] tubercidin-3′,5′-cyclic phosphate, which lacks the electron rich N-7 atom, is equiactive with cAMP as a substrate for PDE and a similar observation has recently been made for cGMP and its 7-deaza analogue.[45] Furthermore, various other cAMP analogues in which other nitrogen atoms have been replaced by carbon have been shown[46] to retain good substrate activity and the idea of identifying individual binding sites was not pursued. Instead the molecular consequences of the overall electronic distribution, the dipole moment, were considered. Thus, enzymes may recognise particular dipoles to orientate the approaching molecule in an approximate fashion, and then initiate the more subtle complexities of secondary binding. A similar approach has been employed recently by Olsson[47] in considering structure-activity relationships in a series of adenosine receptor agonists and antagonists, where both the direction and magnitude of the purine dipole moment appeared to be critical for activity. More generally, it has been proposed[48] that dipole moments could be useful in explaining direct drug-receptor interactions through non-covalent bonding, and this concept may have particular relevance in systems where strong hydrogen-bonding interactions are absent.

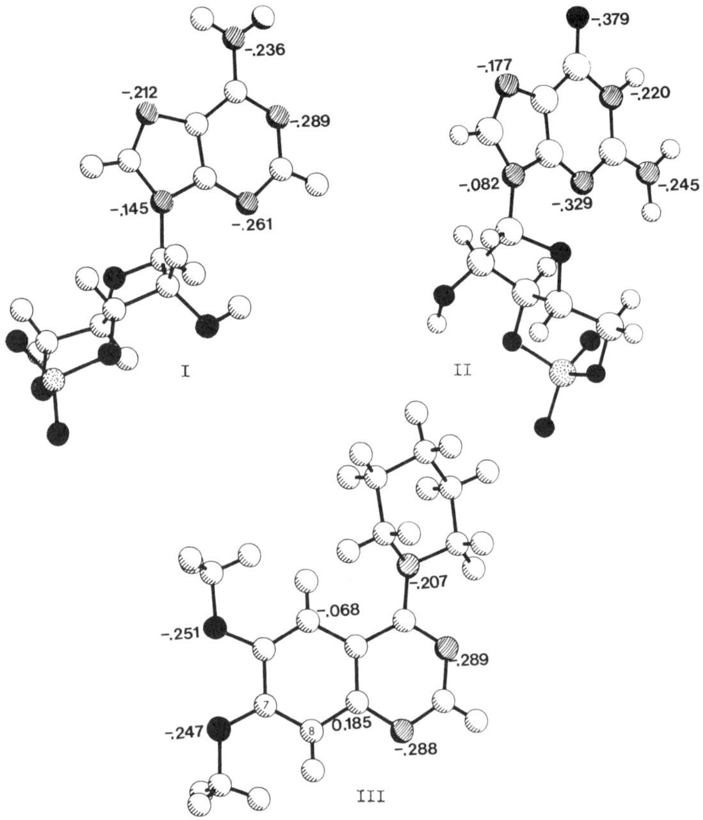

FIG. 6. cAMP (anti form; I), cGMP (syn form; II), and CP-13,509 (III) showing calculated electron densities.

The routine experimental determination of dipole moment magnitude and direction is difficult, but reliable theoretical methods are now available. Thus, for 9-methyladenine and 9-methylguanine, the calculated values were in good agreement with those obtained empirically,[49] though it must also be remembered that nucleotide base dipole moments are strongly influenced by rotation of the ribose moiety around the glycosidic bond.[50] Similar calculations were therefore undertaken with CP-13,509 based on molecular geometry obtained from an X-ray crystal analysis of a closely related quinazoline derivative. This revealed that the piperidine ring and the quinazoline nucleus, are not coplanar but lie at an angle of ca 20° to each other (Fig. 7). This is presumably due

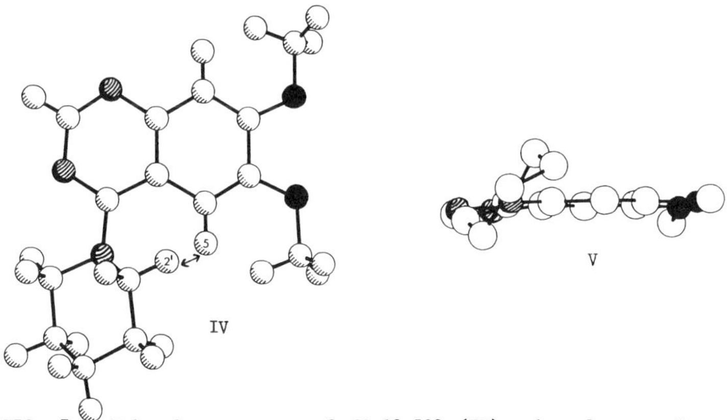

FIG. 7. Molecular geometry of CP-13,509 (IV) taken from an X-ray crystal analysis of a closely related analogue, showing the relative twist of the piperidine ring to the quinazoline nucleus (V).

to steric interactions between the 2′-protons of the piperidine and the 5-proton of the quinazoline. Further evidence that these protons are spatially close is provided by NOE experiments which show that irradiation of the 2′-equatorial protons leads to a 10% signal enhancement for the 5-proton. The calculated dipoles (μ) for CP-13,509 and the parent adenine portion of cAMP have virtually the same magnitude, but when these are aligned as shown (Fig. 8; orientation A), although the quinazoline and purine nuclei might be able to interact with similar binding sites on the enzyme, the piperidine ring is poorly positioned for interaction with any putative enzymic groups which recognise the ribose portion of the natural substrate. However, dipole alignment can still be maintained if the quinazoline is rotated 180° about the dipole axis (orientation B). In this case, the piperidine could clearly be occupying the ribose binding site. In particular, the carbonyl function present in buquineran could be approaching the putative nucleophilic group on the enzyme, possibly acting as a surrogate electrophilic centre in place of the natural phosphate moiety. In considering dipole moment alignment, it must be remembered that the strength of the dipole (inhibitor) - dipole (enzyme) interaction is proportional to the cosine of the angle between the two dipoles. Thus, while maximum interaction occurs when the two dipoles are antiparallel (cos 0°=1), an angle of 30° between the dipoles can still be accomodated (cos 30°=0.86) without the resulting strength of the interaction being severely reduced. By contrast to cAMP, the calculated magnitude of the parent heterocyclic system of cGMP was larger than that of cAMP. However, when the dipoles for CP-13,509 and cGMP are aligned, the piperidine ring is again

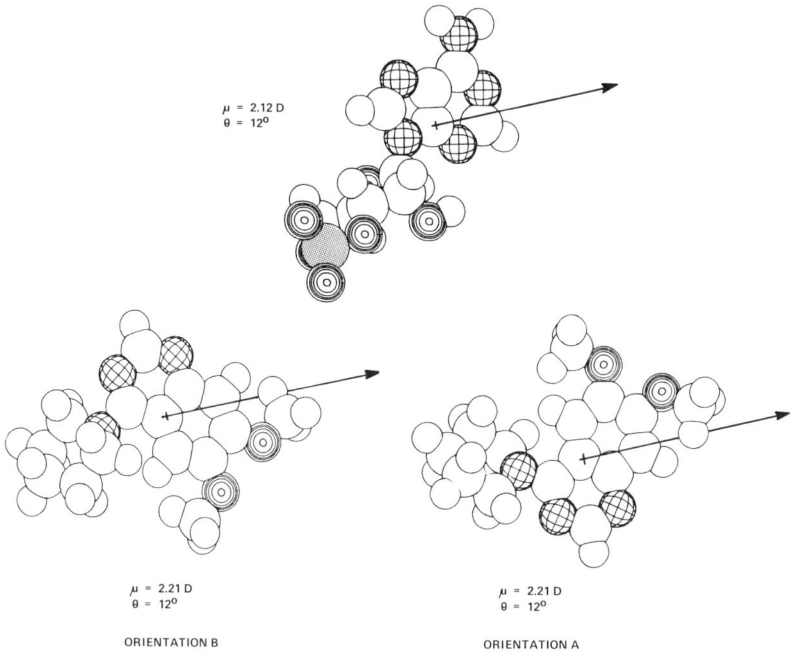

FIG. 8. Spacefill representations of cAMP (anti form) and CP-13,509 with calculated dipole moments (Debye) aligned parallel.

well positioned to interact with the putative ribose binding site of the enzyme (Fig. 9). Therefore inhibition of cGMP PDE might also be expected, and this was generally observed in the series, although selectivity for cAMP PDE was evident in every case.

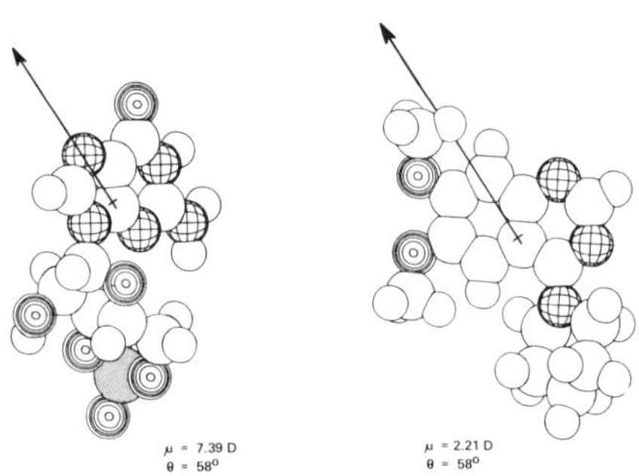

FIG. 9. Spacefill representations of cGMP (syn form) and CP-13,509 with calculated dipoles (Debye) aligned parallel.

4. Alternative heterocyclic derivatives as PDE inhibitors

Although the rationale for enzyme-inhibitor interactions had originally been based on quinazoline derivatives, it may also be applicable to related heterocyclic systems. Therefore, on the basis of our clinical experience with buquineran, we decided to seek an alternative but similar heterocyclic system. The phthalazine nucleus was chosen because the magnitude of its dipole moment was similar to that of the quinazolines, and could presumably also interact with any putative enzymic groups that recognise the purine portion of cAMP. Furthermore, the heterocyclic nucleus (pKa ca 5.0) would be unlikely to protonate at physiological pH. The phthalazine analogue of buquineran was synthesised and showed essentially similar inotropic activity. A series of phthalazine derivatives were therefore synthesised, from which carbazeran (UK-31,557) was chosen for clinical evaluation (Fig. 10). Carbazeran was a

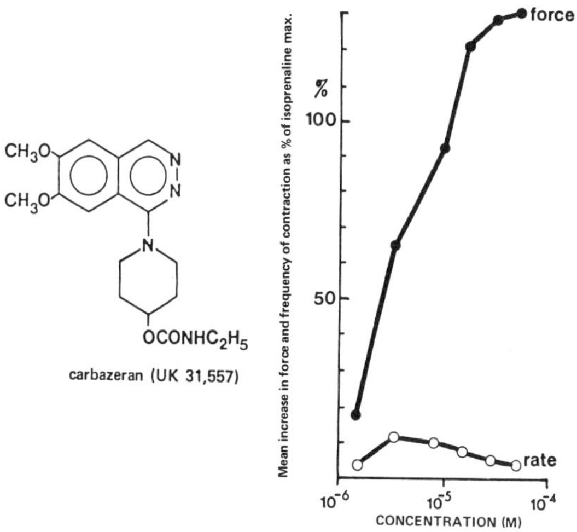

IC_{50} values
cAMP PDE 5μM
cGMP PDE 40μM

FIG. 10. Effect of carbazeran on the force and frequency of contraction in the guinea pig double atrium[16] at various concentrations.

more potent PDE inhibitor than buquineran with a better selectivity ratio, consistent with a moderate increase in inotropic activity. In addition, carbazeran was orally active in the dog (2.5mg kg^{-1} p.o.),[51] producing a marked increase in cardiac contractility with little effect on heart rate. Cardiac stimulant activity has also been demonstrated in man, and most importantly, carbazeran did not prolong the QT interval in the ECG complex,[25] thus satisfying one of our major objectives. However, the full potential of the compound in heart failure could not be exploited since the phthalazine nucleus proved to be a substrate for an enzyme oxidase system,[52] undergoing rapid

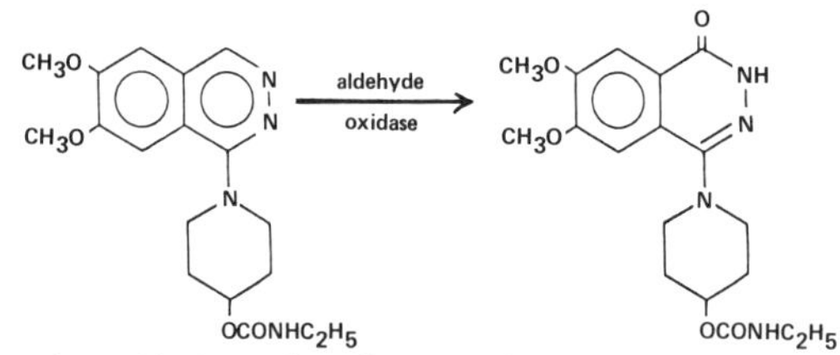

FIG. 11. Enzymic oxidation of carbazeran in man.

metabolism in man (Fig. 11). By contrast, metabolism via this route in rat and dog was not observed and in man appears to be only a minor pathway for related phthalazines such as hydralazine. One obvious way to overcome the problem was to introduce substituents into the 4-position of the phthalazine nucleus, and while this did afford metabolic stability, it could not be combined with the overall cardiovascular profile required.

5. Substitution of the piperidine with alternative groups

We were aware that some quinazoline derivatives were also substrates for this enzyme oxidase system,[53] but 4-piperidino substituted analogues were found to be stable, and effort was redirected towards this structural area. In developing a rationale for enzyme-inhibitor interactions, it has already been proposed that the ureido-function of buquineran could be occupying the cyclic phosphate recognition site. Since this may not be the optimal spatial arrangement, more flexibility was introduced into the system in order to

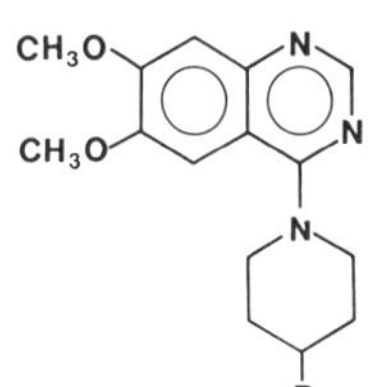

No.	R	inotropic potency vs buquineran
1	CH_3 $-CH_2NCONHC_4H_9$	5.5
2	CH_3 $-CH_2NCONHC_3H_7$	7.5
3	CH_3 $-CH_2CH_2NCONHC_2H_5$	4.1
4	$-CH_2CH_2CON(CH_3)_2$	4.3
5	CH_3 CH_3 $-CHCH_2NCOCH_3$	5.0

TABLE 1. Importance of variation of the piperidine 4-substituent on inotropic potency in the anaesthetised dog[18] (buquineran potency = 1) at doses of $0.05-0.25 mg kg^{-1}$.

optimise the position of the electrophilic group relative to the piperidine ring (Table 1). Inotropic potency was immediately improved when a methylene bridge was introduced (Nos. 1,2) and this was maintained when a two carbon chain was used. For example, the ureido-derivative (No. 3) and the amido-analogues (Nos. 4,5) also demonstrated good inotropic activity in the anaesthetised dog.

In attempting further to improve inotropic and PDE inhibitory potency, we next considered alternative electrophilic groups which might interact with the phosphate binding site of the enzyme in a similar fashion to cAMP. The trigonal planar carbonyl function of the ureido- and amido-derivatives may well be suboptimal since the phosphate group of cAMP is tetrahedral. Various replacements for the phosphate moiety were therefore examined, and the sulphonyl group was considered to be particularly appropriate since it possesses tetrahedral geometry. In addition, adenosine-3′,5′-cyclosulphate[54] in which the charged phosphate group of cAMP is replaced by a neutral cyclic sulphate moiety, is still a good substrate for PDE. Therefore it appears that the negatively charged phosphate group or the neutral sulphate moiety can be accommodated by the enzyme. This geometry was therefore incorporated via a number of sulphonamide derivatives and again good inotropic potency was observed in vivo together with selective inhibition of cAMP PDE in vitro (Table 2). In general N-methylated compounds (Nos. 2,6) were more potent as inotropes than their secondary analogues (Nos. 1,3) and further increases in inotropic activity were obtained when a two carbon bridge was introduced

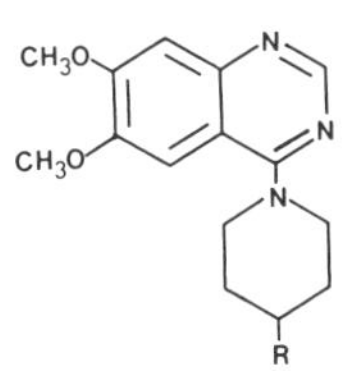

No.	R	Inotropic potency vs buquineran	IC_{50}	
			cAMP PDE	cGMP PDE
1	$-NHSO_2CH_3$	0.5	10.8	55
2	$-N(CH_3)SO_2CH_3$	4.7	8.8	27.0
3	$-CH_2CH_2NHSO_2CH_3$	4.6	3.6	26.0
4	$-CH_2CH_2N(CH_3)SO_2CH_3$	7.0	1.2	3.4
5	$-CH(CH_3)CH_2N(CH_3)SO_2CH_3$	7.3	1.5	7.5
6	$-CH_2CH_2N(CH_3)SO_2CH_3$	8.6	2.5	10.8
7	$-CH(CH_3)CH_2N(CH_3)SO_2C_2H_5$	10.4	1.1	8.2

TABLE 2. Importance of variation of the piperidine 4-substituent on inotropic potency in the anaesthetised dog[18] (buquineran potency = 1) at doses of 0.05-0.25mgkg^{-1} and IC_{50} values (μM) vs beef heart PDE (commercial crude enzyme from Sigma).

between the piperidine and sulphonamide groups (Nos. 3,4,6). Branching of the alkylene chain by a methyl group gave the most active compound of the series (No. 7) and this was also the most potent inhibitor of cAMP PDE. In general, both cAMP and cGMP PDE inhibitory activity were increased in this series but selectivity for the former enzyme was always maintained. Unexpectedly, the two most potent compounds (Nos. 6,7) were only very poorly active in the conscious dog after oral administration (2mg kg^{-1} p.o.), and this may reflect poor bioavailability or rapid metabolic clearance. Although inotropic potency is obviously important, the selection of compounds for further progression depends also on their overall pharmacological profile. Thus, UK-36,327 (Table 1; No. 5) was chosen for development since it is five times more potent than buquineran in both anaesthetised[55] and conscious dogs[56] (oral administration) and has little propensity to increase heart rate either in the whole animal or in in vitro test systems (Fig. 12). cAMP PDE inhibitory potency was also enhanced and enzyme selectivity improved. Furthermore the QT interval of the ECG complex is not affected in the anaesthetised dog or in man, and the compound has demonstrated efficacy in heart failure patients in recent clinical trials.

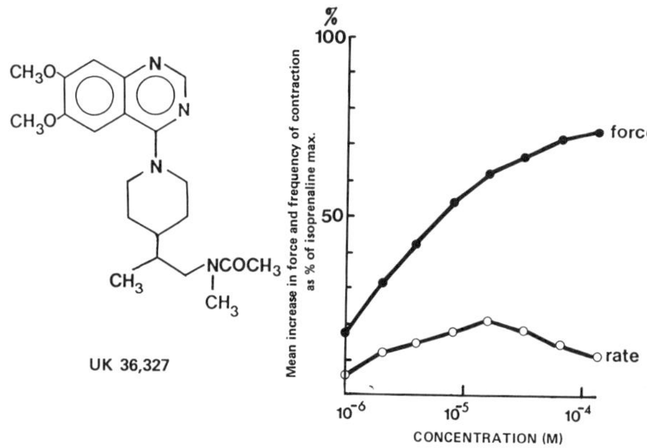

IC_{50} values
cAMP PDE 3.8μM
cGMP PDE 30μM

FIG. 12. Effect of UK-36,327 on the force and frequency of contraction in the guinea pig double atrium[16] at various concentrations.

Finally, in attempting to put our programme into perspective, conscious dog oral activity data after various doses of buquineran, carbazeran and UK-36,327 have been summarised (Fig. 13) confirming their relative inotropic potencies (1:2:5) and demonstrating selective inotropic response with minimal effect on heart rate. All three compounds are potent inhibitors of both cAMP and cGMP PDE but with selectivity for the former enzyme and probably exert their inotropic and vasodilator actions at least in part via this mechanism.

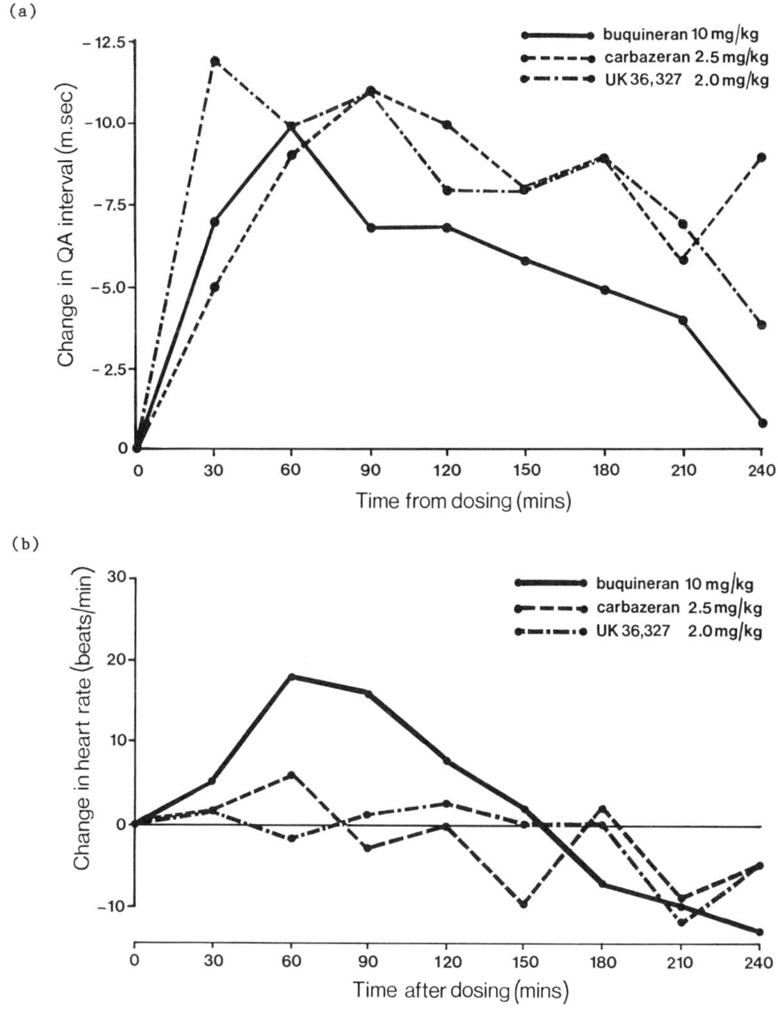

FIG. 13. Effects of buquineran, carbazeran and UK-36,327 at various oral doses on (a) QT interval[57] and (b) heart rate, in the conscious dog.

6. Summary

The work presented here describes SARs for a series of quinazoline and phthalazine derivatives both as cardiac stimulants and PDE inhibitors. Various compounds have been described which are effective inotropes having little intrinsic propensity to increase heart rate, and this should be of major clinical benefit. Some suggestions have been presented as to how cAMP, cGMP and the quinazoline derivatives could interact with phosphodiesterase, and these considerations have influenced and guided our synthetic programmes. In particular, in these unprotonated systems, dipole moment (inhibitor) - dipole moment (enzyme) interaction may be a key recognition event, which could have significance in other structural areas. Finally, substantial improvements in force/rate selectivity and oral activity have been realised over original prototypes, and agents from this mechanistic class may have major potential for the treatment of heart failure.

7. Acknowledgements

The author wishes to express his thanks to colleagues in the Medicinal Biology Department (Drs. K.J. Blackburn, J.R. Joice and colleagues) without whose assistance this work could not have been completed. The theoretical calculations were performed with the assistance of Drs. M.S. Tute and M. Snarey who also provided invaluable help during informal discussions. The NOE NMR measurements were carried out by Dr. M. Kinns.

References

[1] K.T. Weber, Am. J. Med., 1982, 72, 655.
[2] A. Breckenridge, Br. Med. J., 1982, 284, 765.
[3] E.W. Sutherland, G.A. Robinson, R.W. Butcher, Circulation, 1968, 37, 279.
[4] L.H. Opie, Cardiovascul. Res., 1982, 16, 483.
[5] L.H. Opie, Eur. Heart J., 1983, 4, 199.
[6] J.C. Stoclet and C. Lugnier, Med. Chem. Adv. Proc. Int. Symp. - 7th, 1980, 435.
[7] M. Schramm, G. Thomas, R. Towart and G. Franckowiak, Nature, 1983, 303, 535.
[8] M.S. Amer and G.R. McKinney, Life Sciences, 1973, 13, 753.
[9] M.S. Amer and W.E. Kreighbaum, J. Pharm. Sci., 1975, 64, 1.
[10] N. Himori and N. Taira, Naunyn-Schmiedeberg's Arch. Pharmacol., 1976, 294, 34.
[11] J.B. Polson, J.J. Krzanowski, W.H. Anderson, D.F. Fitzpatrick, D.P.C. Hwang and A. Szentivanyi, Biochem. Pharmacol., 1979, 28, 1391.
[12] J.N. Wells and J.R. Miller, Trends in Pharmacol. Sci., 1983, 4, 385.
[13] M.W. Osborne, J.J. Wenger and R.A. Moe, J. Pharmacol. Exp. Ther., 1971, 176, 174.
[14] P. Honerjäger, M. Schäfer-Korting and M. Reiter, Naunyn-Schmideberg's Arch. Pharmacol., 1981, 318, 112. 1
[15] T. Kariya, L.J. Wille and R.C. Dage, J. Cardiovascul. Pharmacol., 1982, 4, 509.

[16] Inotropic response was measured in spontaneously beating guinea pig double atria suspended at 32°C in Kreb's solution under 2g resting tension. Force was recorded isometrically and rate increases were measured by a ratemeter.

[17] Several members of this structural series were examined and all the compounds demonstrated competitive kinetics.

[18] Left ventricular pressure was measured in a sodium pentobarbitone anaesthetised beagle ($30 mgkg^{-1}$ parenteral administration) using a Miller tip pressure transducer in the left ventricle inserted via the carotid artery.

[19] C.T. Alabaster, K.J. Blackburn, J.R. Joice, R. Massingham and P.C. Scholfield, 1977, Br. J. Pharmacol., 1977, 60, 284P.

[20] F. Follath, F. Kersting, G.R.J. Lewis, R.J. Walden, N.N. Woolhouse and C.T. Dollery, Clin. Pharmacol. Ther., 1976, 20, 24.

[21] D.E. Jewitt, P.G. Jackson, D. Kitson and G. Jackson, Br. J. Clin. Pharmacol., 1978, 5, 7.

[22] D.E. Jewitt, P.G. Jackson, M. Monaghan and K. Jennings, Br. J. Clin. Pharmacol., 1978, 5, 13.

[23] I. Hutton, W.S. Hillis, C.E. Langham, J.M. Conely and T.D.V. Lawrie, Br. J. Clin. Pharmacol., 1977, 4, 513.

[24] S.H. Taylor, B. Silke and G.I.C. Nelson, Br. J. Clin. Pharmacol., 1982, 13, 590P.

[25] W.S. Hillis and M. Been, Eur. Heart J., 1982, 3, 97.

[26] D.E. Jewitt, K. Jennings and P.G. Jackson, Am. J. Medicine, 1978, 65, 197.

[27] E.M.V. Williams, Br. Heart J., 1982, 47, 513.

[28] N.D. Goldberg, M.K. Haddox, S.E. Nicol, D.B. Glass, C.H. Sanford, F.A. Kuehl and R. Estensen, Adv. Cyclic Nucleotide Res., 1975, 5, 307.

[29] G. Lowe, Acc. Chem. Res., 1983, 16, 244.

[30] P.J.J.M. Van Ool and H.M. Buck, Eur. J. Biochem., 1982, 121. 329.

[31] M. Landt and L.G. Butler, Biochemistry, 1978, 17, 4130.

[32] F.A. Cotton, E.E. Hazen and M.J. Legg, Proc. Natl. Acad. Sci. USA, 1979, 76, 2551.

[33] B. Springs and P. Haake, Tetrahedron Lett., 1977, 37, 3223.

[34] V. Heinemann and W. Saenger, Nature, 1982, 299, 27.

[35] N. Yathindra and M. Sundaralingam, Biochem. Biophys. Res. Commun., 1974, 56, 119.

[36] G.V. Fazakerley, J.C. Russell and M.A. Wolfe, Eur. J. Biochem., 1977, 76, 601.

[37] A.M. Gronenborn and G.M. Clore, Biochemistry, 1982, 21, 4040.

[38] K. Watenpaugh, J. Dow, L.H. Jensen and S. Furberg, Science, 1968, 159, 206.

[39] K.I. Varughese, C.T. Lu and G. Kartha, J. Am. Chem. Soc., 1982, 104, 3398.

[40] S.S. Tavale and H.M. Sobell, J. Mol. Biol., 1970, 48, 109.

[41] G. Michal, K. Mühlegger, M. Nelboeck, C. Thiessen and G. Weimann, Pharmacol. Res. Comm., 1974, 6, 203.

[42] M. Sundaralingam and J. Abola, J. Am. Chem. Soc., 1972, 94, 5070.

[43] M.E. Druyan, M. Sparagana and S.W. Petersen, J. Cyclic Nucleotide Res., 1976, 2, 373.

[44] J.P. Miller, K.H. Boswell, K. Muneyama, R.KL. Tolman, M.B. Scholten, R.K. Robins, L.N. Simon and D.A. Shuman, Biochem. Biophys. Res. Commun., 1973, 55, 843.

[45] Q-H. Tran-Thi, D. Franzen and F. Seela, Angew. Chem. Int. Ed. Engl., 1982, 21, 367.

[46] J.P. Miller, L.F. Christensen, T.A. Andrea, R.B. Meyer, S. Kitano and Y. Mizuno, J. Cyclic Nucleotide Res., 1978, 4, 133.

[47] R.A. Olsson, "Regulatory Function of Adenosine" Martinus Nijhoff, Boston, 1983, Vol. 2, Chapter 3, P. 33.

[48] E.J. Lien, Z-R. Guo, R-L. Li and C-T. Su, J. Pharm. Sci., 1982, 71, 641.

[49] H. DeVoe and I. Tinoco, J. Mol. Biol., 1962, 4, 500.

[50] H. Berthod and B. Pullman, Biochem. Biophys. Res. Commun., 1972, 46, 125.

[51] C.R. Taylor, J.R.C. Baird, K.J. Blackburn, D. Cambridge, J.W. Constantine, M.S. Ghaly, M.L. Hayden, H.M. McIlhenny, P.F. Moore, A.Y. Olukotun, L.G. Pullman, D.S. Salsburg, C.A.P.D. Saxton and S. Shevde, Am. Heart J., 1982, 102, 515.

[52] B. Kaye, J.L. Offerman, J.L. Reid and S. Hillis, manuscript in preparation.

[53] C. Stubley, J.G.P. Stell and D.W. Mathieson, Xenobiotica, 1979, 9, 475.

[54] E.S. Severin, V.A. Tkachuk and N.N. Gulyaev, Biokhimiya, 1976, 41, 384.

[55] P. Ellis and J.R. Joice and M.J. Randall, Br. J. Clin. Pharmacol., 1983, 15, 125P.

[56] C.T. Alabaster, C.G. Henderson and J.R.C. Baird, Br. J. Clin. Pharmacol., 1983, 15, 124P.

[57] Non-invasive measurement of cardiac contractility as a systolic time interval.

The Cardiac Bipyridines—A New Chemical Entity with Distinct Positive Inotropic and Vasodilatory Activity

By A. A. Alousi*, L. H. Walton, G. Y. Lesher, and A. E. Farah

STERLING–WINTHROP RESEARCH INSTITUTE, RENSSELAER, N.Y. 12144–3493, U.S.A.

The intensive search for new chemical entities with positive inotropic properties has led to the discovery of the cardiac bipyridines.[1-5] The introduction of these bipyridines into clinical practice has stimulated exciting research in the pharmacotherapy of congestive heart failure,[6-11] a disease state in which there is critical reduction in cardiac output.[9] Among the several physiological determinants of cardiac output are inotropy and afterload.[12] The cardiac bipyridines have been shown to improve cardiac output in patients[6-11] and experimental models[3,4,13,14] of heart failure by both enhancing inotropy and reducing afterload. The chemistry, pharmacological actions, and therapeutic indications of the cardiac bipyridines will be briefly reviewed in this article.

<u>Chemical Structure and Name</u>

AMRINONE

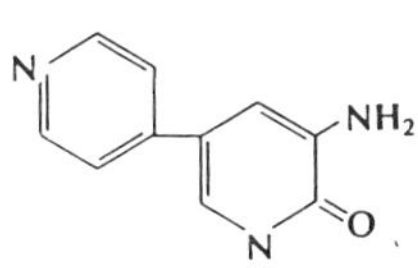

5-Amino [3,4′-Bipyridine] -6(1H)-one

MILRINONE

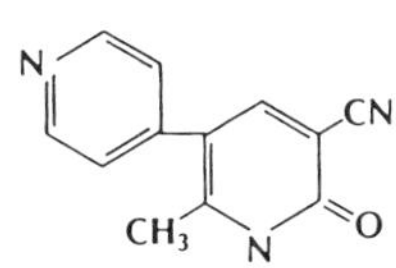

1,6-Dihydro-2-methyl-6-oxo- [3,4′-bipyridine]-5-carbonitrile

FIGURE 1

Chemical Synthesis

The parent cardiac bipyridine, Win 40680, amrinone, has been prepared by several different chemical routes. The most efficient way is outlined in Figure 2. The imino-aldehyde was synthesized from 4-picoline using the Vilsmeier-Haack reaction conditions of Arnold.[15] This aldehyde was condensed with cyano-acetamide to construct the pyridone ring. The initial group is then hydrolysed to an amide which in turn is converted to the

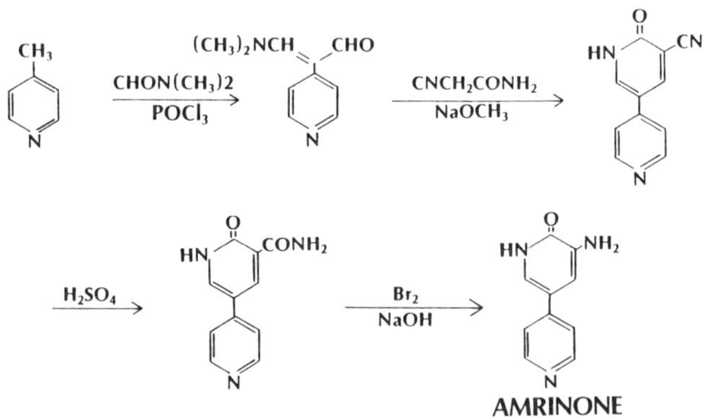

FIGURE 2

amine *via* a Hofmann reaction as shown in Figure 2. Similarly, the preparation of Win 47203, milrinone, starts with 4-picoline as shown in Figure 3. An ester condensation gives the pyridyl acetone[16] which is then condensed with dimethylformamide dimethyl acetal to give the imino-ketone. Finally, the pyridone ring is obtained by condensation with cyanoacetamide (Figure 3).

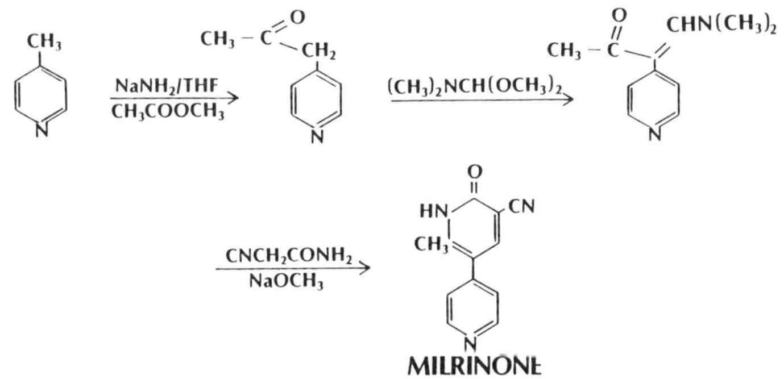

FIGURE 3

Cardiovascular Actions

Effects On The Heart. The cardiac actions of the bipyridines were studied in numerous experimental models including healthy cardiac tissues and experimentally induced heart failure.[3,4,13,14,17-22]

Initially, the positive inotropic and chronotropic activity of amrinone was noted in our *in vitro* primary screening test. In this test we utilize cat or guinea pig spontaneously beating right atria and electrically stimulated papillary muscles.[3,5] The relationship between the positive inotropic and chronotropic effect of a drug can be expressed as the tension-rate index as described below:

$$\text{Tension-rate index} = \frac{\%\Delta \text{in papillary muscle developed tension}}{\%\Delta \text{in right atrial rate}}$$

A drug with a tension-rate index ≥ 3 is usually considered of potential interest as a cardiotonic agent. In our laboratories the relative potency of an inotropic agent is determined by the magnitude of the papillary muscles' inotropic response to

100 μg/ml as compared to that produced by the same concentration of amrinone (110 ± 9.9 % increase in guinea pig papillary muscle developed tension).

Figure 4 illustrates the relative inotropic and chronotropic potency of amrinone and milrinone in isolated guinea pig right atria and papillary muscles. Milrinone, in concentrations of 0.1 to 100 μg/ml, caused 24±2.3 to 246 ± 26.4 % increase in papillary muscle developed tension, a 26 ±2.6 to 234 ± 16.1 % increase in right atrial developed tension and a 7 ± 2.0 to 78 ± 5.0 % increase in right atrial rate. By comparison amrinone in concentrations of 3.0 to 300 μg/ml caused increases in papillary muscle developed tension of 25 ± 2.7% to 187 ± 20.3%, increases in right atrial developed tension of 19 ± 5.8% to 179 ± 14.6%, and increases in right atrial rate of 5 ± 1.5% to 44 ± 5.4%.

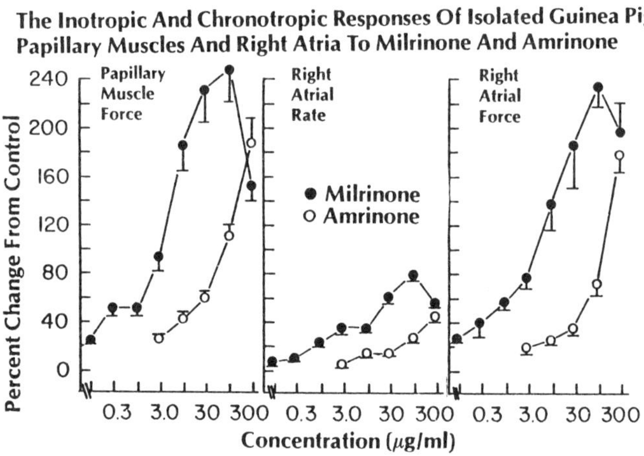

FIGURE 4

In this preparation milrinone was approximately 30 times more potent than amrinone. At a concentration of 10 μg/ml the tension-rate index for amrinone was 4.3; the tension-rate index for milrinone was 5.6. Not only is milrinone more potent than amrinone but the increases in papillary muscle developed tension relative to right atrial rate increases are greater.

Tables 1 and 2 illustrate the structure activity relationships of amrinone and milrinone analogues.

TABLE I

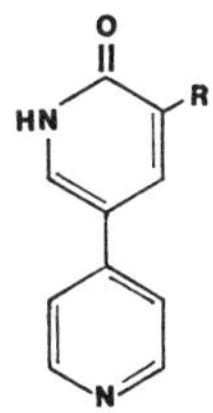

R	RELATIVE POTENCY	R	RELATIVE POTENCY
* NH_2	1	CH_2NH_2	0
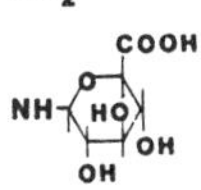	1/2	OH	1
$NHCOCH_3$	1/4	CN	1/2
$NHCH_3$	1	Br	1/2
$N(CH_3)_2$	1	H	1/2

* AMRINONE

Table 1 shows that the N-glucuronide and the N-acetyl derivatives (both metabolites) are somewhat less active than amrinone. The methylamino and dimethylamino derivatives are equal in activity to amrinone (Table 1), but the corresponding derivatives in the milrinone series (Table 2) are less active than the primary amine. There are other discrepancies between these two series; e.g. the nitrile and the 3-unsubstituted analogues in the amrinone series are less potent than some of the other compounds in this series, while in the milrinone series the nitrile and 3-unsubstituted derivatives are among the most potent derivatives. Notice in the milrinone series that larger alkyl groups at the 6-position result in large drops in potency.

TABLE 2

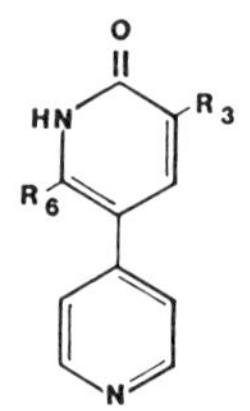

R_3	R_6	RELATIVE POTENCY	R_3	R_6	RELATIVE POTENCY
* NH_2	H	1	o CN	CH_3	30
NH_2	CH_3	30	Br	CH_3	2
$NHCH_3$	CH_3	2	H	CH_3	30
$N(CH_3)_2$	CH_3	1	H	C_2H_5	1
OH	CH_3	10	H	C_3H_7	1

* AMRINONE   o MILRINONE

The onset of the in vitro effects of each of the bipyridines was rapid with a peak effect at 2-3 minutes and a duration greater than 30 minutes. The bipyridines, especially milrinone, seem to bind tightly to cardiac tissues and repeated washes over a period of one hour or longer are necessary to restore tissue responsiveness to subsequent doses.[5] No sign of tachyphylaxis or hypersensitivity was observed after repeated treatment of the isolated tissues with these cardiac bipyridines, and depletion of cardiac norepinephrine by pretreatment of the animal with reserpine did not modify the in vitro effects of amrinone or milrinone.[3,5] Investigation of the effects of the cardiac bipyridines on α-and β- adrenergic, cholinergic and histaminergic receptors in the heart showed no significant stimulation or inhibition of these receptors by either bipyridine.[3,5] Similarly, inhibition of prostaglandin synthetase with indomethacin did not enhance the in vitro tissue response to milrinone.[5]

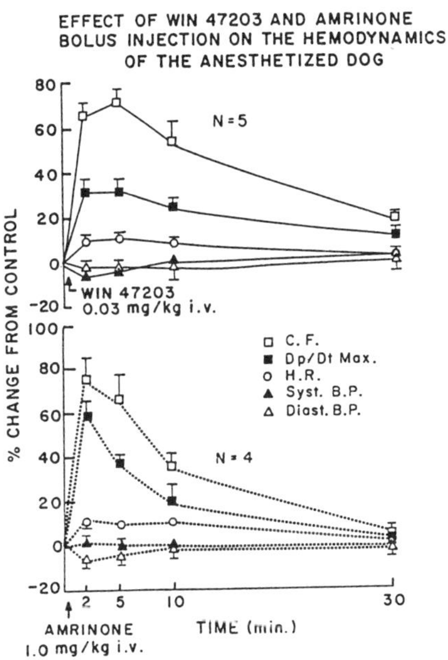

FIGURE 5

In the whole animal, dose-dependent increases in cardiac contractile force and left ventricular dp/dt max were observed after the intravenous administration of amrinone or milrinone.[3,4,20,21] Figure 5 is a time course of the response of the anesthetized dog to single bolus injection of amrinone or milrinone.

To maintain the positive inotropic response, amrinone (10 to 100 μg/kg/min) and milrinone (0.1 to 1 μg/kg/min) were infused intravenously in anesthetized dogs, cats or monkeys for several hours. The onset of action was within 5-10 minutes and the peak effect was reached within 30 minutes. The inotropic and hemodynamic effects were maintained during the infusion period with no sign of tachyphylaxis.[3,4] Similar responses were observed in unanesthetized animals.[3,4]

The cardiac bipyridines are also active when administered orally to unanesthetized dogs. The oral potency of milrinone was about 10 times that of amrinone. Both drugs have rapid onset of action and a total duration greater than 5 hours.[3,4]

Experimental models of heart failure such as the spontaneous or pentobarbital-induced failure in the heart-lung preparation[22], drug-induced failure in anesthetized dogs,[3,4,13] ischemic heart failure preparations [21] and the tamponade-induced failure[20,23] model were used to demonstrate the efficacy of the cardiac bipyridines on the failing heart. In the heart-lung preparation,[22] amrinone caused significant reduction in right atrial pressure and increase in cardiac output. In the propranolol-or verapamil-induced heart failure model in anesthetized dogs, milrinone caused increases in cardiac contractile force, left ventricular dp/dt max, and cardiac output with significant reduction in right atrial and pulmonary artery pressure and in the calculated total peripheral resistance.[4] Figure 6 illustrates the effects of milrinone on propranolol-induced heart failure. Similar results were observed in other experimental models of heart failure.[14,20,23]

The Effect Of Milrinone On Propranolol — Induced Heart Failure In The Anesthetized Dog

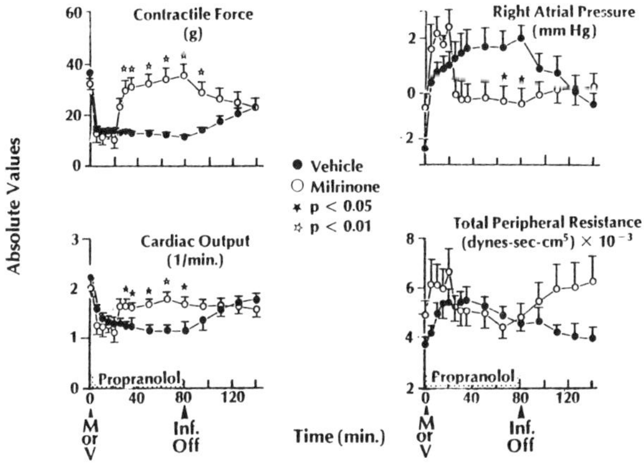

FIGURE 6

<u>Effects on Circulation</u>. In the intact anesthetized animal, the administration of amrinone in doses of 3 to 30 mg/kg i.v. or milrinone in doses of 0.3 to 3 mg/kg i.v. caused significant, dose-related lowering in systolic and diostolic blood pressure and in total peripheral resistance[3,4] suggesting a direct relaxation effect on the smooth muscle vasculatures. This was confirmed later in the isolated denervated canine hind-limb preparation. In this preparation, the intra-arterial injection of minute doses of the bipyridines caused significant reduction in hind-limb perfusion pressure. No difference in the responses was observed between the innervated and denervated preparation indicating the lack of autonomic reflex effect.[24] Misherie <u>et al</u>.[25] demonstrated

the direct relaxing effect by amrinone on isolated aortic and taenia coli strips incubated in vitro and preconstricted with norepinephrine or high K+.

In the Langendorff preparation, perfusion with amrinone led to significant increases in coronary blood flow.[17,26] Eizing et al.,[27] using anesthetized dogs, observed increases in regional blood flow to the renal cortex, spleen, and liver, and decreases in regional flow to intestines, gallbladder, triceps and the brain. Data on the effect of amrinone on myocardial blood flow and endo/epicardial ratio were variable. Grupp et al.[20] demonstrated 30% improvement in endocardial blood flow, while Eizing et al.[27] reported a lowering in endo-/epicardial ratio and Millard et al.[23] observed increases in regional myocardial blood flow without significant changes in transmural blood flow or endo-/epicardial ratio.

Mechanism of Action

The mechanisms involved in the inotropic and vasodilator actions of the cardiac bipyridines have been studied by several investigators.[3,4,25-36] Initial studies in our laboratory indicated that amrinone was not a β-agonist and it did not inhibit Na+, K+ -ATPase.[3] Parker and Harper[29] reported increases in Ca^{2+} influx in dog erythrocytes incubated in low Na+ in the presence of amrinone. He concluded that amrinone stimulates the Na+, Ca^{2+} exchange mechanism in his preparation. In studies with the bioluminescent protein aequorin, Morgan et al.[30] demonstrated modification of the pattern of Ca^{2+} transients produced by amrinone in cat papillary muscle. There was a prolongation of the liberation of Ca^{2+} by the sarcoplasmic reticulum and an increase in the rate of Ca^{2+} sequestration. In isolated smooth muscle strips, amrinone caused relaxation of norepinephrine or K+ contracted muscle. In these preparations, amrinone caused reduction in the norepinephrine-induced increase in Ca^{2+} influx without affecting K+ induced Ca^{2+} influx.[25] Studies from our laboratories indicate that in guinea pig papillary muscle, the increase in muscle developed tension caused by amrinone was dependent on extracellular Ca^{2+} concentration and on the rate of stimulation (Hz). These studies suggest an enhancement of Ca^{2+} influx and/or the stimulation of intracellular Ca^{2+} release by amrinone.[33] In isolated cardiac myocytes, Frangakis et al.[34] demonstrated an increased rate of Ca^{45} exchange in the presence of amrinone. This was dose-dependent and was influenced by extracellular Ca^{2+} concentration. A very high Ca^{2+} concentration in the extracellular fluid (7.5mM) caused inhibition of the inotropic response of the cat papillary muscle to amrinone.[18] In the K+ depolarized guinea pig papillary muscle, amrinone restored muscle contraction.[33]

The slow channel appears to have some role of the inotropic response of cardiac tissue to the bipyridines. Pretreatment of guinea pig papillary muscles with verapamil, at a concentration which reduced initial muscle tension by 90%, caused no significant changes in the dose-response curve to amrinone.[33] However, pretreatment of the papillary muscle with nifedipine or nimodipine caused significant reduction in the inotropic response to milrinone (Alousi et al., unpublished data). In the whole animal, severe heart failure induced with large doses of verapamil was reversed with intravenously administered milrinone.[4] It is obvious that clarification of the exact effect(s) of the bipyridines on Ca^{2+} influx, release and modulation requires further investigation.

Increase in cardiac tissue C-AMP was observed prior to the increased muscle tension caused by the catecholamines.[37] In the presence of the cardiac bipyridines, increases in smooth muscle C-

AMP level was first observed by Meisleri et al.[25] Later, an inhibition of cardiac phosphodiesterase and an increase in cardiac C-AMP level caused by amrinone was also reported.[31,32] This was contrary to previous reports.[3,26] Further investigation of the time course and concentration dependence of the changes in cardiac C-AMP with amrinone and milrinone revealed a different temporal pattern than that observed with the β-agonists as shown in Figure 7.

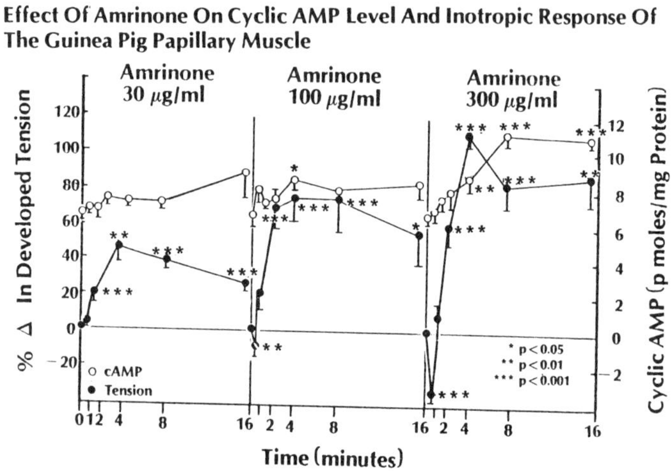

FIGURE 7

Doses of amrinone (30 μg/ml) that caused significant increase in guinea pig papillary muscle developed tension caused no significant increase in C-AMP. Larger increases in muscle developed tension were accompanied by but not preceded by increases in muscle C-AMP. Similar results were obtained with milrinone.[4] The cardiac bipyridines may increase C-AMP secondary to increased muscle tension which may be related to calcium mobilization. To date we have not identified differences in Ca^{2+}

fluxes relative to one bipyridine rather than another (amrinone vs. milrinone). The question of cellular compartmentalization and possible local changes in C-AMP will be addressed in the future.

Therapeutic Indications

Clinical trials designed to test the therapeutic efficacy of amrinone and milrinone have been conducted in patients with New York Heart Association functional Class II and IV congestive heart failure caused by idiopathic cardiomyopathy, ischemic heart diseases or valvular insufficiency. All patients studied were maintained on digitalis and diuretics. Some patients required further management with antiarrhythmics, vasodilators, anti-coagulants and/or other drug regimens. Acute hemodynamic improvement was observed after intravenous bolus or infusion of amrinone or milrinone.[6-11] In patients with left and right heart catheterization, Benotti et al.[6] demonstrated significant increases in cardiac index and left ventricular dp/dt with significant reduction in left ventricular end-diastolic pressure, pulmonary capillary wedge pressure and right atrial pressure after a single bolus dose of amrinone (1.5 to 3.5 mg/kg). Patients with right heart catheterization showed similar changes in cardiac index and a significant reduction in left ventricular pressure and mean endocardial circumferential fiber shortening after bolus and infusion of amrinone.[7] Bolus injection of milrinone caused similar effects on cardiovascular parameters in doses of 25-175 μg/kg.[10,11] The intravenous administration of these cardiac bipyridines caused an almost immediate improvement in cardiac performance that lasted for more than one hour.[6-8] There was no attenuation in the efficacy after sequential administration of the drug or its prolonged intravenous administration.[6,39]

Oral administration of amrinone[38-44] or milrinone[10,11] caused sustained beneficial effects. Amrinone, in doses of 50 to 300 mg t.i.d., caused significant increases in cardiac index and stroke volume with significant reduction in pulmonary wedge pressure, mean right atrial pressure and systemic vascular resistance.[33-44] The administration of milrinone (10 to 50 mg total daily dose) caused similar sustained effects.[10,11] Weber and others have reported that long-term therapy with the cardiac bipyridines produced significant improvement in exercise tolerance time and the quality of life with a reduction in numbers and/or doses of concomitant medications.[41,43,44]

Side Effects

Amrinone. In more than 500 patients treated with amrinone the incidence of reduction in platelet count was approximately 15%. The reduction in count was reversible and was not accompanied by haemorrhage or petechia. A similar incidence of gastrointestinal disturbance was observed. In most instances the gastric irritation was reduced or eliminated with dose-adjustment, antacids, and/or administration with meals. Hepatic enzyme changes and cardiac arrhythmias were also noted in a few patients; 2% experienced abnormal hepatic function, and 3% were reported to have arrhythmias.

Milrinone. No serious side effect has been observed to date with milrinone. Clinical studies were initiated in October of 1982.

Summary

In summary, the cardiac bipyridines, amrinone and milrinone, represent a new class of chemical agents with inotropic/vasodilatory activity. They exhibit a unique pharmacological profile and mechanism of action. Although there is no well-defined structure-activity relationship among the bipyridines tested, substitution of a methyl group for the H in the C_6 position in the structure of amrinone led to a significant increase in the potency of the compounds.

Clinical experiences with amrinone have resulted in useful therapeutic applications; however, the drug is not free from troublesome but reversible side effects. To date no significant side effects have been observed with milrinone. Extensive clinical evaluation of the latter should determine the therapeutic value of this class of compounds in the treatment of congestive heart failure.

BIBLIOGRAPHY

1. A. Alousi, A. Farah, G. Lesher and C. Opalka, Federation Proceedings, 1978, 37, 914.
2. A. Farah and A. Alousi, Life Sciences, 1978, 22, 1139.
3. A. Alousi, A. Farah, G. Lesher and C. Opalka, Circulation Research, 1979, 45, 666.
4. A. Alousi, J. Canter, M. Montenaro, D. Fort and R. Ferrari, J. Cardiovascular Pharmacology, 1983, 5, 792.
5. A. Alousi, G. Stankus, J. Stuart and L. Walton, J. Cardiovascular Pharmacology, 1983, 5, 804.
6. J. Benotti, W. Grossman, E. Braunwald, D. Davolos and A. Alousi, New England J. Medicine, 1978, 299, 1373.
7. T. LeJemtel, E. Kung, E. Sonnenblick, H. Ribner, M. Matsumoto, R. Davis, W. Schwartz, A. Alousi and D. Davolos, Circulation, 1979, 59, 1098.
8. K. Weber, V. Andrews, J. Janicki, J. Wilson and A. Frishman, American J. Cardiology, 1981, 48, 164.
9. S. Siskind, E. Sonnenblick, R. Forman, J. Scheuer and T. LeJemtel, Circulation, 1981, 64, 966.
10. C. Maskin, L. Sinoway, B. Chadwick, E. Sonnenblick and T. LeJemtel, Circulation, 1983, 67, 1065.

11. A. McDowell, D. Baim, J. Cherniles, T. Bekele and E. Braunwald, J. American College of Cardiology, 1983, 1, 675.

12. E. Braunwald, "Heart Disease", W.B. Saunders Company, Philadelphia, (1st.Ed. 1980) Vol. 1, Chapter 13, p.453.

13. A. Alousi, D. Fort, Circulation, 1979, 60, II-16.

14. J. Jentzer, T. LeJemtel, E. Sonnenblick and E. Kirk, American J. Cardiology, 1981, 48, 75.

15. Z. Arnold, Collect. Czech. Chem. Commun., 1963, 28, 863.

16. S. Raymond and R. Levine, J. American Chemical Society, 1960, 82, 472.

17. G. Onuaguluchi and R. Tanz, J. Cardiovascular Pharmacology, 1981, 3, 1342.

18. S. Rendig, E. Amsterdam and D. Mason, Circulation, 1981, 64, IV-70.

19. J. Rosenthal and Ferrier, J. Pharmacology and Experimental Therapeutics, 1982, 221, 188.

20. I. Grupp, G. Grupp, N. Fowler, M. Gabel, A. Alousi and R. Millard, Federation Proceedings, 1980, 39, 976.

21. A. Nusrat, D. Tepper, J. Hertzberg and R. Aronson, Clinical Research, 1981, 29, 228.

22. E. Kabela, D. Michaels and E. Felice, Archivos del Instituto de Cardiologica de Mexico, 1979, 49, 892.

23. R. Millard, G. Dube, G. Grupp, I. Grupp, A. Alousi and A. Schwartz, J. Molecular and Cellular Cardiology, 1980, 12, 647.

24. A. Alousi, A. Helstosky, Federation Proceedings, 1980, 39, 855.

25. K. Meisheri, R. Palmer and C. Van Breemen, European J. Pharmacology, 1980, 61, 159.

26. J. Azari and R. Huxtable, European J. Pharmacology, 1980, 67, 347.

27. S. Einzig, G. Rao, M. Pierpoint and J. White, Canadian J. Physiology and Pharmacology, 1982, 60, 811.

28. A. Schwartz, I. Grupp, G. Grupp, C. Johnson, P. Berner, E. Wallick and K. Imai, Circulation, 1979, 60, II-16.

29. J. Parker and J. Harper, J. Clinical Investigation, 1980, 66, 254.

30. J. Morgan, N. Lee, J. Blinks, Federation Proceedings, 1980, 39, 854.

31. P. Honerjaeger, M. Schaefer-Korting and M. Reiter, Naunyn-Schmiedeberg's Archives of Pharmacology, 1981, 318, 112.

32. M. Endoh, S. Yamashita and N. Taira, J. Pharmacology and Experimental Therapeutics, 1982, 221, 775.

33. A. Alousi, G. Stankus and J. Stuart, Circulation, 1982, 66, II-309.

34. C. Frangakis, K. Lasher and A. Alousi, Circulation, 1982, 66, II-57.

35. H. Adams, J. Rhody and J. Sutko, Circulation Research, 1982, 51, 662.

36. P. Siegel, G. Morgan and C. Sweet, Federation Proceedings, 1982, 41, 1310.

37. G. Robinson, R. Butcher, I. Oye, H. Morgan and E. Sutherland, Molecular Pharmacology, 1965, 1, 168.

38. T. LeJemtel, E. Keung, H. Ribner, R. Davis, J. Wexler, M. Blaufox and E. Sonnenblick, American J. Cardiology, 1980, 45, 123.

39. R. Malacoff, G. Curfman, J. Wynne, J. Neill and E. Braunwald, American J. Cardiology, 1980, 45, 433.

40. L. Siegel, E. Keung, S. Siskind, R. Forman, H. Flinberg, J. Strom, D. Efstathakis, E. Sonnenblick, S. Leuri, S. Karaffa, L. DeBoer and T. LeJemtel, Circulation, 1981, 63, 838.

41. K. Weber, V. Andrews, G. Kinasewitz, J. Janicki and A. Fishman, American Heart J., 1981, 102, 569.

42. J. Chesebro, C. Harrison and S. Deets, American J. Cardiology, 1981, 47, 491.

43. J. Benotti, L. Lesko and J. McCue, J. Clinical Pharmacology, 1982, 22, 425.

44. C. Maskin, R. Forman, N. Klein, E. Sonnenblick, T. LeJemtel, American J. Medicine, 1982, 72, 113.

Acknowledgment

The authors are grateful to the staff of SWRI and the clinical invistigators for providing some of the information in this article.

The Use of Interactive Computer Graphics in the Study of Molecular Structure and Intermolecular Interactions

By A. C. T. North*, A. J. Geddes, and E. A. Potterton

THE ASTBURY DEPARTMENT OF BIOPHYSICS, UNIVERSITY OF LEEDS, LEEDS LS2 9JT, U.K.

1. Introduction

Interactive computer display systems provide a powerful new tool for chemists in assessing the optimum conformations of molecules, in studying the disposition of functional groups on the surfaces of molecules, in exploring possible modes of interaction between two or more molecules and hence in the selection or design of molecules in order to achieve desired reactivities.

Recent years have seen very rapid advances in the hardware available for interactive computer graphics. As with all such recent developments with computers, these have led both to the availability of more powerful systems and to reductions in price, so that it is now practicable to plan for an individual research group, or even an individual scientist, to have its own system; at the same time the increasing range of hardware means that there is no unique way of achieving a desired objective and any configuration that appears to be the most suitable today may well be superseded next year. This continual change in hardware impedes the interchangeability of software, which poses a very real dilemma - interchangeable software demands either standardised hardware or relatively powerful computers that can run machine-independent programs written in high level languages; against this must be set the pressure for individuals to have their own systems which, for economic reasons, must be based on micro-computers which only function efficiently when programmed in a machine-dependent way. It is imperative that any choice of systems takes into account the cost of software implementation; this may well lead to different decisions being reached in commercial organizations (which must cost their manpower fully) and academic institutions (in which the proper objective of training man-power often leads to useful software in the process). For the various reasons described above the computer display system that we have developed leads us to strategies that are not necessarily the only ones appropriate, for example, in how we divide operations between our display host computer and the university "mainframe" to which it is linked.

1.1 Merits of Computer Graphics as a Method for Studying Molecular Structure.
A very high proportion of the human brain is devoted to the processing of visually-derived information and our well-developed spatial senses have long been exploited in the use of molecular models. The advent of computer graphics has enormously enhanced the means of communication between man and computer, allowing him to see model representations directly rather than by having to construct models from coordinates output by the computer in numerical form. The most obvious advantage of computer-based models is their precision and flexibility - flexibility in two senses: first, the geometrical parameters of the model may be varied readily; second, the form of representation on the screen may be varied, for example, allowing a protein molecule to be represented as:

(i) a skeletal model showing the bonds joining all of the atoms
(ii) a simplified model showing the course of the polypeptide backbone
(iii) a "space-filling" model showing the shape of the molecular surface.

In addition to the direct substitution of computer graphical models for mechanical models, computer graphics offers further facilities. One of these is the ability to superimpose representations of different structures so that they can be directly compared in a manner that is impossible with "real" models. A second is the ability to couple with the display of molecular structure the calculation of properties that can be related to molecular geometry. If the calculations are relatively simple, it may be possible to carry them out in "real-time"; in practice, this means that, if the calculation can be completed within the time interval between successive refreshing of the display screen (of the order of 1/50 sec.), there is no apparent delay between a change in molecular geometry and the presentation of the derived property. Other properties may take longer to evaluate (and may perhaps require the use of a computer more powerful than the display's host) in which case the operator will have to wait between changing the geometry of a molecule and seeing the effect of the change. There is no doubt of the advantage accruing from "real-time" calculations and, hence, of the desirability of increasing computer power so that they can be achieved.

Some structurally-dependent properties may be represented by numerical values (e.g. the conformational energy of a molecule or the frequency of a proton NMR signal) but others may well be represented pictorially on the screen (for example, the electrostatic potential on the molecular surface, which can be shown by appropriate colouring of the surface features). The use of colour, now becoming much more widely available, has a far more than cosmetic effect on computer displays. The "professional" user of a display can become adept at interpreting a display, for example, recognizing different types of atom by their context. Increasingly, though, display systems will become just one of the many tools available to the research chemist and he will rightly expect the nature of each atom or functional group to be made clear and, just as with conventional models, colour is an ideal aid. Colour can be used to distinguish between two different molecules, e.g. a drug and its receptor site, such a means of distinction becoming essential when molecular surfaces are being shown. Or it can be used to distinguish between two or more molecules that are being superposed to compare them. Yet again, as indicated above, it can be used to represent properties such as the variation of electric potential over the surface of a molecule.

Finally, while such applications may not be strictly interactive, the visual representation of molecular structures on the display screen may give insight into molecular phenomena that would not be attainable from inspection of numerical output from the computer. For example, several groups have carried out simulations of molecular dynamics; by showing the results as a series of pictures representing the structures at successive intervals of time, it is possible to distinguish a variety of vibration modes and to see concerted movements of the type required for a substrate molecule to move into the active site of an enzyme.

2. Basic Features of Molecular Graphics System

2.1 Geometrical Relationships. Molecules are made up from fragments of defined geometry, appropriately linked together. The construction of molecules in the computer and manipulation of their conformations require only two basic operations, translation and rotation in space.

Translation is accomplished by changing the origin of the coordinate system -

$$x' = x + x_o$$
$$y' = y + y_o$$
$$z' = z + z_o$$

where the prime symbols indicate the new coordinates and $x_o y_o z_o$ represent the coordinates of the old origin in the new system.

Rotation is accomplished by the set of transformations:

$$x' = a_{11}x + a_{12}y + a_{13}z$$
$$y' = a_{21}x + a_{22}y + a_{23}z$$
$$z' = a_{31}x + a_{32}y + a_{33}z$$

where a_{11}, etc., represent the elements of a rotation matrix, which can be set up so as to yield rotations about any direction in space; such directions may include, say, the vertical axis of the display screen (thus allowing the molecule to be rotated in order to give a 3-dimensional effect) or any chemical bond in the molecule.

It is useful to be able to change the scale of the representation in order to fit it on to the screen and this may be accomplished by multiplying each of the coefficients a_{11}, etc., by a scale factor S. The operations of scaling, translation and rotation can then be combined into the transformations:

$$x' = S(a_{11}x + a_{12}y + a_{13}z + x_0)$$
$$y' = S(a_{21}x + a_{22}y + a_{23}z + y_0)$$
$$z' = S(a_{31}x + a_{32}y + a_{33}z + z_0)$$

In matrix notation, this involves the multiplication of each coordinate vector by the elements of a 3 x 4 matrix.

A slight elaboration, in which a 4 x 4 matrix is used (termed homogeneous coordinates) has some conveniences. Our objective here, however, is not to discuss advanced programming procedures (for which, see refs. 1 and 2) but rather to show that the basic manipulative procedures of computer graphics are essentially very simple; nevertheless if the display consists of a large number of points, the total number of arithmetic operations required to generate a new picture may be very large and this obviously may limit the rate at which new representations can be generated and displayed.

In our own current display system, these operations are carried out in software within the host computer. In advanced commercial systems, such transformations are often carried out by special purpose hardware within the display itself. This is clearly faster, but may have a drawback; if it is desired to carry out conformation-dependent calculations of properties (e.g. internal energies) within the host computer, then the host requires access to the transformed coordinates, which may not be readily available for return from the display. A possible compromise may be to use special purpose hardware within the host to provide transformed coordinates for use by both host and display[3].

2.2 Ordering of Coordinate Lists. Obviously if a translation or rotation operation is imposed on all of the coordinates in the list, then the molecule will be moved as a rigid body. If the operations are applied to some but not all of the coordinates, then one molecular fragment will be moved relative to the rest. This procedure is used to change molecular conformations by varying the torsion angle about single bonds - atoms to one side of the bond are left unchanged and those on the other side are moved by using a rotation matrix designed to give rotation about an axis coinciding with the bond. A rotation of the whole molecule is termed a GLOBAL rotation, one of part of the molecule is termed a PARTIAL rotation.

The ease with which molecular representations may be adjusted depends partly on setting up an appropriate logical structure for the data and procedures for assembling molecular structures from fragments. This is in fact by far the most difficult aspect of molecular graphics software. For example, a polymeric molecule such as a protein consists of a main chain, each unit of which carries a side chain. A rotation about a bond within the main chain causes all of the atoms to one side of that bond to move; on the other hand, rotation about a bond in the side chain causes only the atoms towards the extremity of the side chain to move.

One way of forming an appropriate data structure is to form two separate lists. The first contains all of the atomic coordinates - three, together with a type designator, for each atom. The second is an explicit list of bonds containing at least the following four pieces of information for each bond:

$$n_1 \quad n_2 \quad r \quad t$$

where n_1 and n_2 give the number of the atom at each end of the bond (i.e. the bond joins atom n_1 to atom n_2); r gives the total number of atoms beyond n_2 that would be affected by a partial rotation about that bond ; t indicates the type of the bond, which can be used to instruct the program how the bond should be drawn (e.g. a light line for a single bond, a heavy line for a double bond) and also to indicate whether rotation about the bond is allowed or not (e.g. rotations are not allowed about double bonds or bonds within aromatic rings).

Whatever logical data structure is used, it must be capable for our purposes of allowing molecules to be assembled from fragments, allowing two or more molecules to be moved independently and indicating to the computer whether atoms are within one molecule or within different molecules so that internal energies and interaction energies may be properly calculated.

2.3 Hardware and Program Control. Our hardware comprises:

PDP 11/45 host computer, with interfaces to:

- Hewlett-Packard CRT display
- keyboard
- 2 banks of 8 control knobs
- "bit-pad" tablet
- graph plotter for hard copy
- University Amdahl computer

Our experience has shown that, for interactive display work, it is most convenient for the various manipulative operations to be effected by analogue input devices. We find that rotatable control knobs give a very natural way of effecting rotations and translations of the molecule. The control of the programs is through a "menu" of options which are presented on the display screen; the desired option is chosen by typing in a code letter on the keyboard or by using the bit-pad to point to the required entry on the "menu".

The display tube is of the calligraphic type, i.e. the beam is moved at constant brightness along each straight line of the picture in turn. (In the alternative, raster-scan approach, the beam is tracked over the entire screen in parallel lines as in a TV set, and is brightened up whenever it crosses a feature in the object; at present, calligraphic displays allow a higher resolution in a complex picture).

2.4 Molecular Mechanics Calculations. In considering the optimum conformations of molecules and assessing interactions between two or more molecules, the crudest parameter is the elimination of close contacts between non-bonded pairs of atoms. This involves a simple calculation of the distances between non-bonded atoms and comparison with the sum of their Van der Waals radii. Any pairs that are too close can be flagged on the display. When a range of conformations is being explored, for example by partial rotations of pendant groups, it is unnecessary to recalculate the distances between pairs of atoms whose relative positions are unaffected by the change in conformation. Again, appropriate data structures can ensure that time is not wasted, and that distances are only calculated between pairs of atoms whose relative positions are liable to change.

More refined than the simple detection of "bumps" is the use of an appropriate potential function such as $V = \frac{a}{r^6} - \frac{b}{r^{12}}$; to this may be added terms for electrostatic and hydrogen-bond interactions. Obviously, the more complex the functions, the more time-consuming are the calculations and present computer speeds restrict what can be done in "real-time" to a fairly small (e.g. ∿200) number of pairwise interactions. This may nevertheless be sufficient to permit real time monitoring of the docking of a small substrate in the active site of an enzyme until an approximate fit is achieved.

A full determination of the minimum energy conformation of a molecule involves additional energy terms including those concerned with torsion angle variation, inter-bond angle bending and bond stretching, using a self-consistent set of parameters. Such a full relaxation cannot usefully be carried out in an interactive fashion but it is instructive to monitor the course of the search for the minimum energy conformation by transmitting the conformation after each cycle (or series of cycles) to the display. It may be helpful to store the sequence of conformations so that they can be shown on the screen at a more appropriate speed for the nature of the movements to be seen more clearly.

2.5 Representation of Molecular Surfaces. While the experienced worker can mentally "clothe" a skeletal representation and not be deceived by the apparent large spaces between atoms, a space-filling representation undoubtedly gives a much clearer impression of the molecular surface. With conventional models, space-filling components have the undesirable property of hiding the underlying covalent connections of the molecule and making interior features invisible. The advantage of computer graphics models is that the surface can be shown in a transparent way or "switched off" when not required.

The "chicken-wire" representation of molecular surfaces, developed from the representation of electron-density contours in crystallographic maps[4], has been widely used. It is derived by:

(i) defining a fine 3-dimensional grid of points

(ii) evaluating the electron-density due to the molecule at each point of of the grid or, more simply, setting to 1 each grid point that falls within the Van der Waals radius of an atom, leaving 0 at each point outside

(iii) using a contour-drawing routine to depict the boundary between the 0 and 1 values in the figure field or at an appropriate low value of density.

Contouring in three dimensions yields a surface comprising nodes joined by short straight line segments. An advantage is that it can be defined by the same type of data structure as a molecule, with the nodes and lines corresponding to atoms and bonds; the same transformation routines can therefore be used to rotate and translate the surface.

An alternative method of depicting surfaces is by means of an array of dots[5,6]. One advantage is that the dots may more easily be coloured according to the type of atom to which that part of the surface belongs than can the lines of the "chicken-wire" representation.

Representation of surfaces, particularly by the dot method, greatly increases the number of features that must be displayed compared with skeletal models and it is therefore only practicable on very high performance display systems.

3. Simulation of Drug-Receptor Docking

It is well known that the affinity of a drug for its receptor site, exemplified by the affinity of a substrate molecule for the active site of an enzyme, depends upon a complementarity of surface shape, balancing of charged groups, formation of hydrogen bonds between appropriate donors and acceptors, and juxtaposition of hydrophobic surface patches. While it is readily apparent when an approximate fit has been achieved, it is not practicable in general to simulate the docking of a drug into a receptor site from a distance, partly because some movement of the side chains of the receptor protein may be needed to give access to the binding site, partly because it is difficult to evaluate the different types of interaction on an absolute scale. Not the least problems are making proper allowance for dielectric effects and evaluating the energy concerned with de-solvation.

A two-stage strategy seems to be appropriate. In the first stage, interactive molecular graphics is used to identify a suitable binding site and to manoeuvre the drug molecule into an apparently suitable position. This may require some adjustments to the conformation of the drug molecule and of groups of the receptor. The use of real-time energy calculations can be of great assistance in assessing the best position of binding and in verifying that conformational changes to drug and to receptor are energetically permitted. In the second stage, an automatic energy minimisation procedure may be used to adjust the conformations and relative positions of drug and receptor in order to determine the optimal mode of binding and the energy associated with it.

We used this approach to predict the mode of binding of trimethoprim (TMP) to the enzyme dihydrofolate reductase (DHFR) of E.coli[7]. Previously reported crystallographic studies of the binding of other inhibitors were used to identify the active site and important functional groups. The TMP molecule was then moved manually into position so that it made satisfactory contacts and hydrogen-bond interactions, real-time energy calculations being used to monitor its position. The resulting position of TMP, its internal conformation and the conformations of side chains of the enzyme were then subjected to an energy minimization procedure, revealing that the most favourable binding was achieved when TMP was itself in a low energy conformation with $\tau_1 \cong 180°$ and $\tau_2 = 60\text{-}90°$. The predicted geometry of the complex (Figure 1) was confirmed subsequently by crystallography[8].

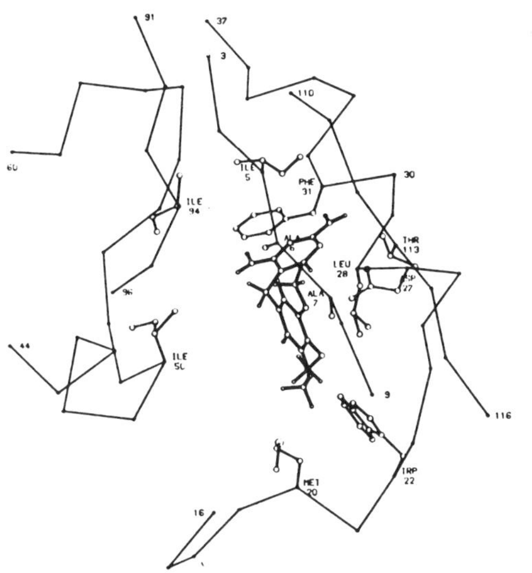

Figure 1. Preferred mode of binding of TMP to E.coli DHFR as deduced by computer graphics. The Cα backbone of the enzyme is shown, together with those side chains that interact with the inhibitor.

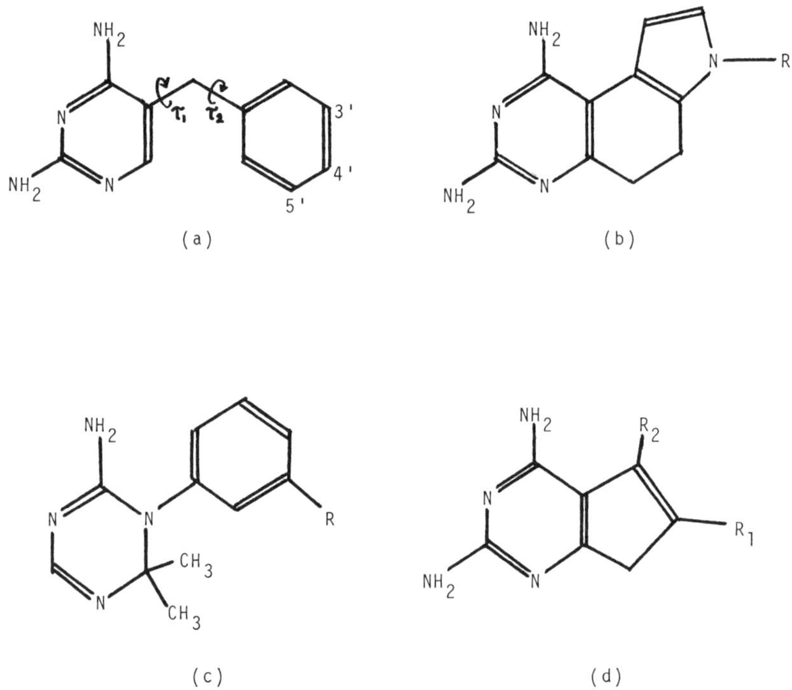

Figure 2. Inhibitors of DHFR

(a) benzylpyrimidine, showing the torsion angles that define its various conformations and the positions at which groups have been added in the various derivatives
(b) pyrroloquinazolines
(c) 5-phenyltriazines
(d) a novel series of compounds. R_1 = trimethoxybenzene is predicted to have enhanced binding over R_1 = H. R_2=methyl or ethyl are predicted to bind more strongly than large groups or R_2 = H.

We then proceeded to assess the binding of a number of TMP derivatives[9] (Figure 2a). We calculate slightly less favourable binding for the 6-methyl and 7-methyl substituted derivatives, mainly because the substitution prevents the inhibitor from adopting the preferred binding conformation with $\tau_1 \cong 180°$ and additionally, in the case of the R isomer of 7-methyl TMP, because of steric hindrance of the added methyl group with the cofactor. In a series of derivatives substituted at the 3', 4' and 5' positions of the phenyl ring, there is a satisfactory agreement (Figure 3) between our calculated enthalpies of binding and the experimentally observed inhibition constants, K_i [13,14].

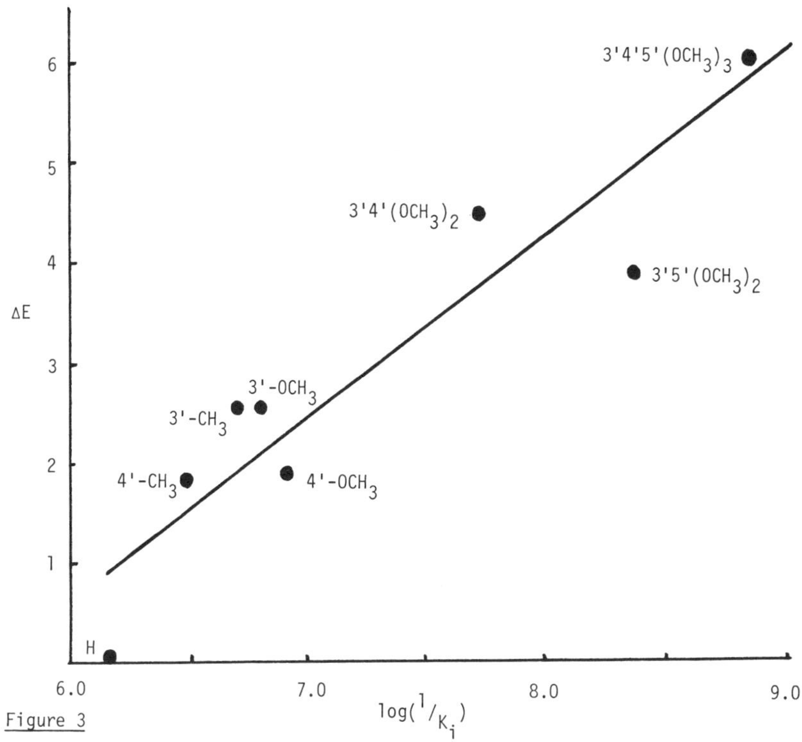

Figure 3

In a comparison of the binding to L.casei DHFR of three pyrrolquinazolines (Figure 2b) our calculated difference in binding energy between R = H and R = CH_3 is within the limits of error but we calculate a significant gain in binding energy when R = H is replaced by R = C_6H_5, which agrees well with the experimental data.

We have also modelled the binding of triazines (Figure 2c) to DHFR. We predict a substantial gain in binding energy when R = H is replaced by

R = $OCH_2C_6H_5$, again in agreement with the large increase in inhibition that has been reported. Any substituents on the terminal phenyl ring (cf. Baker's antifol) would be exposed to solvent, thus not materially affecting binding, whereas substitution of the $OCH_2C_6H_5$ fragment at the para position of the central phenyl ring seems to be precluded by close contacts with the enzyme.

The above work has shown that molecular graphics methods are able to place members of a family of compounds in the correct rank order of binding. They enable us to design inhibitor molecules that would bind more strongly by increasing the enthalpic contributions to free energy arising through optimum complementarity of shape between inhibitor and binding site, well-directed hydrogen bonds, etc. A second approach to increasing the binding energy is to minimize the loss of entropy enforced upon the inhibitor when it fits into the binding site; this may be achieved by designing the inhibitor with flexibility restricted close to the conformation required for optimum binding. An example of such a DHFR inhibitor is shown in Figure 2d.

4. Superposition of Molecules

The ability of a molecular graphics system to show two or more molecules superimposed (drawn in different colours or distinguished in other ways) is of great value in allowing detailed comparisons, for example of corresponding enzymes from different species. Such comparisons may assist in establishing the mechanism of activity and in the design of species-specific inhibitor compounds. (Use of the approach in comparing drug molecules is discussed in the next section).

The method may also be used to derive a model of a protein of unknown 3-dimensional structure on the basis of a related protein whose 3-dimensional structure is known.

It is by now well established that protein molecules belong to families, the similarity of whose members arises from an evolutionary derivation from a common ancestor. The similarity is often preserved much more clearly in the 3-dimensional conformation than in the amino acid sequence itself. When a family relationship is suspected, it is possible to build a 3-dimensional model of the "unknown" protein by folding its backbone as closely as possible to that of the known related protein. Favourable signs that the model is essentially correct include:

(a) substitutions of amino-acids in the interior of the protein should maintain its predominantly hydrophobic nature;

(b) changes, particularly in the interior, should generally be of a coordinated and compensatory nature, e.g. substitution of a small by a large side chain would be accompanied by a reverse change in a spatially neighbouring group;

(c) insertions or deletions of amino-acid residues should occur at surface bends and loops.

This method has been used to predict the 3-dimensional structure of the hormone relaxin[10]. Chemical sequence studies had shown that pig relaxin showed strong similarities to pig insulin, each molecule being composed of two chains and containing three disulphide bridges. Although many of the amino-acids in the two molecules are different, the strong similarities, particularly of the conformationally important bridges, suggested that relaxin might well have an insulin-like fold. A model was therefore constructed and found to be satisfactory by the criteria described above. Subsequently, the amino-acid sequences of relaxins from several more species have become available and appear to be consistent with the core of the molecule having an insulin-like fold[11] (Figure 4a). Of particular interest is a characteristic patch of polar side chains on one part of the molecular surface in all of the known relaxins (Figure 4b). This occurs in a region distinct from the parts of the surface of insulin believed to be responsible for insulin activity; it may well represent the receptor-binding site of relaxin.

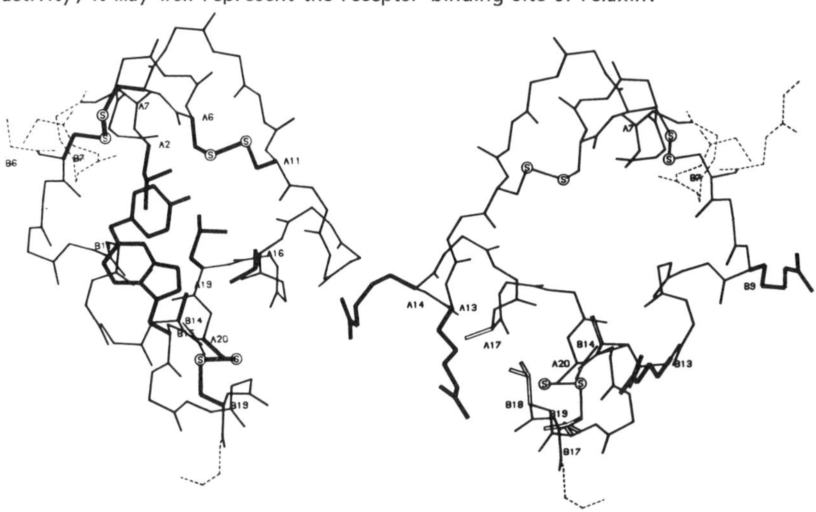

Figure 4(a) Residues forming the molecular core of the rat relaxin model.
(b) Residues in the putative binding site; surface side chains common to pig, rat and shark relaxins are shown by thick lines.

A similar approach has been used to construct a model of the enzyme renin[12], which has a closely homologous amino-acid sequence to several other acid proteinases, some of whose 3-dimensional structures have been found crystallographically. The availability of this model should be of significant help in the design of renin inhibitors, of potential value as anti-hypertensive agents.

5. Combination of Molecular Superposition and Energy Minimization

Comparison of drug molecules of similar activity has been used to establish their common features. Examination of the common features of a series of drugs may then be used to build up a picture of the receptor site to which they are complementary.

The conformations of small molecules are normally derived from crystallographic studies in which the precise geometry adopted may be influenced by crystal lattice forces. Moreover, the conformation of a drug molecule, when liganded to its receptor site, may be changed slightly from the absolute minimum energy conformation.

In trying to deduce the disposition of functional groups required for binding from the structures of two or more drug molecules, we need to find molecular conformations that are as similar as possible but still of low energy. This may be achieved by adjusting the conformations of the molecules in order to minimize an expression that is the sum of three terms: the first term is the least squares distance between the corresponding atoms of the equivalent functional groups; the other terms are the internal energies of the two molecules. The terms are added together, with suitable weighting factors[15].

Although such optimization procedures may be used only on a relatively powerful computer (probably not the display host) it is often valuable to follow the course of events in successive cycles of optimization by feeding the conformations generated at the end of each cycle to the display system. This may well assist in the development of the most effective strategy and is one example of the way in which a visual form of output from the computer can give far greater insight into the behaviour of a system or process than a numerical output.

6. Acknowledgments

It is a pleasure to acknowledge the contributions of my colleagues, particularly M. J. McCall and E. E. Eliopoulos, and financial support from the U.K. Science and Engineering Research Council and the British Diabetic Association.

References

1. W.M.Newman and R.F.Sproull. "Principles of Interactive Computer Graphics" (2nd ed.), McGraw-Hill, 1979.
2. J.D.Foley and A.Van Dam. "Fundamentals of Interactive Computer Graphics", Addison Wesley, 1982.

3. A.C.T.North, N.P.E.Gammage and C.A.Pinches. J.Molecular Graphics, 1983, 1, 52.
4. C.D.Barry and A.C.T.North. Cold Spring Harbor Symposia on Quantitative Biology, 1971, XXXVI, 577-584.
5. R.Langridge, T.E.Ferrin, I.D.Kuntz and M.L.Connolly, Science, 1981, 211, 661-666.
6. L.H.Pearl and A.Honegger. J.Molecular Graphics, 1983, 1, 9-12.
7. A.A.Chabot, A.J.Geddes, A.C.T.North and E.A.Potterton, Acta Crystallogr., 1981, A37, C340.
8. D.J.Baker, C.R.Beddell, J.N.Champness, P.J.Goodford, F.E.A.Norrington, D.R.Smith and D.K.Stammers, FEBS Lett., 1981, 126, 49-52.
9. E.A.Potterton, A.J.Geddes and A.C.T.North in "Pteridines and Folic Acid Derivatives" (ed.Blair), de Gruyter, 1983.
10. N.Isaacs, R.James, H.Niall, G.Bryant-Greenwood, G.Dodson, A.Evans and A.C.T.North. Nature (London), 1978, 271, 278-281.
11. G.G.Dodson, E.E.Eliopoulos, N.W.Isaacs, M.J.McCall, H.D.Niall and A.C.T.North. Int.J.Biol.Macromol., 1982, 4, 399-405.
12. T.Blundell, B.L.Sibanda and L.Pearl, Nature(London), 1983, 304, 273-275.
13. C.Hansch, R-C.Lai, J.M.Blaney and R.Langridge, J.Med.Chem., 1982, 25, 777-784.
14. D.P.Baccanari, S.Daluge and R.W.King, Biochemistry, 1982, 21, 5068-5075.
15. G.J.Willis, Ph.D.Thesis, University of Leeds, 1980, and unpublished work of A.J.Geddes and E.E.Eliopoulos.

Systematic Search of Conformational Space: Use and Visualization

By G. R. Marshall*, N. Van Opdenbosch, and J. Font

WASHINGTON UNIVERSITY SCHOOL OF MEDICINE, BOX 8101, 660 SOUTH EUCLID AVENUE, ST. LOUIS, MO. 63110, U.S.A.

One difficulty common to multivariable problems is the presence of multiple minima. Conformational hyperspace shares this problem, and energy minimization gives answers dependent on the starting conformation. Many approaches to this problem have been suggested; all are a compromise between the desire to determine the global minima and the computational price necessary to rest assured that the deepest minima found is, in fact, global. When faced with results from a Monte Carlo simulation, the question of sufficient sampling is paramount. When molecular dynamics are used. then the issue of sufficient equilibrium and simulation in the time domain must be addressed. When one is dealing with thermodynamic properties, then these concerns have been sufficiently explored that acceptable guidelines can be found. When one is considering potentially unique solutions to a set of constraints in conformational hyperspace, such guidelines are difficult to assess. For this reason, systematic exploration of the parameter space at a known increment offers the security of knowing the possible error between what is calculated and what may be correct. For example, systematic search of the possible conformations of a molecule at one degree increments of the torsional variables would clearly identify the conformer with the desired property ($\pm$ 1°).

Locus and vector maps. One problem facing the systematic examination of conformational space is an appropriate metric for dealing with the problem under consideration. One is often overwhelmed with conformers, often expressed as values of torsional variables, and one needs to compare properties of different molecules which do not share these variables. As molecules diverge from congeneric series, torsional rotations of one molecule have no relevance to others in non-congeneric series. One must, therefore, transform the results to an appropriate metric for comparison. One method which has been found useful is the use of locus plots, if an appropriate frame of reference such as an important group, or partial pharmacophore, can be established across the non-congeneric series.

In a series of congeners, the problem is straightforward. For example, in the case of biogenic amines, the aromatic ring can be used to orient the set of analogs and the locus of possible positions of the nitrogen monitored and compared (Figure 1).

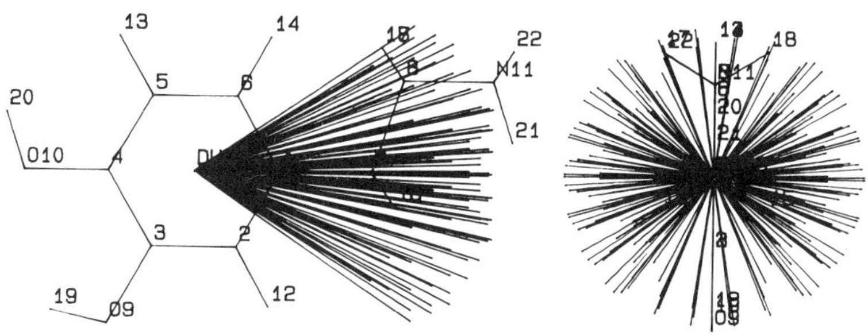

Figure 1. Locus of sterically allowed positions of the nitrogen atom of dopamine superimposed on the molecule. Vector from ring center indicates common reference point.

For each sterically allowed conformation at the given increment of the search, the x,y,z coordinates of the nitrogen are recorded with the orientation of the ring being fixed in the x,z plane and its centroid at the origin of the coordinate system (x=0,y=0,z=0). Alternatively, one may be interested in the possible interaction of the lone pair of the nitrogen with a hypothetical site on the receptor. By recording the locus of the bond comprising the nitrogen to lone pair vector, one can visualize the possible binding sites relative to the common frame of reference, the benzene ring (Figure 2).

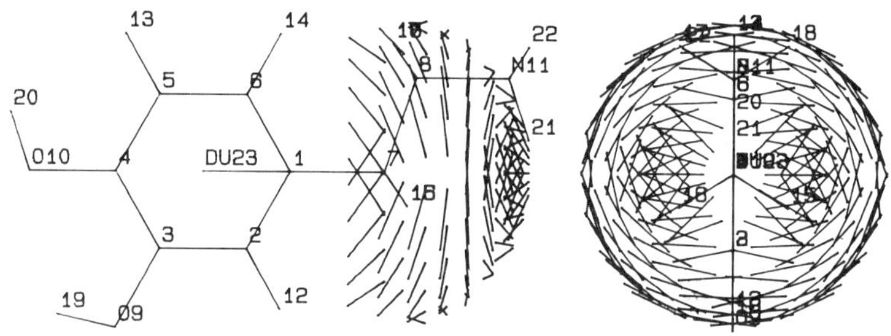

Figure 2. Vector map of the possible orientations of the lone pair of nitrogen superimposed on dopamine.

One area of interest has shown this technique to be of particular use. Modification of peptides to render them less susceptible to proteolytic digestion often involves modification of the amide bond. One is attempting to retain the appropriate conformation which generates the desired biological response, generally in the absence of information concerning the specific conformation. It is desirable, therefore, that the amide bond modification be able to mimic the set of conformers available to the amide as thoroughly as possible so that the modified peptide will be able to assume the biologically active conformation. It should be obvious that no overlap in conformations between the analog and the parent compound should lead to inactivity. The measure of overlap between the analog and the parent compound is a measure of conformational mimicry. In the case of the amide bond, let us consider several candidate modifications. In Fig. 3 the geometrical differences between the amide, the retroamide and the trans double bond are emphasized.

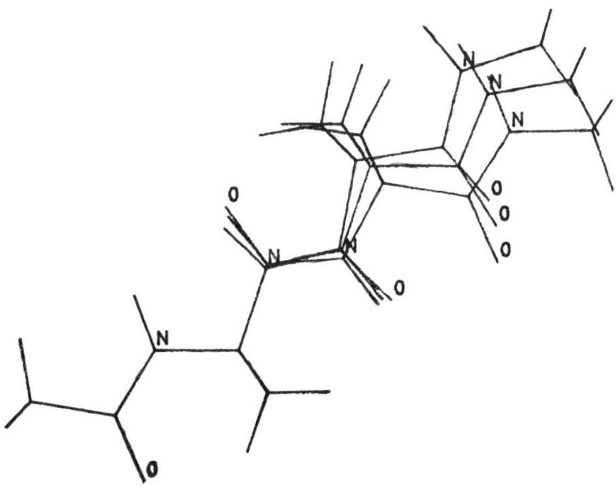

Figure 3. Superposition of acetyl-alanyl-methylamide and the retroamide and trans double bond analog to emphasize geometrical changes due to amide bond modification.

The N-terminal acetyl groups of acetyl-alanyl-alanyl-methylamide and of the two analogs with the central amide bond modified are aligned and used as a means of orientation. Slight changes in bond length and bond angles associated with the modifications result in divergence of the C-terminal portion of the molecules. These, as well as changes in the size of the atoms attached to the peptide chain in the analogs, result in major differences in conformational possibilities.

In order to monitor the similarities, we chose (1) to record the vector corresponding to the carbon alpha - carbon beta of the second alanyl residue. We first allowed the conformations available to the parent dipeptide to be explored as indicated in Fig. 4.

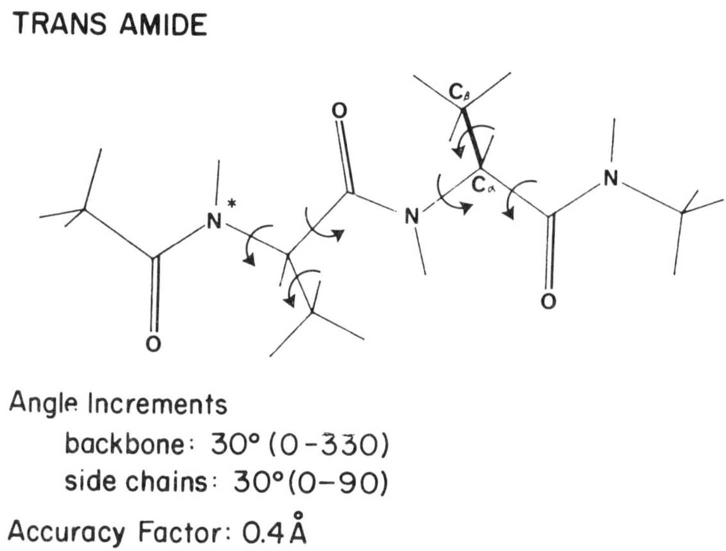

Figure 4. Outline of systematic search of the conformational space of the dipeptide which generated the vector maps in Fig. 5.

The amide bond was then modified and the vectors available to the analog which corresponded to those seen for the parent amide recorded. The sets for the amide , the trans double bond and the retroamide are seen in Fig. 5.

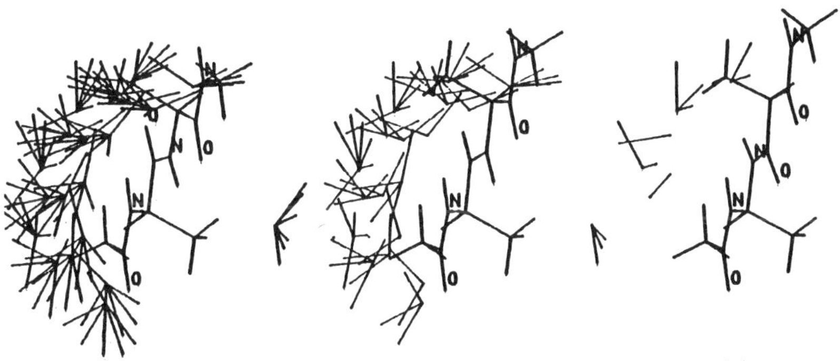

<u>Figure 5.</u> (A.) Vector map for normal <u>trans</u> amide. (B.) Vector map for <u>trans</u> double map resulting from intersection with A. (C). Vector map for retroamide bond resulting from intersection with A.

The percent of conformational mimicry for the <u>trans</u> double bond is approximately 66% and 8% for the retroamide. One could argue that statistically one has a two-thirds chance of retaining biological activity with the <u>trans</u> double bond and only one-fifth chance with the retroamide. A cursory review of the peptide literature would suggest that these data give a rough approximation to the observed success of these modifications. Other modifications of synthetic interests are the thioamide with approximately 56% overlap and the reduced amide bond with 100% overlap. Elimination of the planarity associated with the amide bond in the latter modification increases the conformational flexibility considerably and entropic differences may dominate. While such a ranking may be useful prior to development of a synthetic approach, the utility of the analogs in determining the receptor-bound conformation is inversely proportional to their conformational mimicry. The recent report by Roques <u>et al.</u> (2) on a retroamide analog of thiorphan which retains activity has considerable information content and can be used to ascertain the receptor-bound conformation of thiorphan when bound to enkephalinase.

Orientation maps and receptor mapping. When one is considering the binding of small molecules to proteins, multiple binding modes are clearly possible and can be considered if sufficient data is available to force such an elaborate hypothesis. In the case of the opiod mu receptor, Humblet and Marshall (3) have elaborated a more complicated model of that originally proposed by Portoghese and Fries (4) with a receptor site (modelled by an oxygen) which can interact with, at least, two binding sites for nitrogen as well as two binding sites for the phenyl rings.

The importance of this example is the use of a receptor group for overlap in the pharmacophoric sense rather than functional groups in the binding molecules. The studies on inhibitor binding to dihydrofolate reductase by x-ray crystallography show mutiple binding modes with lack of correspondence of assumed identical groups in the inhibitors. The binding groups in the receptor, however, stay in approximately the same positions. This suggests that the use of orientation maps and extended functional groups to include those to which the compound binds hypothetically may be of use. An example would be to analyze the data on the two inhibitors of angiotensin converting enzyme, captopril and enalopril, to see if an idea of the geometry of the enzyme active site could be deduced. One could either look for correspondence of groups A, B, and C from each of the two compounds as shown in Fig. 6 or extend these groups to include the hypothetical active site groups to which they bind.

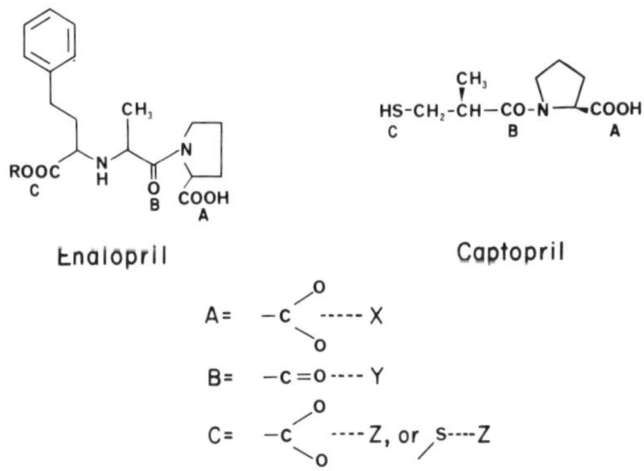

Figure 6. Structure of two inhibitors of the proteolytic enzyme, angiotensin converting enzyme, with group extensions to represent active site components.

Illustrated are a positively charged group X at an optimal salt bridge distance from group A, a hydrogen-bond donor group Y such as an amide nitrogen to interact with the carbonyl group B, and a group Z to represent the hypothesized zinc atom in the active site to which the sulfhydryl and carboxylate groups C bind. A more relevant approach would be to look for three-dimensional alignment of groups X, Y, and Z from each of the active analogs, in this case, captopril and enalopril. Such a correspondence can be found by independent minimization of each molecule with the further constraint that groups X, Y, and Z occupy the same coordinates in space for each active analog. This approach will answer the question of the existence of conformers available to the analogs with appropriate disposition of groups X, Y, and Z. Unfortunately, the question of the uniqueness of the solution is not addressed by minimization, similar to the question of local versus global minima. The alignment found will, therefore, depend on the starting point and relative weights of force constants. Systematic search of conformational space with recording of the relative positions of groups X, Y, and Z for each sterically allowed conformer allows generation of multidimensional orientation (relative distance) maps (5). Intersection of maps from each active analog allows identification of the set of orientations common to the analogs, i. e. all possible arrangements of groups X, Y, and Z consistent with the hypotheses. In the case of angiotensin converting enzyme, the two compounds captopril and enalopril are not sufficiently different to determine a unique three-dimensional arrangement of X, Y, and Z. Others inhibitors of angiotensin converting enzyme are rapidly being published and should supply sufficient data to determine such a unique arrangement.

The reasons for the desirability of this information should be fairly obvious. Determination of the pharmacophore, or active site binding requirements, provide a basis for the comparison of molecules not belonging to the same congeneric series. If one can understand the orientation of molecules when binding to a site, then a crude, but useful, description of the site can be obtained. Two published examples will serve to illustrate this point. Numerous amino acid analogs had been tested as competitive inhibitors for the methionine binding site of S-adenosyl:ATP transferase. Using the alpha amino acid portion of the molecule as the frame of reference, the volumes of active, rigid, carbocyclic analogs were combined by Sufrin et al. (6) to give an indication of the volume available at the binding site. Several closely related analogs did not bind, presumably because of negative steric interactions. Each of these analogs required unique volume not required by any active analog, which presumably represented the volume required in part by the receptor. More recently, the question of a common mode of action of the convulsant, picrotoxinin, and a series of gamma-butyrolactone analogs with both convulsant and anticonvulsant activity was examined by Klunk et al. (7). Analysis of the structure-activity data independently gave a similar picture for the two structurally different set of compounds assuming that a lactone ring formed the basis of molecular recognition by the receptor. Further study allowed the development of a classification scheme which could distinguish between agonist and antagonist based on the presence of steric volume in a particular region of space relative to the pharmacophore, the lactone ring. These studies on compounds with minimal flexibility provide the impetus to develop such ideas for more generally interesting flexible molecules.

As an example of the use of the orientation map approach with flexible molecules, a study in progress (8) on the tripeptide hormone, thyroliberin or thyrotropin releasing hormone, Glp-His-Pro-NH2, can be used (Fig. 7). Conventional modification studies suggested that the carbonyl of the pyroglutaminyl ring (Glp), the planar aromatic ring of His, and the C-terminal carboxamide group were candidates for groups A, B, and C. Analysis of the orientation maps available to six selected analogs (Fig. 7) led to only three possible pharmacophoric patterns for these groups. For each pattern, some variability in the distances AB, AC and BC was allowed. This gives rise to several conformers capable of presenting the particular pharmacophoric pattern. In each case, the conformer family could be represented by the midpoint of the torsional angle range with a variation in torsional angle. The concept of conformer family is a useful one which allows the grouping of conformers based on contiguous positions in torsional angle space. The family basically represent conformers sharing the same potential well in which distortion of normal geometry is not necessary for conversion between conformers. Semirigid analogs have been designed (9) capable of conformations compatible with

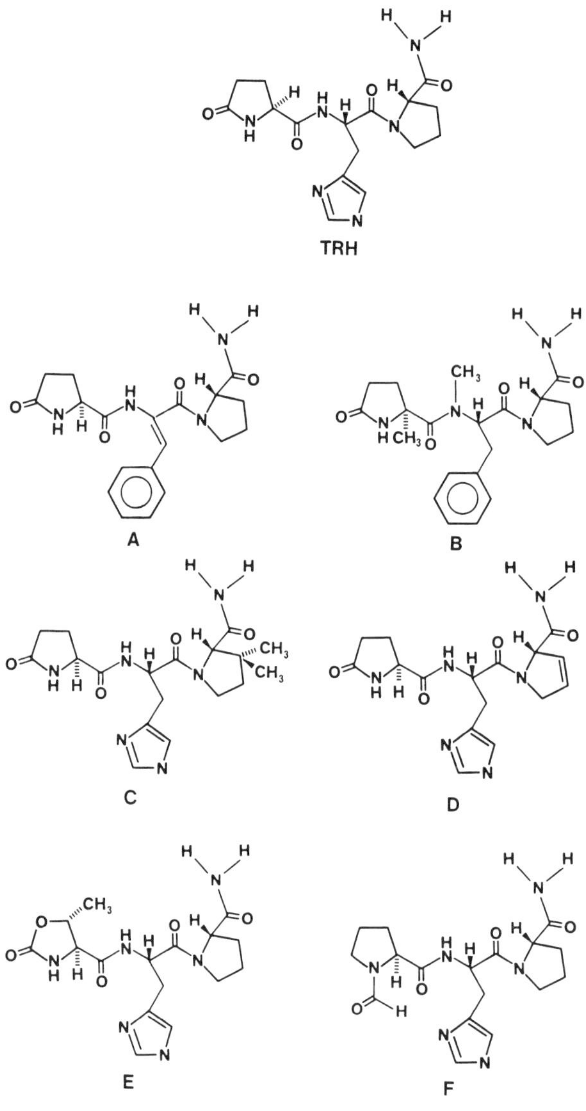

Figure 7. Thyrotropin releasing hormone and six analogs used in analysis leading to three candidate pharmacophores.

only one of the three patterns and are currently being prepared synthetically. Activity of one of these analogs would uniquely determine the receptor-bound conformation of TRH.

Conclusion. The ability of computational chemistry to explore conformational space and present properties of interest of the set of sterically allowed conformations has proven to be of general utility. Transformation of the coordinate data into relative distances has proven a powerful tool for analysis of complex comparisons. Locus plots of the possible positions of atoms, possible directions of atomic vectors, possible directions of group dipole moments, et cetera, can be quite instructive if an appropriate reference frame can be established for the series. Examples from real structure-activity problems have shown how these techniques can be applied to give new insight. Future applications will surely increase as these computational techniques become more readily available.

References.

1. Marshall, G. R., C. Humblet, N. Van Opdenbosch and J. Zabrocki. In: Peptides:Synthesis Structure Function. D. Rich and E. Gross, eds. Pierce Chemical Co., 1981, p. 669.
2. Roques, B. P., E. Lucas-Soroca, P. Chaillet, J. Costenin and M. Fournie-Zaluski. Proc. Natl. Acad. Sci. USA 80: 3178 (1983).
3. Humblet, C.and G. R. Marshall. Ann. Rep. Med. Chem. 15: 267 (1980).
4. Fries, D. S. and P. S. Portoghese. J. Med. Chem. 19: 1155 (1976).
5. Marshall, G. R., C. D. Barry, H. E. Bosshard, R. A. Dammkoehler and D. A. Dunn. In: Computer-Assisted Drug Design. E. C. Olson and R. E. Christofferson, eds. ACS Symposium 112, Washington, D. C., 1979, p. 205.
6. Sufrin, J. A., D. A. Dunn and G. R. Marshall. Mol. Pharmacol. 19: 307 (1981).

7. Klunk, W. E., B. L. Kalman, J. A. Ferendelli and D. F. Covey. Mol. Pharmacol. 23: 511 (1983).

8. Font, J., A. Hortmann and G. R. Marshall, work in progress.

9. Marshall, G. R. In: X-ray Crystallography and Drug Action. A. S. Horn and C. J. De Ranter, eds. Clarendon Press, Oxford, 1984, p. 484.

The Use of Molecular Graphics in the Design of Enzyme Inhibitors

By A. Kröhn

ROCHE PRODUCTS LIMITED, P.O. BOX 8, WELWYN GARDEN CITY, HERTS. AL7 3AY, U.K.

In recent years, it has become increasingly clear that the probability of finding truly novel therapeutic agents by random screening is very low. The escalating cost of the development of new drugs and the more stringent conditions imposed by regulatory and patent authorities has given a new impetus to find more rational approaches to drug design.

During the past decade, important advances have been made in molecular pharmacology and medicine. Many new potential drugs are now tested on isolated receptors and as inhibitors of key enzymic processes giving good quality biological data without the masking effects of drug metabolism and tissue distribution. Physico-chemical methods have undergone equally rapid development. Using X-ray crystallography and nmr techniques, it is now possible to study the structures of large biomolecules and their interactions with drugs in considerable detail. Minicomputers and visual displays have also become common in many chemical laboratories. X-ray crystallographers were the first to realise the great potential of these new devices for visualisation of their data in terms of molecular structures. Medicinal chemists and biochemists, interested in the action of drugs at the molecular level, have more recently adopted computer graphics for molecular modelling.

At first, the computer modelling tended to copy the Kendrew models traditionally used by peptide and protein chemists. Later, new molecular graphics techniques such as matching and docking of molecules, tracing loci of important atoms, generation and manipulation of molecular surfaces and volumes have been devised. Colour has become an important factor for identification of individual atoms and for highlighting the local properties of molecules (e.g. electron densities or lipophilic character).

The hardware used for computer graphics has evolved rapidly in the last five years. We started modestly in the mid-1970's exploring the possibilities of a PDP 11/40 minicomputer interfaced to a VT11 display. This system has been gradually upgraded over the years and our present system consists of a VAX 11/750 processor, a Megatek 7000 black and white display and a Ramtek 6211 low resolution colour raster graphics terminal.

The development of good computer software is a time consuming task (we have spent about 5 man years to develop our present system). Several software packages are now available, but we feel that the software should be written to satisfy the needs of the chemists rather than to be imposed on them from an outside source. However, specialised computer programs for energy calculations or plotting are now readily obtainable and can be incorporated into individual graphics systems.

Crystallographic data are centralised in two important data bases: the Brookhaven Protein Data Bank and the Cambridge Crystallographic Data Centre. Most pharmaceutical houses are now aware of the need to generate crystallographical data of molecules internally to give them proprietory knowledge of selected biochemical processes of medicinal interest.

The inhibition of enzymes has become the most fruitful area of rational drug design. Various enzymic processes have been clearly linked to disease states and the relative ease of testing compounds in *in vitro* enzymic assays has contributed significantly to the successful discovery of new inhibitors. In our work, we have assumed that biologically active molecules have functional as well as purely structural features. While functional groups interact directly with enzymes, structural parts are responsible for holding these groups in the correct geometrical orientation. Modification of functional parts has already led to the discovery of new types of inhibitors (e.g. suicide substrates, transition-state analogues etc.). In our studies we have concentrated on the modification of structural parts of inhibitors as well as the addition of groups able to interact with an enzyme.

The most amenable cases are those where crystal structures of the enzyme and also the enzyme complexed with various inhibitors are known. However, in the majority of cases crystal structures are not known and design of inhibitors has to rely on models based on structures of related enzymes ("protein extension") or on an "active analogue approach".

The elastase family of enzymes are serine proteases which degrade the elastin fibres in the injured connective tissue. In healthy organs the activity of elastases is checked by the presence of natural inhibitors: for example α-1-antitrypsin and α-2-macroglobulin. Patients suffering from pulmonary emphysema, a disease caused by loss of the elasticity of the lungs, have low concentrations of these inhibitors. Their depletion has usually a genetic origin or can be caused by the environment, especially by heavy smoking.

INHIBITORS

Natural: α_1-antitrypsin

α_2-macroglobulin

Synthetic:

(i) irreversible: phenylmethanesulphonyl fluoride

peptide chloromethylketones

(ii) reversible: H.Ala.Ala.Ala.OH [1]

Ac.Pro.Ala.Pro.Ala.OH [1]

(iii) crystal structure:

Pr-Ala-Pro-ethylamide [2]

Pr-Ala-Pro-cyclopentylamide [2]

Ac-Pro-Ala-Pro Ala-CHO [2]

TFA-Lys-Ala-NH$C_6H_5CF_3$ [3]

We have selected porcine pancreatic elastase (PPE, EC 3.4.4.7), the crystal structure of which is known, as a model for the more relevant human granulocyte elastase (HGE, EC 3.4.2.11), but the testing of inhibitors has been carried out against both enzymes in vitro[4].

Only small changes in the conformation of PPE have been observed upon complexation with these inhibitors allowing us to simulate the docking of the inhibitors into the active site without undue concern about a possible induced fit.

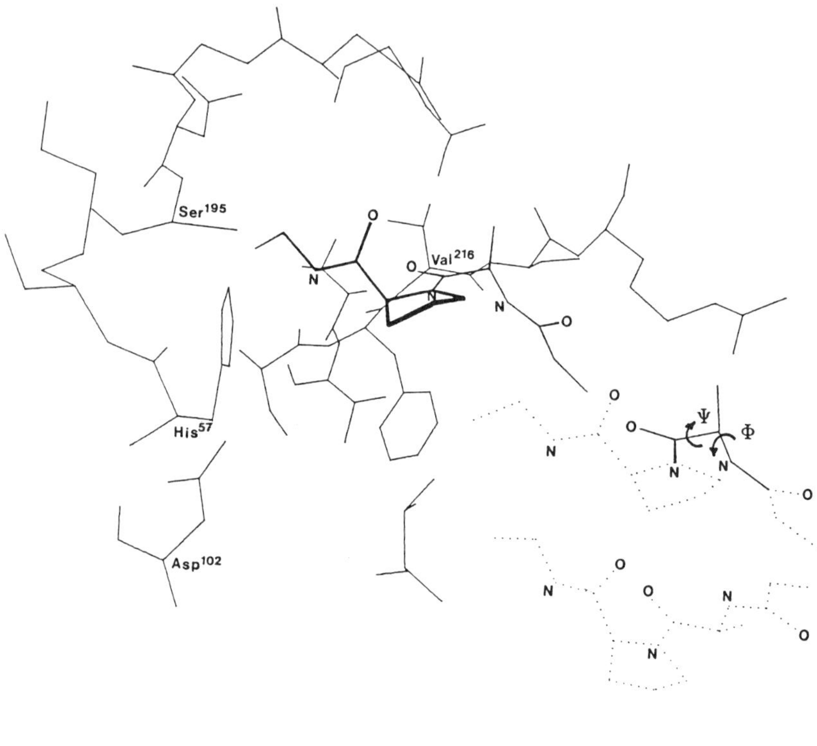

Figure 1 The docking of Pr-Ala-Pro-ethylamide to the surface of PPE. (The hydrogen bonds on each site of Val^{216} have been optimised and unfavourable contacts with the enzyme excluded).

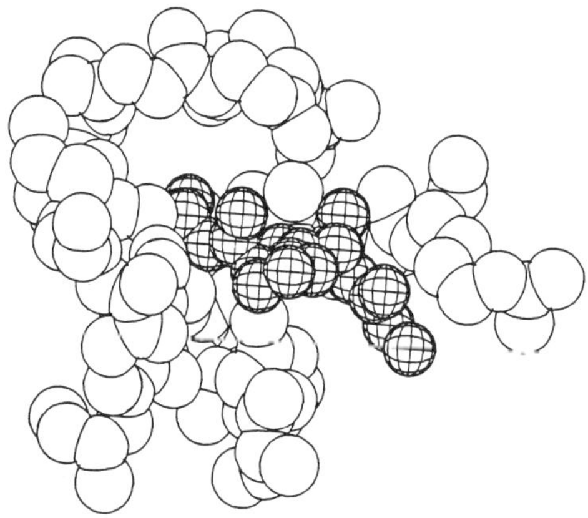

Figure 2 Space filling model of the PPE and Pr-Ala-Pro-ethylamide (the inhibitor is cross hatched).

The conformation of bound inhibitors could be precisely determined and the cavities forming the active site studied in detail.

By overlaying of the crystal structures of bound inhibitors we have been able to relate their modes of binding with respect to each other and also to the possible binding of the substrate.

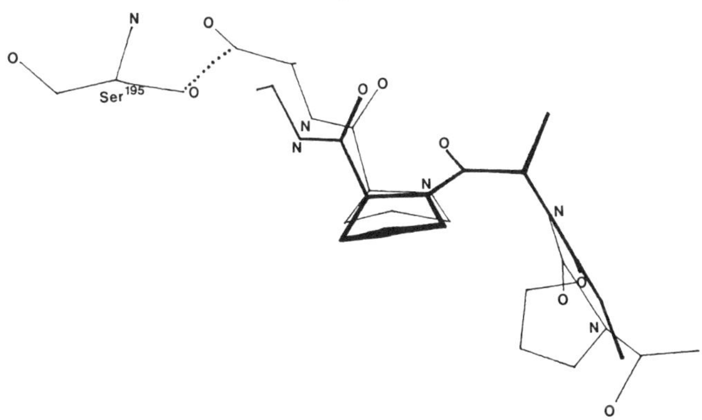

Figure 3 Comparison of the crystal structures of Pr-Ala-Pro-ethylamide (tapered) and Ac-Pro-Ala-Pro-Ala CHO (both bound to the enzyme).

However, substantial differences exist between the human enzyme and the enzyme isolated from porcine pancreas. To take full advantage of molecular graphics in designing useful therapeutic agents it will be necessary to determine the crystal structure of HGE.

Like many other pharmaceutical companies ROCHE has an active interest in the development of new dihydrofolate reductase (DHFR, EC 1.5.1.4) inhibitors. A large variety of species use DHFR for one carbon transfer, an important step in the initiation of protein synthesis and biosynthesis of nucleic acid bases.

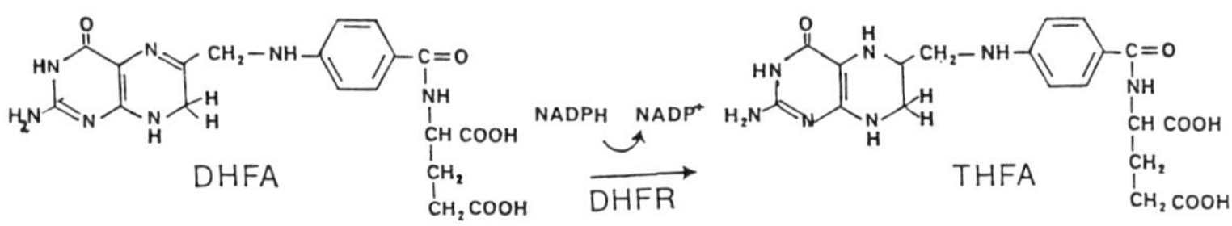

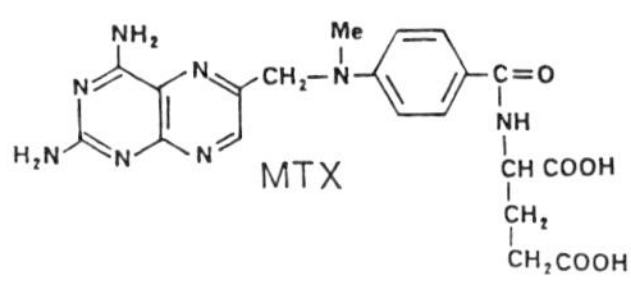

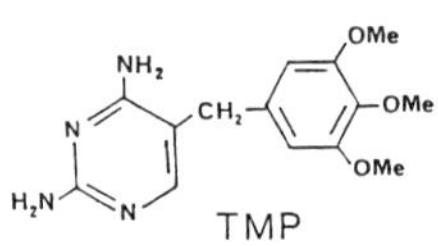

Compounds active against the human enzyme, such as methotrexate (MTX), are used as anticancer agents, primarily due to their ability to slow down DNA synthesis in fast growing cancer cells. Trimethoprim (TMP) on the other hand is a highly selective inhibitor of bacterial DHFR and in combination with sulfamethoxazole is used as the successful antibacterial drug BACTRIM®.

Many molecular graphics laboratories became interested in the inhibition of DHFR in the late 1970's when the crystal structures of <u>E.coli</u> and <u>L.casei</u> DHFR complexed with MTX and NADPH were published[5,6].

We have used computer graphics to replace MTX with TMP on the active site of DHFR so that the diaminopteridine and diaminopyrimidine rings of MTX and TMP respectively were in the same orientation towards the nicotinamide ring of NADPH. The phenyl ring of TMP was placed in the cavity previously occupied by the corresponding ring in MTX while the non-bonded interactions were optimised.

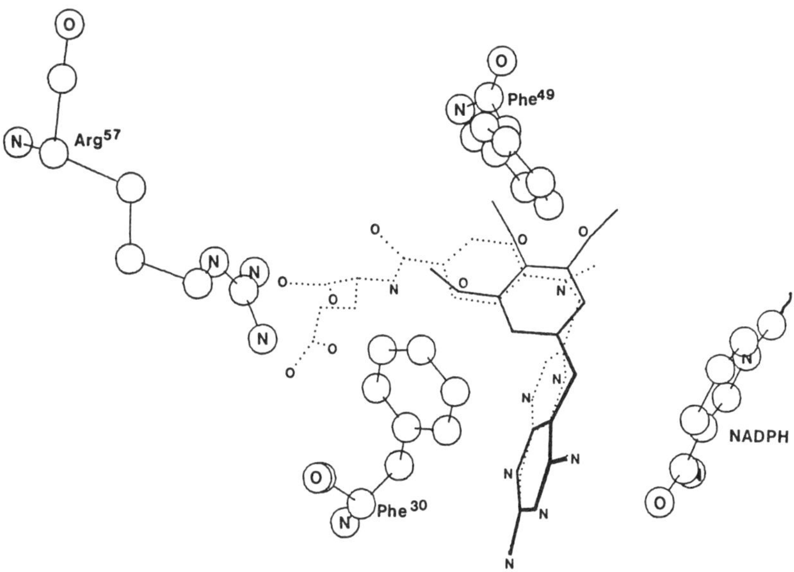

Figure 4 The matched conformations of MTX (dashed lines) and TMP on the surface of L.casei DHFR.

When the crystal structure of TMP complexed with E.coli DHFR became available[7], it was gratifying to find this structure to be practically identical with that predicted by molecular graphics.

Close examination of the enzyme - TMP complex has indicated that an empty channel exists between the methoxy group of TMP and basic moiety of Arg^{57}. We have synthesised a series of m-substituted carboxyalkyl derivatives of brodimoprim and tested them against bacterial DHFRs.

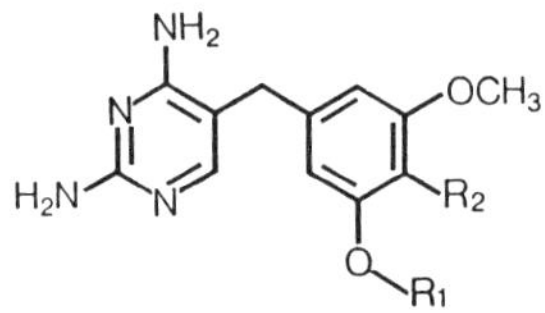

R_1	*E. coli* DHFR K_i (nM)		*L. casei* K_i (nM)
	R_2[8]= OCH_3	R_2= Br	R_2= Br
CH_3	1.37	0.603	11.3
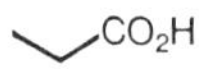	2.6	1.590	14.2
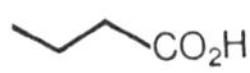	0.37	0.585	2.71
	0.035	0.032	0.893
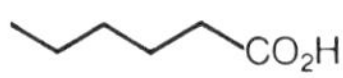	0.066	0.040	0.203
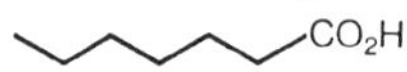	0.024	0.012	0.383
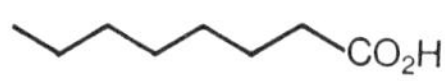	0.050	0.029	0.580

Similar conclusions have been reached by the Wellcome Group working directly with TMP derivatives[8].

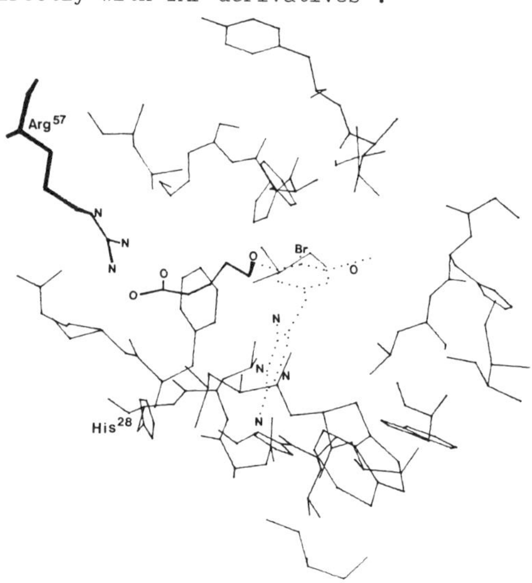

Figure 5 The model of the brodimoprim derivative interacting with Arg^{57} DHFR.

In our hands new brodimoprim derivatives have increased anti-DHFR activity as well as the necessary selectivity towards bacterial enzymes. Unfortunately these compounds do not apparently penetrate bacterial cell walls, so that perhaps some form of pro-drug approach will be needed to obtain activity *in vivo*.

There is considerable interest in investigating the potential of the renin-angiotensin system for controlling blood pressure. Angiotensin converting enzyme (ACE, EC 3.4.15.1) transforms the decapeptide angiotensin I to the powerful vasoconstrictor angiotensin II. ACE is a zinc containing protease similar to carboxypeptidase A, thermolysin and penicillopepsin. On the basis of the 2D model of ACE, the Squibb group of researchers has been able to develop a clinically useful inhibitor captopril[9].

Asp–Arg–Val–Tyr–Ile–His–Pro–Phe–⌇–His–LeuOH   Angiotensin I

↓ ACE

Asp–Arg–Val–Tyr–Ile–His–Pro–PheOH   Angiotensin II

\+ His–LeuOH

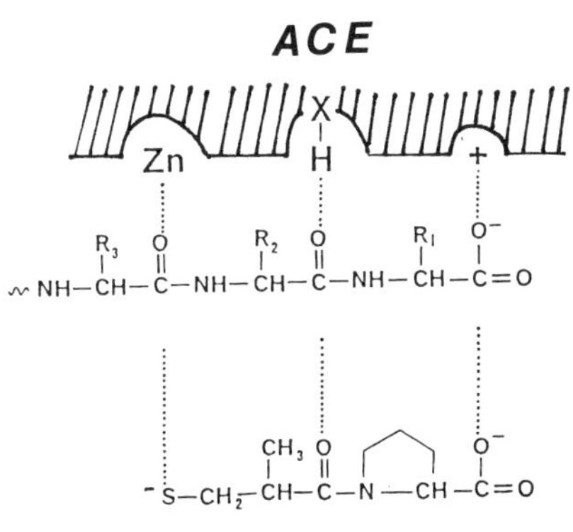

captopril

It has been shown that at least three groups, the thiol, the carbonyl of the amide bond and the carboxyl, are essential for high biological activity. Our aim was to design and synthesise conformationally restricted analogues of captopril in which the binding groups have been fixed in various geometrical arrangements[10]. The major conformer of captopril in solution as well as in crystal had the amide bond in the <u>trans</u> conformation[11,12]. The thiol group is freely rotating about the CH-CO and CH_2-CH- bonds and can be described by the locus of this atom. However, the high energy conformations, particularly those where the methyl or the thiol groups are eclipsed with the proline ring, can be excluded, since it is unlikely that they will be found in the enzyme-inhibitor complex.

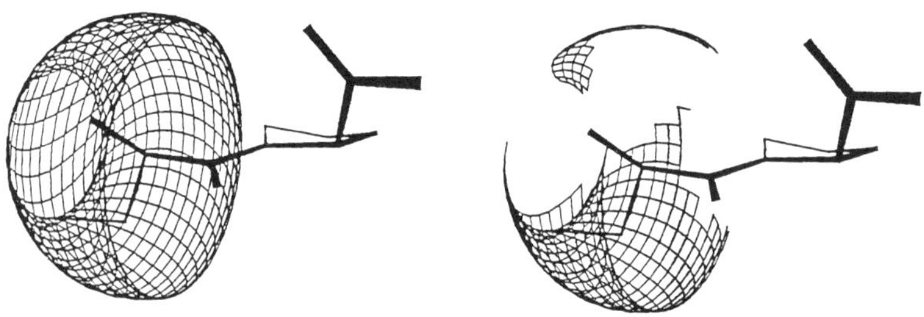

(a) (b)

Figure 6 The locus of the thiol group of captopril with freely rotating single bonds (a). Position of the thiol group for conformations of captopril with total internal energies < 20 kcal/mol (b).

Three classes of bicyclic inhibitors have been initially selected to mimic different conformations of captopril.

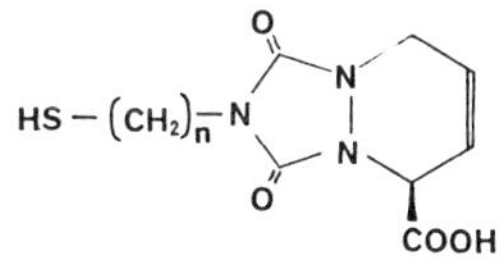

n	I_{50}
1	0.7
2	0.48
3	6

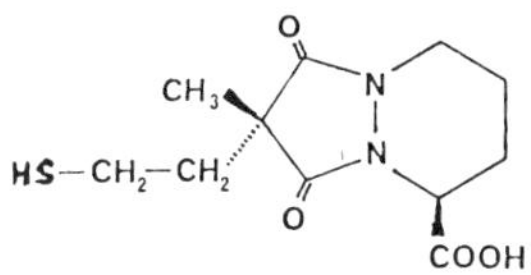

I_{50} = 0·1

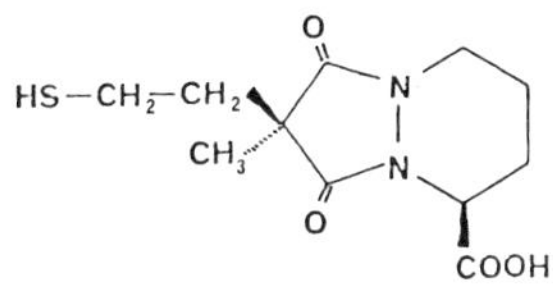

I_{50} = 43

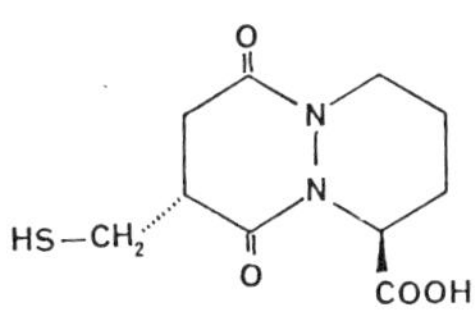

I_{50} = 0·038

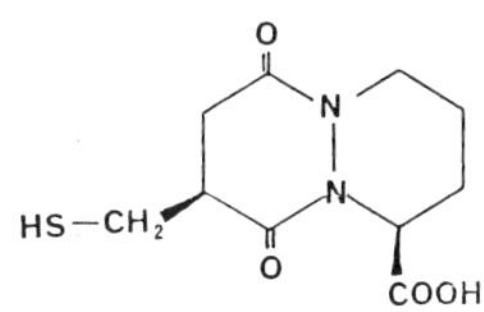

I_{50} = 0·1

I_{50} (μM)

The expansion of the proline ring and the introduction of the second nitrogen had only a small effect on the biological activity of these compounds but simplified considerably the chemical synthesis. In all bicyclic analogues the ring systems are rigid, fixed by the set of two amide bonds and strongly favouring the axial position of the carboxyl group to relieve so-called "A" strain.[13] This rigidity of the ring systems has been confirmed in solution, by detailed nmr studies[12].

For each bicyclic system the amide bond can be matched with that of captopril and then the locus of the sulphur atom can be traced by allowing free rotation about single bonds.

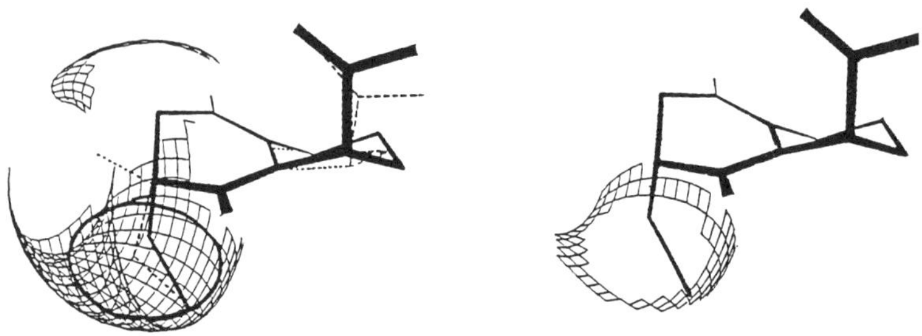

(a) (b)

Figure 7 Matched molecules of captopril and its bicyclic analogue (a); the common locus with S atoms within 0.5Å of each other (b).

The common locus for each pair can be defined, and knowing the in vitro activity in each bicycle, different activities can be assigned to different parts of the captopril locus.

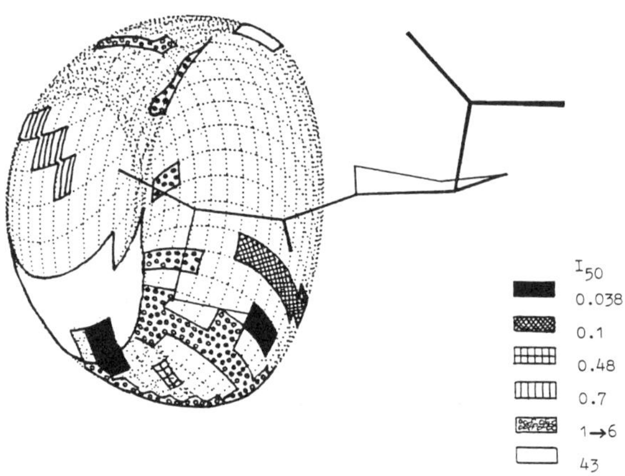

Figure 8 Activities (I_{50}µM) assigned to the parts of the captopril locus.

Assuming standard geometry for the Zn ...... S-CH_2 fragment and the fact that the torsion angle Zn ...... S-CH_2-$CHCH_3$ is likely to be trans in the captopril-ACE complex, for steric reasons, the following picture of the catalytic site can be proposed.

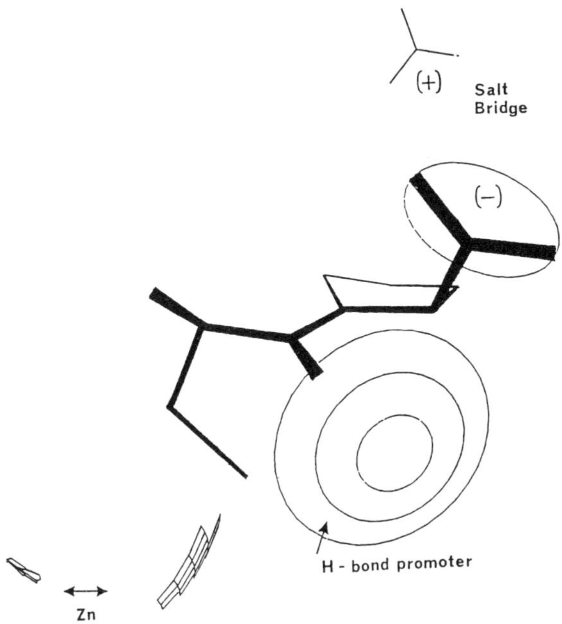

Figure 9 Orientation of captopril for binding to the postulated catalytic site of ACE.

It is probable that the carboxy group binds to a positively charged centre, either $\overset{+}{N}H_3$ or arginine NH-CH(NH_2)$\overset{+}{N}H_3$. The carbonyl group is able to interact with the donor group of the enzyme in the region indicated by the part of the sphere shown. The expected positions of the zinc atom are shown by a part of the mesh plot approximately trans to the carboxy group. In this way, the main components of the catalytic site of ACE can be mapped out. This story of ACE inhibition is far from complete. We have several very potent new classes of non-thiol ACE inhibitors, one of which is currently being evaluated in man.

The design and development of modern therapeutic agents is a multidisciplinary effort. The following scientists have made significant contributions to this work: Drs C H Hassall, J J Daly, W H Johnson, I Kompis, G Lawton, W J Lloyd, C J Moody, J E Pearson, N A Roberts, R Then and Dr W A Thomas.

1. L Sawyer, D M Shotton, J W Campbell, P L Wendell, H Muirhead, H Watson, R Diamond and R C Ladner, J.Mol.Biol., 1978, 118, 137.

2. L Sawyer, private communication.

3. D Hughes, L C Sieker, J Bieth and J Dimicoli, J.Mol.Biol., 1982, 162, 645.

4. C H Hassall, W H Johnson and N A Roberts, Biorganic Chem., 1979, 8, 299.

5. D A Matthews, R A Alden, J T Bolin, S T Freer, R Hamlin, N Xuong, J Kraut, M Poe, M N Williams and K Hoogsteen, Science, 1977, 197, 452.

6. D A Matthews, R A Alden, J T Bolin, D J Filman, S T Freer, R Hamlin, W G J Hol, R L Kisliuk, E J Pastore, L T Plante, N Xuong and J Kraut, J.Biol.Chem., 1978, 253, 6946.

7. D J Baker, C R Beddell, J N Champness, P J Goodford, F E Norington, D R Smith and D K Stammers, FEBS Letters, 1981, 126, 49.

8. L F Kuyper, B Roth, D P Baccanari, R Ferone, C R Beddell, J N Champness, D K Stammers, J G Dann, F E A Norrington, D J Baker and P J Goodford, J.Med.Chem., 1982, 25, 1122.

9. M A Ondetti, B Rubin and D W Cushman, Science, 1977, 196, 441.

10. C H Hassall, A Kröhn, C J Moody and W A Thomas, FEBS Letters, 1982, 147, 175.

11. M Fujinaga and M N G James, Acta Cryst., 1980, B36, 3196.

12. C H Hassall, A Kröhn, C J Moody and W A Thomas, J.Chem.Soc., Perkin Trans. 1, 1984, 155.

13. Y L Chow, C J Colon and J N S Tam, Canad.J.Chem., 1968, 40, 2821.

The Design of Triazole Fungicides

By A. F. Marchington

ICI PLANT PROTECTION PLC, JEALOTTS HILL RESEACH STATION, BRACKNELL, BERKS. RG12 6EY, U.K.

Let us begin with a fundamental question. What are the research departments of the major drug and crop protection companies trying to do? They are trying to invent small, biologically active molecules whose effects have commercial worth or advantage. In that case, what makes a molecule biologically active? Usually a molecule possesses activity primarily because it binds to an active site on a biological macromolecule, most commonly a three-dimensional protein structure. In the past, the discovery and subsequent development of biological activity has been done in three conceptual ways: (a) large scale empirical screening; (b) close analogue chemistry; (c) the more rational design of biologically active molecules at the molecular level. Although of greater financial potential, the great difficulty with the latter approach has always been our stark inability to answer several crucial questions concerning the binding of a small molecule to its protein receptor. For instance, what do (a) the substrate and (b) the active-site of the receptor actually look like as the one approaches the other. The free energy changes associated with the removal of a molecule from its solvent sheath are to some extent amenable to experimental evaluation, but the exact nature and geometry of the recognition process, collision and chemical binding of a small substrate to a protein is still largely a mystery.

In molecular terms, how one molecule appears to another, whether it be the substrate or the binding site, is really two questions.

Where are the nuclei?. This is not just a question of equilibrium shape as measured by n.m.r., x-ray or neutron spectroscopy, but also concerns what possible shapes the molecule can assume as it interacts with its partner; in general, what flexibility it possesses. Flexibility is clearly a property of both small molecules and the protein binding sites.

<u>Where are the electrons?</u>. This question too can only be studied experimentally for molecules in equilibrium and in a roughly homogeneous environment such as a crystal or in solution. What we really want to know is how the distribution of these electrons around the nuclei determine the likelihood of effective collision and how they then behave during the interaction. Since molecules interact most strongly at their accessible surfaces, it is important to know what these surfaces look like.

Advances in theoretical methods and computer technology mean that both these questions can now be answered using a computer and any number of easily obtained programs. Having obtained the answer to our problem theoretically however, there is a further difficulty. How can these often complex molecular properties be displayed. This is really the crux of the matter for the world's crop protection companies. It is a fact of life that the scientists trained to make molecules will not be influenced by those trained to design them unless the proposed rationale can be seen to be obvious. Computer Molecular Graphics provides this link between a chemist's intuition and the vast array of chemical, physical and biological information. As an example of the use of computer graphics and theoretical methods this paper describes a study in the design of the triazole fungicides, for instance the ICI compound diclobutrazol ('Vigil') and the Bayer compound triadimefon ('Bayleton'). The Eli Lilly material triarimol ('Elancocide') though not a triazole fungicide is equivalent in its mode of action. This general class of fungicide is now attracting wide commercial interest in both the crop protection and pharmaceutical industries as inhibitors of fungal ergosterol biosynthesis (Figure 1).

<u>Figure 1</u>. Triazole fungicides

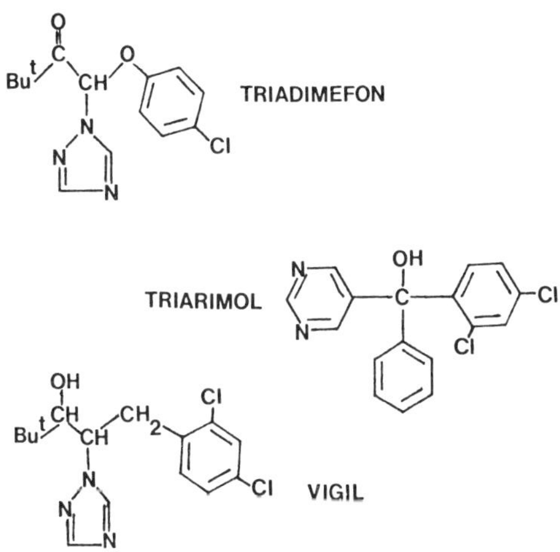

This design work can be divided into three stages :-

1. Assembly of biochemical, physical and biological information mostly from the literature, but also experiment, to construct a crude two dimensional picture of the site of action of these compounds.

2. A computer graphics facility was then used with available crystal data, molecular orbital and molecular mechanics calculations, infra-red and n.m.r. studies to construct a three dimensional model of the target enzyme active site (a cytochrome P-450) designed specifically to accommodate both the natural substrate (24-methylene-24,25-dihydrolanosterol) and these known antagonists in their minimum or low energy forms.

3. This model was then used to suggest new structure-activity relationships and contribute towards novel fungicide design.

The Qualitative Enzyme Model

It has been shown (1) that the triazole fungicides inhibit the 14α -demethylation of 24-methylene-24,25-dihydrolanosterol, the ergosterol precursor. This is a crucial step in ergosterol biosynthesis which has to be completed before a number of other steps can begin more or less in parallel. This inhibition is brought about by the compounds binding to the heme prosthetic group of the cytochrome P-450 oxidase enzyme system which catalyses this transformation. When added to a rat liver cytochrome P-450 preparation, for instance, an unmistakable Type II Soret difference spectrum is produced indicating that the triazole 4-nitrogen coordinates to the heme ferric ion which maintains its ferric (Fe^{3+}) low spin resting state. In so doing the antagonist has to displace the natural sixth liquid of the heme which is probably a water molecule (2) or possibly an imidazole group derived from a protein histidine. The other axial ligand, below the heme plane, is believed to be a cysteine sulphur as first suggested by Murakami and Mason (3).

The 14α -demethylation of dihydrolanosterol proceeds in three main stages with the two intermediates - the alcohol, 5α -lanost-8-ene-3β,32-diol, and the aldehyde, 3β-hydroxy-5α -lanost-8-en-32-al, being tightly protein bound. The cytochrome P-450 is the component of the enzyme system required to initiate oxidation of the 14α -methyl group, but not of that responsible for the subsequent oxidation steps required for its elimination as formic acid (4). This initial oxidation also seems to be directly inhibited by the alcohol and aldehyde metabolities.

Consequently, in computer modelling the antagonism of these heme binding fungicides it seemed necessary to consider only the first oxidation of the parent lanosterol to the 14-methyl alcohol. It is only in the last few years that a plausible mechanism for this oxidation has been suggested (5). On binding the substrate the ferric porphyrin is converted from low to high spin due to the displacement of the high field sixth ligand. Studies with spin labelled substrates have shown that this substrate binding site places the bound molecule very close to the iron (6). This complex is then reduced and as Fe^{2+} can then bind molecular oxygen. Further one-electron reduction yields a species which is less well defined but corresponds to the hypothetical state $[Fe^{3+}\ O_2^{2-}]$ which has all the electron equivalents required for methyl hydroxylation, water production and regeneration of the ferric resting state. This final step, however, requires an effector molecule - a free acylating group, provided in bacterial hydroxylase by the carboxy terminal tryptophan, or the pemultimate glutamine of putidaredoxin. This acyl group is responsible through a peracyl group of generating the final iron-oxene intermediate.

Figure 2 shows the crude two dimensional model of the P-450 active site. In designing an inhibitor for this process there are three central features to consider :-

1. The heme prosthetic group available for complexation.

2. The hydrophobic substrate binding site specific for lanosterol. Indeed a recent paper by Dus (7) implicates two binding sites for various cytochrome P-450's - one for substrate and the other for nascent product, and both with activated thiol groups.

3. The occurrence of hydrophilic groups in an otherwise grossly hydrophobic environment. The porphyrin propionate side chains as suggested by Peterson et al. (8) and the acyl effector group could both intervene between the bound substrate and the plane of the heme. There is also the possibility of hydrogen bonding with the displaced histidine (if present) and also a general polar interaction with the polar interface which exists by virtue of the enzyme sitting in a membrane.

The task now was to locate the natural substrate and the flexible inhibitors in a three-dimensional computer model of the enzyme site to examine if interactions with these features could go some way to providing plausible structure-activity relationships.

Computer Modelling Of Enzyme Site

The crystal structures of the protoporphyrin IX and the lanosterol nucleus were obtained directly by computer link to the Crystal Structure Search and Retrieval (CSSR) library provided by the SERC on the Edinburgh Dec 10 Computer. The approach and interaction of the lanosterol with the iron-oxene system was then modelled on the graphics screen. Ideally, one might prefer to model some transition state for the reaction of the oxene with the 14-methyl group. However, since the intermediary alcohol could well be an inhibitor for this enzyme, the alcohol ground state geometry was chosen with an iron-oxygen distance of 1.9Å and a carbon-oxygen-iron angle of 130°. These values are those obtained theoretically by Loew for an iron-carbene system

Figure 2. Model enzyme site

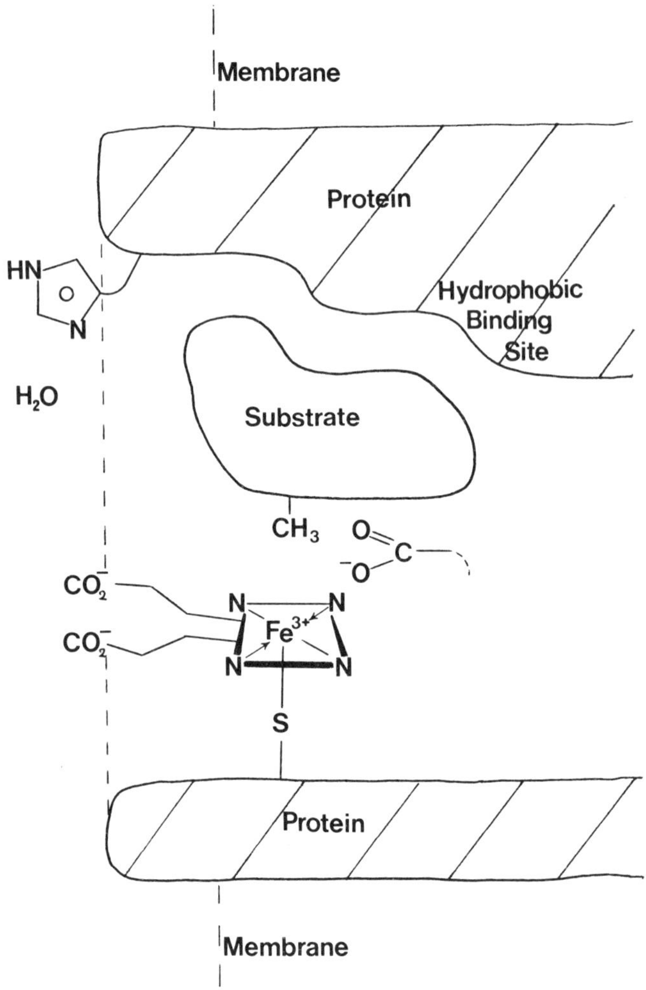

(9). There are now three single bonds :- iron-oxygen, oxygen-carbon and carbon-carbon about which the bound lanosterol can exercise internal rotations. The computer graphics facility could now be used to investigate the possible orientations of the lanosterol relative to the porphyrin ring and calculate sumultaneously, by molecular mechanics, the total internal energy of interaction.

Figure 3. Lanosterol on P-450 porphyrin ring

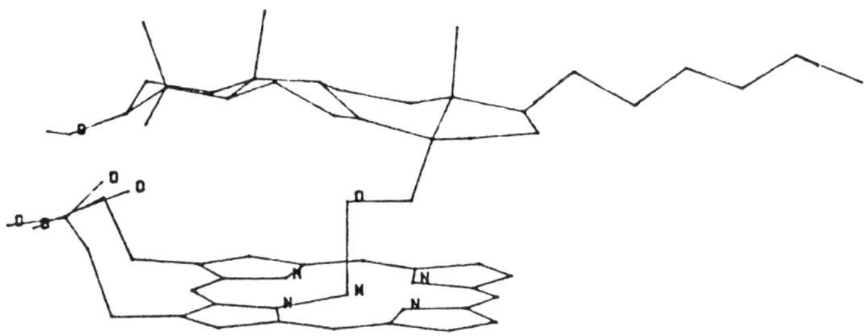

Figure 3, for example, places the lanosterol so that the 3β hydroxyl polar group lies over the propionate side chains. To reduce the complexity of this picture one can now replace the lanosterol structure by a surface canopy to represent the extent of the hydrophobic substrate binding site. There is also the facility to code this surface to signify the electronic properties of the substrates such as their electron density, electrostatic potential, or HOMO/LUMO values. Theoretical work of this type is currently suggesting quite remarkable complementarity of electron properties between bound substrates and protein binding sites (10).

The Shapes of Bound Antagonists

At the beginning of this study the crystal structures of specimen antagonists were unknown. Theoretically the task of calculating all the low energy shapes for just one molecule of interest is considerable. A complete global minimisation for a typical triazole fungicide eg. diclobutrazol with five axes of rotation, sampled at 30° intervals, involves a quarter of a million individual calculations. Even with a large computer this severely degades the quality of calculation which can be done at each point. A strategy was used, therefore, which attempted to reduce this number to a manageable level. Firstly, a crude molecular mechanics method based on Van der Waals contacts was used to eliminate from a full conformational search all those shapes which are sterically too high in energy to be considered for further analysis. All the remaining steric minima were then analysed using semi-empirical molecular orbital methods and subject to a single full ab initio calculation to obtain the absolute minimum energy conformation. The calculated structure for RR-diclobutrazol, for example, agrees very well with the crystal structure as determined by Branch and Nowell (11). Good agreement between calculated and x-ray was also observed for the less fungicidally active isomer (RR-) of triadimenol (12). The calculated structure for both diclobutrazol and triadimenol (Bayer) also seemed consistent with measured n.m.r. coupling constants.

Hydrogen bonding. In setting up these calculations the hydroxyl proton was placed so as to be unavailable for possible hydrogen bonding with the 2-N of the triazole. Theoretically it is well known that extraordinary lengths have to be undertaken to account for this phenomenon properly, even for simple molecules. It seemed more sensible to calculate the other energy contribution theoretically but to look for the formation of internal hydrogen bonding in a dilution experiment in the infra-red. Intra-molecular hydrogen bonds are not observed in the available crystal structures. Infra-red dilution studies show internal hydrogen bonds in both diclobutrazol diastereoisomers but in neither the active RS-, SR- or the less active RR-, SS- triadimenol. In short, the presence of an intra-molecular hydrogen bond between the hydroxyl group and the triazole 2-nitrogen does not in itself relate directly to activity.

Figure 4. Comparison of diclobutrazol and lanosterol in P-450 active site.

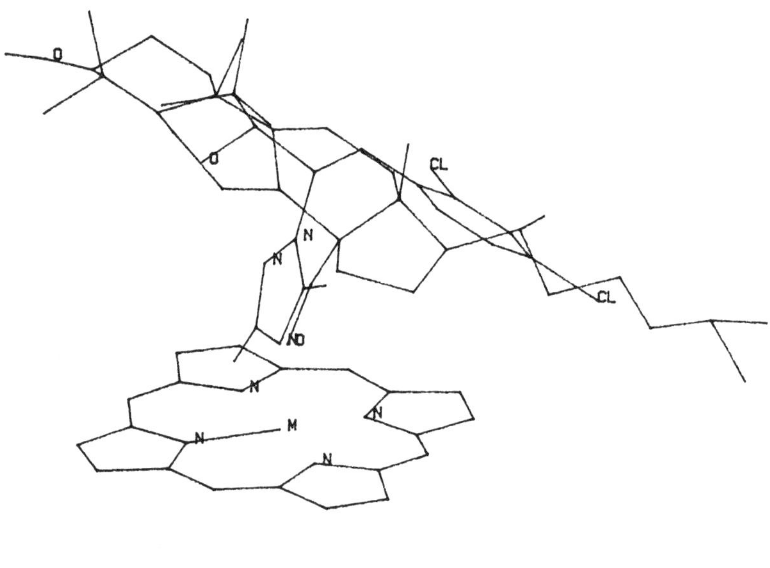

A Comparison of the Antagonists with the Natural Substrate

As an example of the techniques, Figure 4 shows a comparison of the fungicidally active RR- diclobutrazol with the natural substrate lanosterol. The sterol C-32 alcohol is chelated to the iron porphyrin. The three central features of the model cytochrome P-450 can be elucidated. The hydrophobic binding site, the polar region between this hydrophobic region and the heme plane, and a common complexation to the porphyrin iron.

Three features might be noted :

1. The hydrophobic substrate binding site consists of three volumes :-

   a) a region corresponding to the lanosterol A ring which terminates in a polar group, the 3β hydroxyl. The inhibitor makes no use of this space in the enzyme cleft.

   b) a bulky volume occupied by the inhibitor tertiary butyl group and in part by the sterol 6β methyl. It might be expected therefore that extension of the t-butyl group other than onto the A ring would reduce activity. In fact, *in vitro* activity has been shown to be highly sensitive to the size of the this lipophilic moiety.

   c) a deep cavity into which the lanosterol molecule protrudes its side chain and the triazole fungicide projects the benzyl group. This suggests that the benzyl group could be greatly extended, which again agrees with *in vitro* data. The *para*-phenylbenzyl compound for example, shows good activity.

2. The antagonist hydroxyl function lies at a distance relative to the heme group which would make it a candidate for hydrogen bonding to either a heme propionate side chain or an effector acyl group. More generally the polar hydroxyl and the triazole 2-N could mark the interface with polar protein, membrane phospholipid head groups or solution. This agrees very much with the model proposed by Peterson *et al*. (8) for the 5-*exo* hydroxylation of *d*- and *l*-camphor in mammalian cytochrome P-450, and is also consistent with the relationship they noted from steroid metabolism by cytochrome P-450, between the position hydroxylated and its relation to a polar funtional group.

3. The triazole group binds perpendicularly to the heme group and gauche to the iron-nitrogen bonds in the porphyrin plane.

Structure and Activity

Figure 5 now summarises what the model requires for *in vitro* anti 14-demethylase activity. A gauche conformation *is* required between the polar function and the iron chelating group leading to restrictions on the substitution pattern at A, B, C and D. Logically this leads to the possibility of other substitution patterns which will achieve this gauche conformational requirement with groups of the right kind. Substitution at A and B, for example, leaving C and D as hydrogen yields a series of compounds which have all the correct requirements for activity and yet are different in overall appearance.

Figure 5. Model P-450 14-demethylase inhibitor

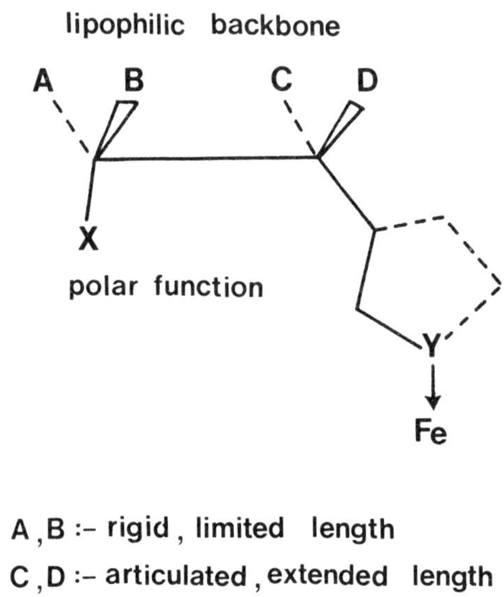

In the search for novel and highly active fungicides or anti-mycotics, active site antagonism is only one of the myriad of processes which the synthetic chemist has to consider. At best then, this approach provides only part of the solution. However, the power of computer graphics in drug and pesticide design lies not so much in the answers it provides but more in the questions it suggests.

Literature cited

1 P. Gadher, E. I. Mercer, B. C. Baldwin and T. E. Wiggins, Pesticide Biochemistry and Physiology, 1983, 19, 1.

2 B. W. Griffin and J. A. Peterson, J. Biological Chemistry, 1975, 250, 6445.

3 K. Murakami and H. S. Mason, J. Biological Chemistry, 1967, 242, 1102.

4 G. F. Gibbons, C. R. Pullinger and K. A. Mitropoulos, Biochem. J., 1979, 183, 309.

5 S. C. Sligar, K. A. Kennedy and D. C. Peerson, Proc. Natl. Acad. Sci. USA, 1980, 77, 1240.

6 J. Pirrwitz, D. Schwarz, H. Rein, O. Ristau, G. R. Janig and K. Ruckpaul, Biochim. et Biophysica Acta, 1982, 708, 42.

7 K. M. Dus, Xenobiotica, 1980, 12, 745.

8 J. A. Peterson, D. H. O'Keeffe, J. Werringloer, R. E. Ebel and R. W. Estabrook, in "Microenvironments and Metabolic Compartmentation", P. A. Srere, R. W. Estabrook, Eds. Academic: New York, 1978 p. 433.

9 G. H. Loew, J. Am. Chem. Soc., 1980, 102, 3655.

10 P. K. Weiner, R. Langridge, J. M. Blaney, R. Schaefer and P. A. Kollman, Proc. Natl. Acad. Sci. USA, 1982, 79, 3754.

11 S. K. Branch, N. H. Anderson, R. S. Loeffler, A. F. Marchington and I. N. Nowell, S.C.I. Symposium: Ergosterol Biosynthesis Inhibitors, Reading UK, 1983.

12 T. Spitzer, J. Kopf and G. Nickless, Cryst. Struct. Comm., 1982, 11, 325.

The Anthracycline Antibiotics: Some Synthetic Endeavours

By R. J. Stoodley

DEPARTMENT OF ORGANIC CHEMISTRY, THE UNIVERSITY, NEWCASTLE UPON TYNE NE1 7RU, U.K.

Daunomycin (1a) was the first clinically effective member of the anthracycline class of antibiotics, being used principally in the treatment of acute leukemia.[1] Its 14-hydroxy derivative (1b), known as adriamycin, possesses important therapeutic advantages[1] and, currently, is the leading anticancer drug, capturing ca. 20% of the market.[2] Adriamycin (1b) plays a significant role in the treatment of acute leukemia, breast cancer, Hodgkin's disease, lymphomas and sarcomas. However, as with all cancer-chemotherapy agents, its use is accompanied by unpleasant side-effects, the most serious of which is dose-relating cardiotoxicity. 4-De-O-methyldaunomycin (1c), known as carminomycin, represents a further natural anthracycline of clinical value.[3]

The synthesis of anthracyclines is a subject of considerable current interest.[4-7] In addition to providing a stimulus for the development of new strategies and methodology, it has afforded structurally modified anthracyclines that would be inaccessible from the natural compounds. Up to now, all syntheses of anthracyclines have involved three basic operations: preparation of the aglycone, preparation of the glycone, and the coupling of these moieties. This strategy has obvious flexibility, enabling anthracyclines modified in the anthracyclinone moiety, the sugar moiety, or both moieties to be assembled. Examples of such modified anthracyclines, which possess chemotherapeutic improvements over their parents and are undergoing clinical evaluation, are 4-demethoxydaunomycin (1d),[6] 4-demethoxyadriamycin (1e),[6] 4'-epiadriamycin (2a)[7] and 4-demethoxy-4'-epiadriamycin (2b).[6]

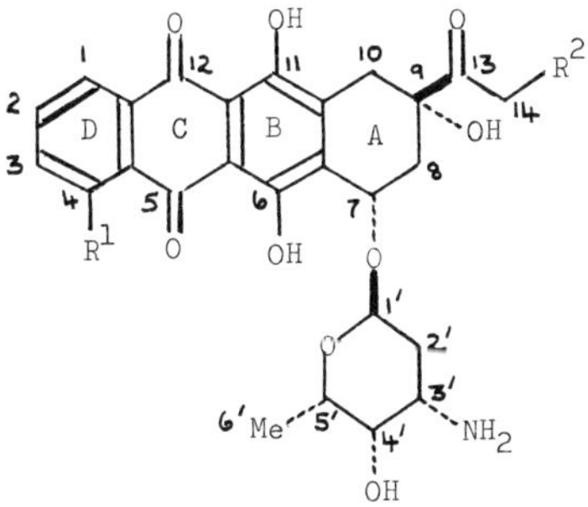

(1)

a; R^1 = OMe, R^2 = H

b; R^1 = OMe, R^2 = OH

c; R^1 = OH, R^2 = H

d; R^1 = R^2 = H

e; R^1 = H, R^2 = OH

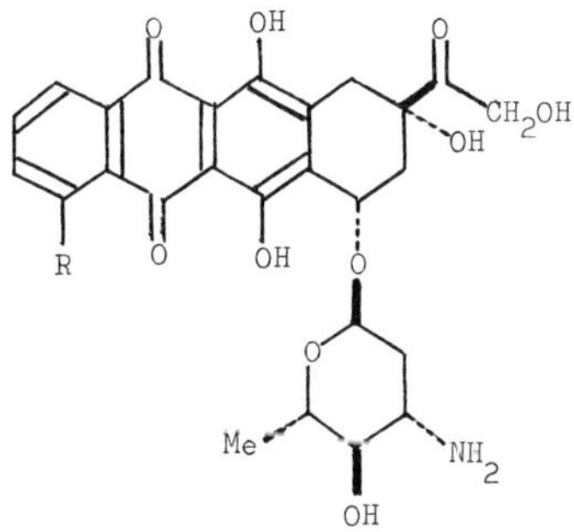

(2)

a; R = OMe

b; R = H

In this discussion, I would like to focus attention upon 4-demethoxy-daunomycin (1d), an orally effective compound that has improved anticancer properties and reduced cardiotoxicity over adriamycin (1b). If its clinical trials are successful, this unnatural anthracycline is likely to be required in kilogram quantities.

4-Demethoxydaunomycin (1d) is currently prepared from (+)-4-demethoxy-daunomycinone (3a) and L-daunosamine (4).[6,8] To effect the coupling reaction, it is necessary to convert the sugar into the glycal (5), the chloride (6a), or the chloride (6b). After the glycosidation step, the O- and N-protecting groups are removed and the anthracycline (1d) is isolated as its hydrochloride. Although the required α-glycosides are formed preferentially or even exclusively, the overall yields for the (3a) ⟶ (1d) transformation are not high [ca. 29% for the route involving the chloride (6b)[8]] and chromatographic purification steps are required.

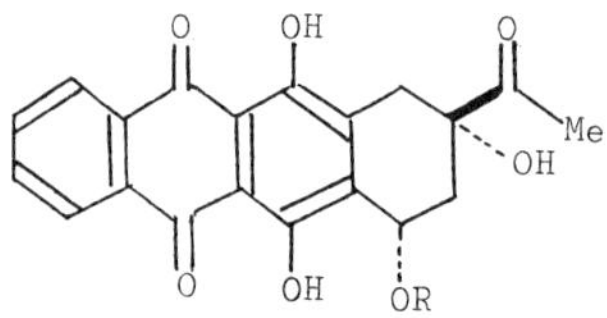

(3)

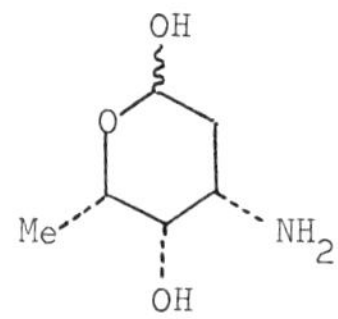

(4)

a; R = H

b; R = Me

c; R = $COCF_3$

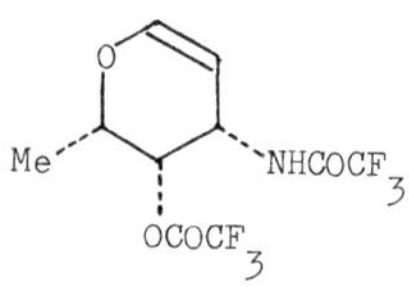

(5)

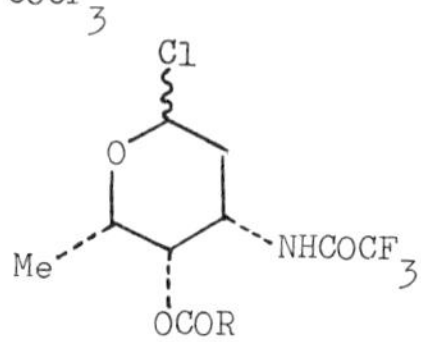

(6)

a; R = CF_3

b; R = $C_6H_4NO_2$-p

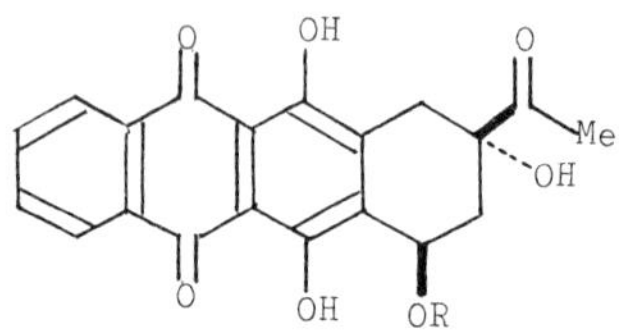

(7)

a; R = H

b; R = Me

c; R = $COCF_3$

Whereas several syntheses of L-daunosamine (4) and its derivatives have been reported,[5,9-14] only three syntheses of (+)-4-demethoxydaunomycinone (3a) are known.[6,8,15] By current standards, the last-mentioned compound represents a modest synthetic target. Nevertheless, it presents two significant challenges relating to diastereocontrol and enantiocontrol, which will now be considered.

Although 4-demethoxy-7-epidaunomycinone (7a) and its 7-O-alkyl derivatives can be converted into 4-demethoxydaunomycinone (3a) by a trifluoroacetolysis - solvolysis sequence, the transformations are inefficient and require a chromatographic purification step.[6,16] For example, when treated with trifluoroacetic acid followed by sodium hydrogen carbonate, the compound (7b) gave 4-demethoxydaunomycinone (3a) in 44% yield.[16] In part, the inefficiency is associated with the thermodynamics of the trifluoroacetolysis reaction, in which the trifluoroacetate (3c) is favoured over its 7-epimer (7c) to the extent of ca. 4 : 1.[16] Since the majority of syntheses of 4-demethoxydaunomycinone (3a) rely upon such equilibration reactions, they are inevitably inefficient.

In fact, in spite of the large number of synthetic endeavours involving the anthracyclinone (3a), the diastereocontrol challenge has rarely been accepted. One exception, reported by Krohn and Tolkiehn,[17] is shown in Scheme 1. The ketone (10), prepared from the cycloadduct of naphthazarin (8) and the diene (9) by hydrolysis, was attacked by ethynylmagnesium bromide from the least-hindered side to give, following aerial oxidation under basic conditions, predominantly the diastereoisomer (11). After elaboration of the D-ring, the resultant anthracyclinone (12) was subjected to the action of mercury(II) oxide and sulphuric acid to afford a ca. 20 : 1 mixture of 4-demethoxydaunomycinone (3a) and its 7-epimer (7a). Following recrystallisation, the overall yield of the compound (3a) was ca. 30% [based upon the dienophile (8)].

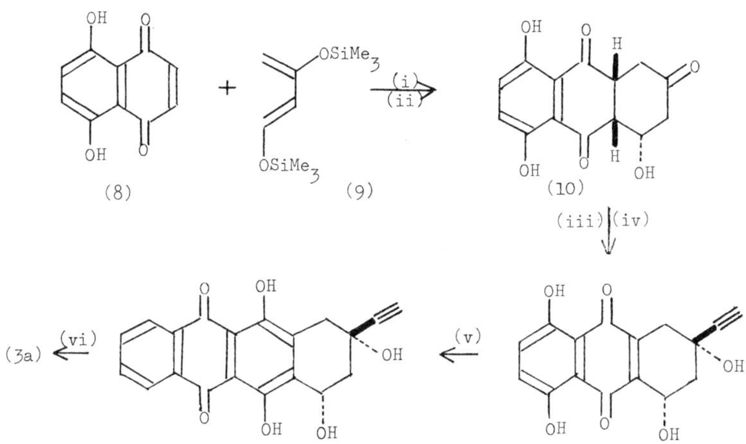

Steps: (i) 20°C (ii) H^+ (iii) HC≡CMgBr (iv) NaOH-O_2

(v) Δ -CH_2:CH·CH:CH(OAc), NaOH-O_2 (vi) HgO - H_2SO_4

Scheme 1

A diastereocontrolled synthesis of 4-demethoxy-7-O-methyldaunomycinone (3b), developed in the author's laboratory,[18] is summarised in Scheme 2. The stereostructure of the compound (15) was confirmed by X-ray analysis, indicating that the cycloaddition had occurred by way of the least-hindered endo transition state and that epimerisation of a bridgehead hydrogen atom had not accompanied the hydrolysis. As in Krohn's case, the diastereocontrol was achieved by steric effects, the acetylide attacking the ketone (16) from the least-hindered face to give almost exclusively the acetylenic alcohol (17). It is noteworthy that the sequence, which afforded the compound (3b) in 22% overall yield [based upon the dienophile (13)], involved no chromatography.

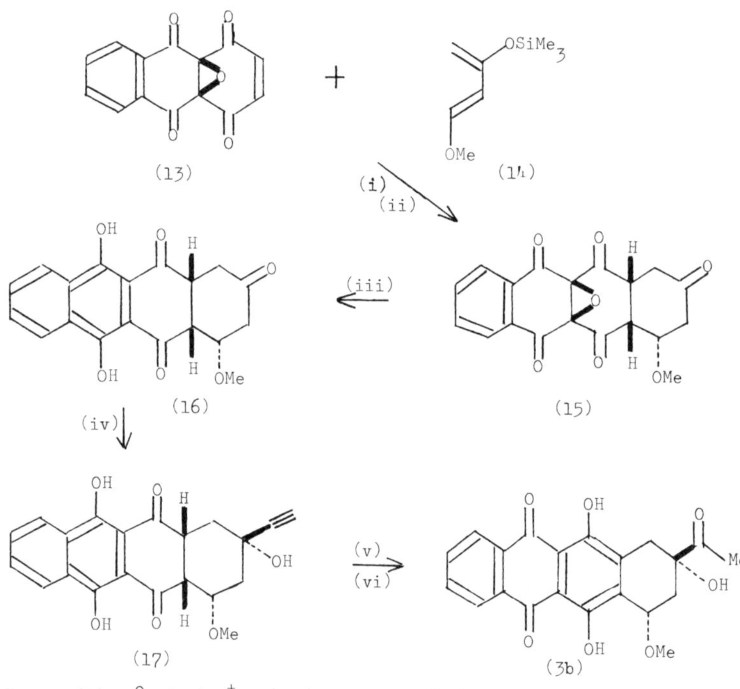

Steps: (i) 20°C (ii) H^+ (iii) $Na_2S_2O_4$ (iv) HC≡CMgBr

(v) $Pb(OAc)_4$ (vi) $HgO-H_2SO_4$

Scheme 2

The synthesis of the BCD-ring synthon (13) is deserving of comment, since quinizarin (18) - a tonnage chemical used in the dyestuff industry - is employed as the starting material. Oxidation of quinizarin (18) with lead(IV) acetate afforded the diquinone (19), which was converted into the dienophile (13) by the action of m-chloroperoxybenzoic acid; the overall yield of the compound (13) was 50%.[19] It is noteworthy that, in the synthesis, the dienophile (13) acts as an operational equivalent of the quinizarin tautomer (20). Such a species is inaccessible from quinizarin (18) and, furthermore, although the diquinone (19) is a powerful dienophile, its internal double bond is more reactive towards Danishefsky's diene (14).[20]

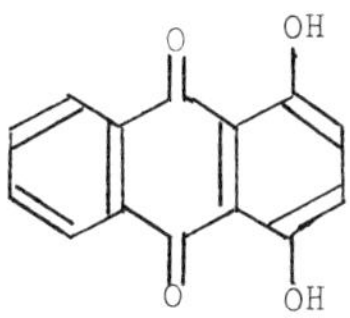

(18)

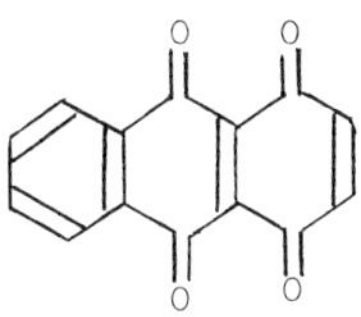

(19)

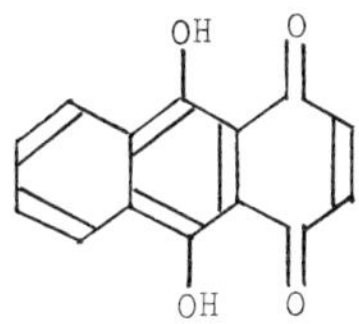

(20)

With this background information, I now wish to address the enantiocontrol challenge presented by (+)-4-demethoxydaunomycinone (3a). Although the compound has been prepared from resolved precursors in two studies,[6,8] the only asymmetric synthesis is that of Terashima's group.[15] The optically pure AB-ring synthon (25) was obtained from the precursor (21) by the route outlined in Scheme 3. A new reducing agent, prepared from lithium aluminium hydride, (-)-N-methylephedrine and N-ethylaniline, was devised to effect the initial reduction of the ketone (21). The derived alcohol (22) possessed an enantiomeric excess of 92% and, after recrystallisation, was obtained in an optically pure state. Epoxidation of the allylic alcohol (22) gave a 19 : 1 mixture of the oxirane (23) and its diastereoisomer. The mixture was then subjected to the action of lithium aluminium hydride and Fetizon's reagent to yield, after recrystallisation, the optically pure AB-ring synthon (25) in ca. 42% overall yield.

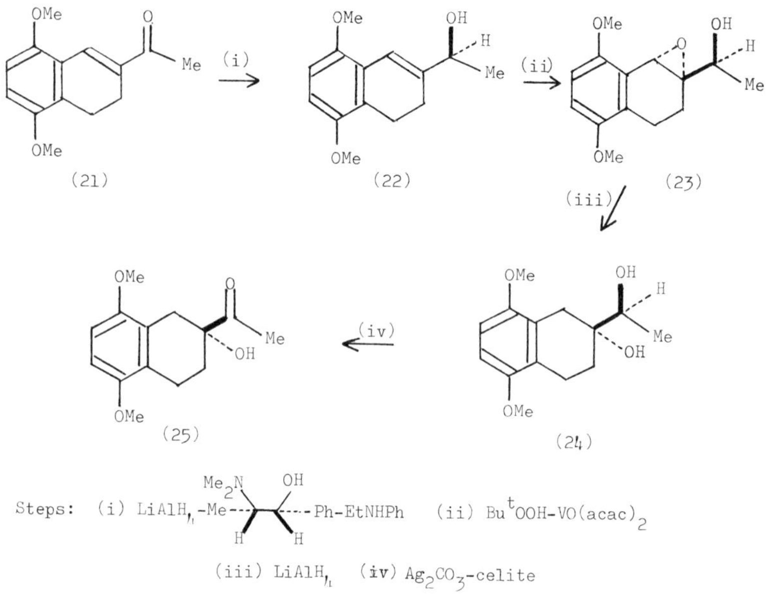

Scheme 3

The transformation of the hydroxy-ketone (25) into (+)-4-demethoxy-daunomycinone (3a), illustrated in Scheme 4, used methodology pioneered by Wong and his co-workers and refined by Arcamone's group. It should be noted that, under the Friedel-Craft conditions, the optically pure ketone (25) reacted with the anhydride (26) with concomitant de-O-methylation and partial racemisation to give the tetracycle (27) with an enantiomeric excess of 71%. Several recrystallisations were required to produce the optically pure tetracycle (27) which was then converted, via its methylated dioxolan derivative (28), into a mixture of the compounds (3b) and (7b). The synthesis was completed by a trifluoroacetolysis - ammonolysis sequence, the inadequacy of which has already been mentioned. The overall yield of (+)-4-demethoxy-daunomycinone (3a) was ca. 10% [based upon the AB-ring synthon (25)].

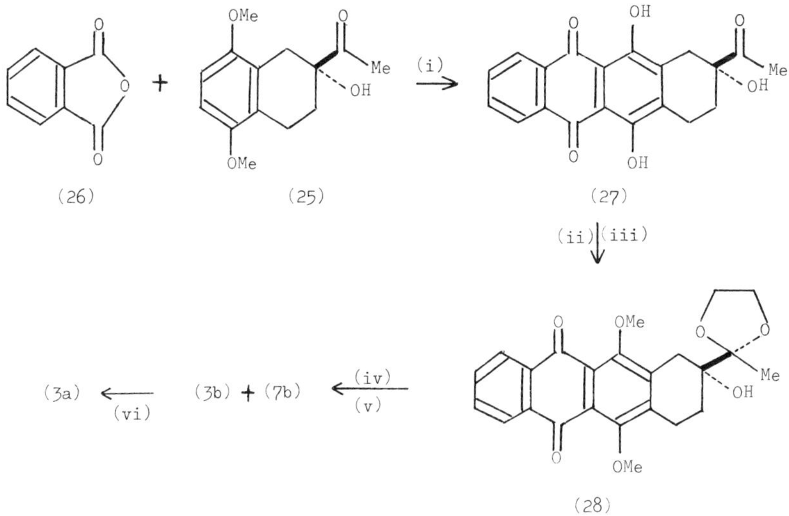

Steps: (i) Δ -$AlCl_3$-NaCl (ii) Me_2SO_4-K_2CO_3 (iii) $(CH_2OH)_2$-H^+

(iv) NBS-hν, MeOH, H^+ (v) $AlCl_3$ (vi) CF_3CO_2H, NH_3

Scheme 4

In the hope of introducing enantiocontrol into the synthesis shown in Scheme 2, we decided to prepare an analogue of Danishefsky's diene (14), in which the methyl group was replaced by an enantiomerically pure ligand. To be of value, such a diene had to satisfy four important requirements. First, it should easily be prepared from readily available and inexpensive starting materials. Second, it should undergo the cycloaddition reaction with the dienophile (13) to give the desired cycloadduct. Third, its ligand should be compatible with the reagents to be used in the subsequent steps of the synthesis. Fourth, its ligand must be removable from the 7-O-substituted 4-demethoxydaunomycinone (or a presursor) without cleavage of the C(7)-O bond.

These considerations led to the selection of the compound (29) as the target diene. It is noteworthy that if the cycloaddition reaction and functionalisation steps were to be successful, a new strategy for the synthesis of anthracyclines would emerge. Furthermore, if the diastereoselection turned out to be the opposite of that required, there would be the possibility of remedying the situation by the use of a wide range of other sugars, including those of the L-series.

The synthesis of the diene (29) was relatively straightforward. Thus when the bromide (30) (readily prepared from D-glucose) was added to a dimethyl sulphoxide solution of the salt (31) (easily prepared from acetone, ethyl formate and sodium methoxide), the enone (32) was obtained. Under the conditions devised by Danishefsky to prepare the diene (14) from trans-4-methoxybut-3-en-2-one, the enone (32) was transformed into the diene (29). Although the overall yield of the diene (29) was only ca. 26% [mainly due to a competing reaction in which the bromide (30) underwent an elimination of hydrogen bromide], no chromatography was involved in the sequence and both the enone (32) and the diene (29) were nicely crystalline compounds.[21]

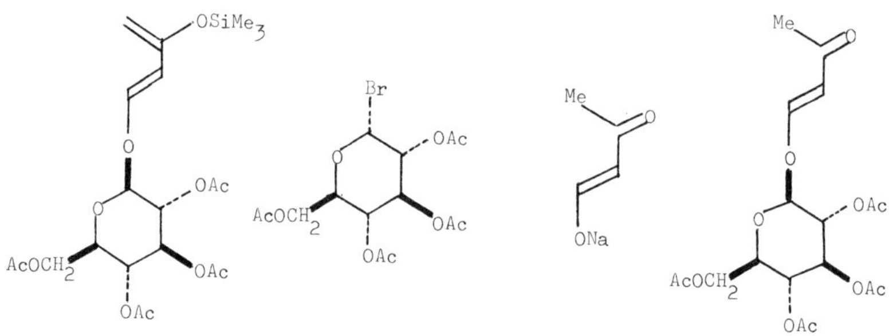

(29) (30) (31) (32)

When the diene (29) was treated with the BCD-ring synthon (13) and the crude cycloadduct subjected to acidic hydrolysis, only one diastereoisomeric ketone, i.e. either (33) or (34), was isolated. The homogeneity of the sample, obtained in 57% yield after recrystallisation [based upon the diene (29)], was established by 360 MHz n.m.r. spectroscopy. Irrespective of the stereostructure of the ketone, this was an exciting result since it implied that the sugar-based diene (29) had undergone the cycloaddition reaction at only one of its two diastereotopic faces. That the ketone did in fact possess the required geometry, i.e. (33), was inferred on the basis of subsequent chemistry. Thus when subjected to the action of zinc and acetic acid, the compound (33) was converted into the leuco-derivative (35) which reacted with ethynylmagnesium bromide followed by lead(IV) acetate to give the anthracycline (36). By a hydrolysis-hydration sequence, the last-mentioned compound was transformed into (+)-4-demethoxydaunomycinone (3a), which possessed an optical rotation close to that recorded in the literature. The overall yield of (+)-demethoxydaunomycinone (3a) was 22% [based upon the diene (29)], and, significantly, no chromatography was involved in the synthesis.[21]

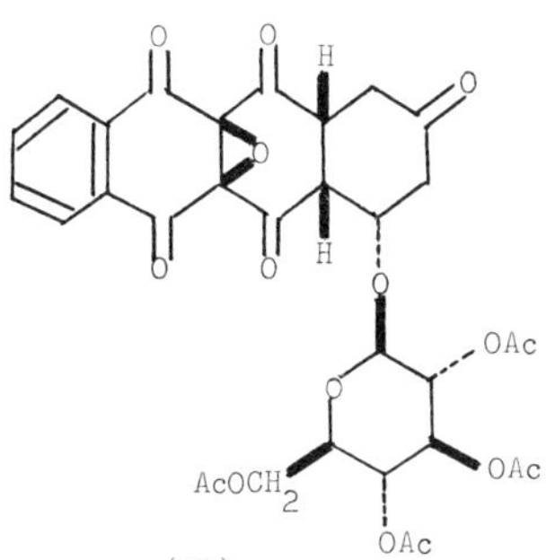

(33)

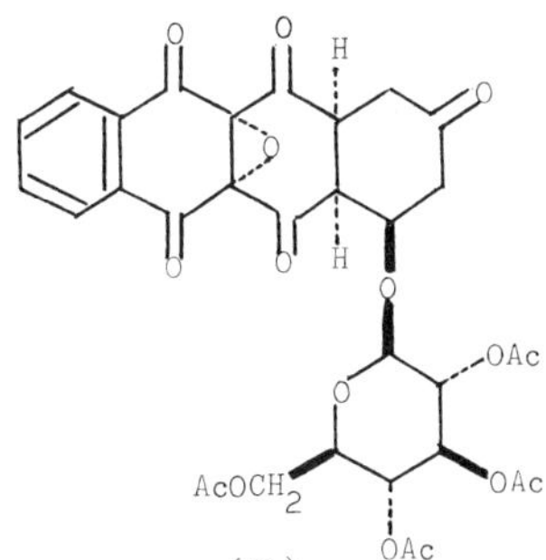

(34)

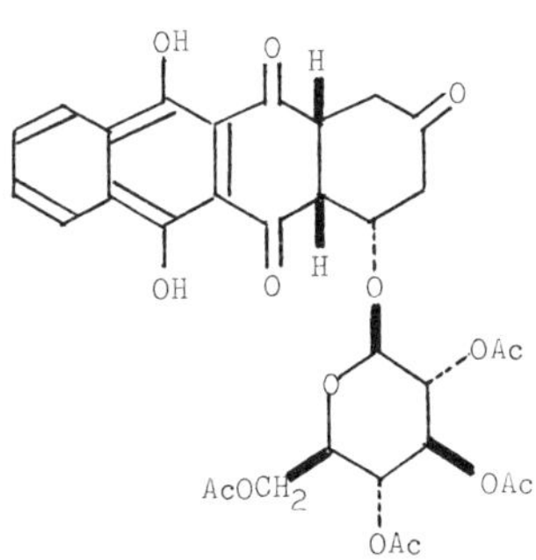

(35)

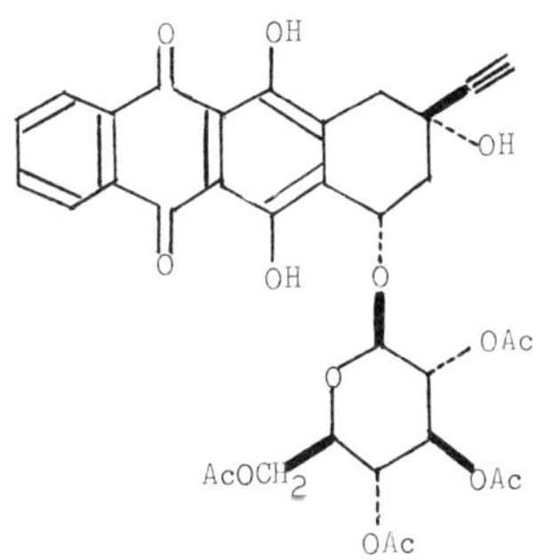

(36)

The origin of the remarkable diastereofacial reactivity exhibited by the sugar-based diene (29) is an intriguing mechanistic issue. In general, acyclic dienes incorporating an enantiomerically pure ligand have had a poor record of diastereodifferentiation. A notable exception, however, is Trost's diene (37) which, for example, reacts with juglone in the presence of boron triacetate to give the cycloadduct (38) exclusively.[22] It has been suggested that the compound (37) adopts the conformation (39) for the cycloaddition reaction, in which a charge-transfer interaction exists between the phenyl group and the diene moiety.

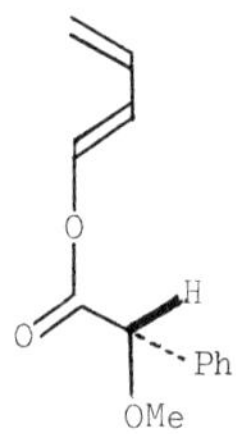

(37)

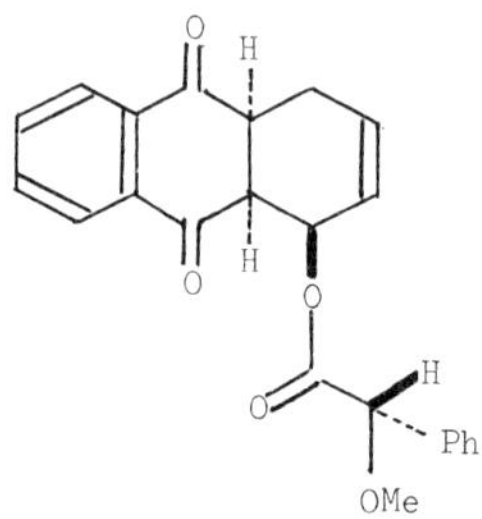

(38)

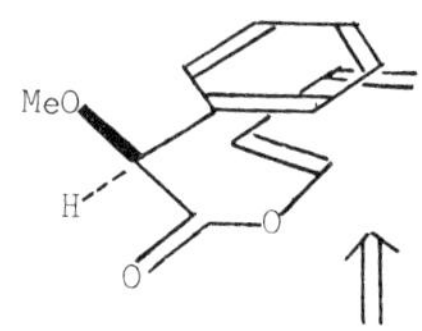

(39)

In the case of the sugar-based diene (29), it is evident that the transition state (40), leading to the cycloadduct (41), must be substantially favoured over its diastereoisomeric counterpart. How does the sugar moiety exert this dramatic effect? X-Ray crystallography and n.O.e.-difference spectroscopy have provided some information regarding the conformation of the diene (29) and it is tempting to extrapolate this to interpret the diastereodifferentiation. The X-ray structure (42) reveals that the CH_2:C($OSiMe_3$)·CH:CH·O·C moiety is planar and that the C(1)-O(ring) bond of the sugar is at 87° with respect to the C-O bond of the CH:CH·O moiety. That a similar conformational situation prevails in deuteriochloroform solution is suggested by n.O.e.-difference spectroscopy. Thus irradiation of the anomeric proton caused a 13% enhancement of the oxygenated methine proton and a 4%

enhancement of the non-oxygenated methine proton of the CH:CH·O group; the anomeric proton was enhanced to the extent of 6% and 3% when the former and latter methine protons were respectively irradiated.

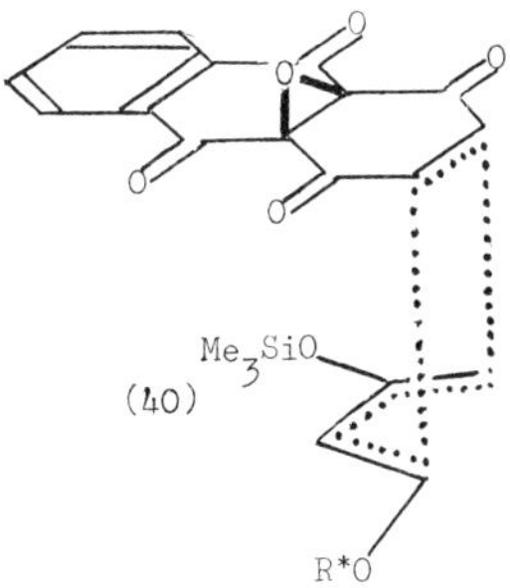

(40)

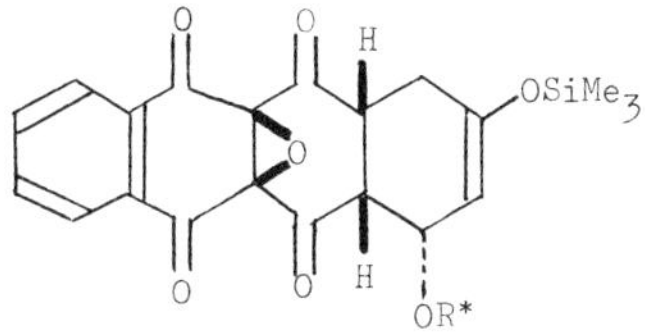

(41)

R* =

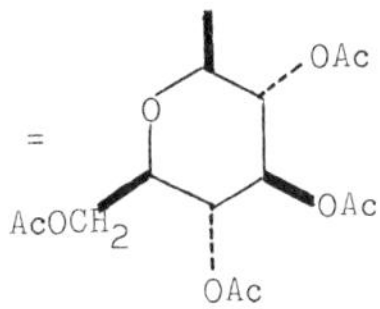

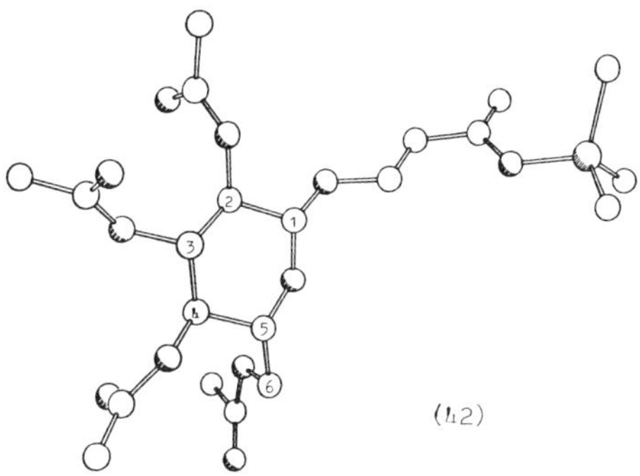

(42)

On the basis of the available evidence, it is enticing to invoke that the diene (29) reacts by way of the conformation (43), in which the pertinent feature is a near-orthogonal arrangement of the C(1)-O(ring) bond of the sugar and the C-O bond of the CH:CH·O moiety. The dienophile (13) then adds from the least-hindered face, avoiding the generation of a <u>cis</u> co-planar 1,3-interaction with the C(1)-O(ring) bond of the sugar. Presumably, the orthogonal geometry, which is a manifestation of the <u>exo</u>-anomeric effect,[23] is stabilised by the feedback of the lone-pair p-orbital electrons on the anomeric O atom to the antibonding σ^*-orbital of the C(1)-O(ring) bond of the sugar.

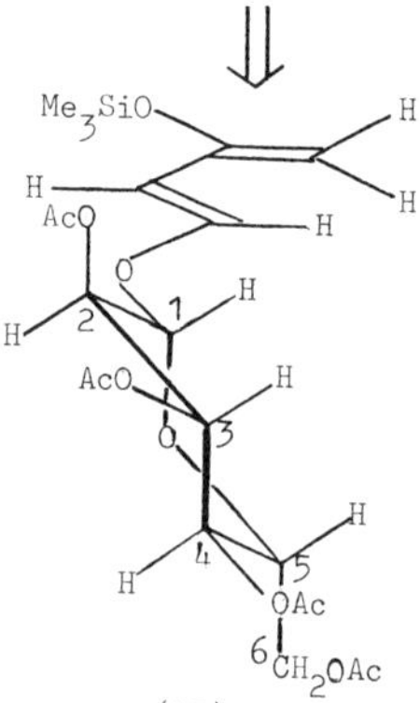

(43)

Acknowledgements

The reduction of ideas to practice rests upon the skills and contributions of one's co-workers. Drs. Philip Harland and Ramesh Gupta were responsible for the work involving the sugar-based diene and I would like to thank them most warmly for their efforts. My thanks are also due to Dr. David Williams, of Imperial College, for the X-ray analyses, and to Dr. Ian Sadler, of Edinburgh University, for the n.O.e.-difference spectra. The financial support of the University of Newcastle upon Tyne and the S.E.R.C. are also gratefully acknowledged.

References

[1] F. Arcamone, "Doxorubicin Anticancer Antibiotics", Academic Press, New York, 1981, Chapter 1.

[2] S. Penco, La Chimica e L'Industria, 1983, 65, 359.

[3] Ref. 1, Chapter 8.

[4] T.R. Kelly, "Annual Reports in Medicinal Chemistry", Academic Press, New York, 1979, Vol. 14, Chapter 28.

[5] Ref. 1, Chapter 2.

[6] Ref. 1, Chapter 7.

[7] Ref. 1, Chapter 6.

[8] M.J. Broadhurst, C.H. Hassall and G.J. Thomas, J.Chem.Soc.,Perkin Trans. I, 1982, 2249.

[9] I. Dyong and R. Wiemann, Chem.Ber., 1980, **113**, 2666.

[10] G. Fronza, C. Fuganti and P. Grasselli, J.Chem.Soc.,Chem.Commun., 1980, 442.

[11] A. Grugnola, P. Lombardi, C. Gandolfi and F. Arcamone, Gazz.Chim.Ital., 1981, **111**, 395.

[12] P.M. Wovkulich and M.R. Uskokovic, J.Am.Chem.Soc., 1981, **103**, 3956.

[13] C. Fuganti, P. Grasselli and G. Pedrocchi-Fantoni, Tetrahedron Lett., 1981, **22**, 4017.

[14] C. Fuganti, P. Grasselli and G. Pedrocchi-Fantoni, J.Org.Chem., 1983, **48**, 909.

[15] N. Tanno and S. Terashima, Chem.Pharm.Bull., 1983, **31**, 821.

[16] J.S. Swenton, D.K. Anderson, D.K. Jackson and L. Narashinhau, J.Org.Chem., 1981, **46**, 4825.

[17] K. Krohn and K. Tolkiehn, Tetrahedron Lett., 1978, 4023.

[18] D.A. Jackson and R.J. Stoodley, J.Chem.Soc.,Chem.Commun., 1981, 478.

[19] M. Chandler and R.J. Stoodley, J.Chem.Soc.,Perkin Trans. I, 1980, 1007.

[20] T.R. Kelly, R.N. Goerner, Jr., J.W. Gillard and B.K. Prazak, Tetrahedron Lett., 1976, 3869.

[21] R.C. Gupta, P.A. Harland and R.J. Stoodley, J.Chem.Soc.,Chem.Commun., 1983, 754.

[22] B.M. Trost, D. O'Krongly and J.L. Belletire, J.Am.Chem.Soc., 1980, **102**, 7595.

[23] R.U. Lemieux, Pure App. Chem., 1971, **25**, 527.

Natural Product Synthesis *via* π-Allyltricarbonyliron Lactone Complexes

By G. D. Annis, E. M. Hebblethwaite, S. T. Hodgson, A. M. Horton, D. M. Hollinshead, S. V. Ley*, C. R. Self, and R. Sivaramakrishnan

DEPARTMENT OF CHEMISTRY, IMPERIAL COLLEGE, LONDON SW7 2AY, U.K.

Some years ago we became interested in the synthesis of δ-lactone natural products typical examples of which are shown below.

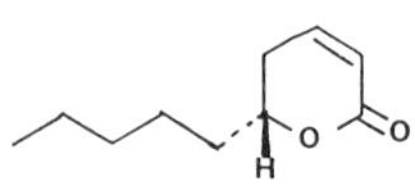

Massoia lactone

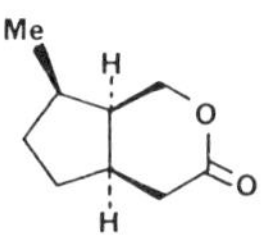

Boschnialactone

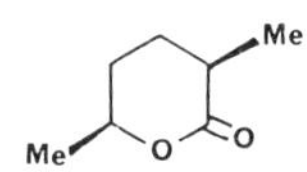

Carpenter bee pheromone

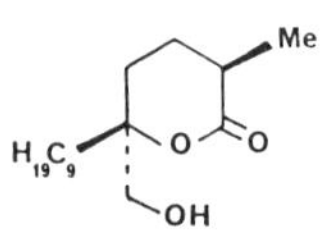

Malyngolide

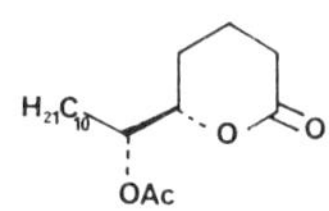

Mosquito oviposition attractant

These molecules have interesting biological properties often as pheromones while others, such as massoia lactone, first isolated from massoibark oil and used as a native medicine in New Guinea, was shown to be a skin irritant and produce systolic standstill in frog heart muscle. Interestingly, massoia lactone has also been isolated from the mandibular glands of two species of formicine ants. Malyngolide on the other hand, which is a natural product isolated from the blue green algae Lyngbya majuscula, was shown to have antibiotic activity against mycobacterium smegmatis and Micrococcus pyogenes.

In an effort to develop a synthetically interesting approach to these molecules we emnbarked on a programme designed to effect the conversion of dienes to δ-lactones.

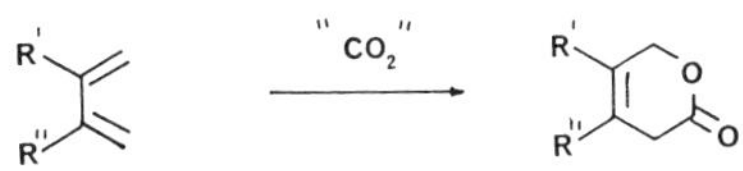

While synthetic equivalents for the addition of CO_2 are known they are not especially regioselective. We chose to investigate the use of π-allyltricarbonyliron lactone complexes as potential precursors for lactones. It had been shown that simple vinyl epoxides reacted with coordinatively unsaturated iron carbonyl species to give crystalline stable π-allyltricarbonyliron lactone complexes.[1] We speculated that these complexes upon oxidation might well undergo carbon-carbon bond formation with corresponding elimination of the iron species to produce either δ- or β-lactones depending on which bonds are formed.[2]

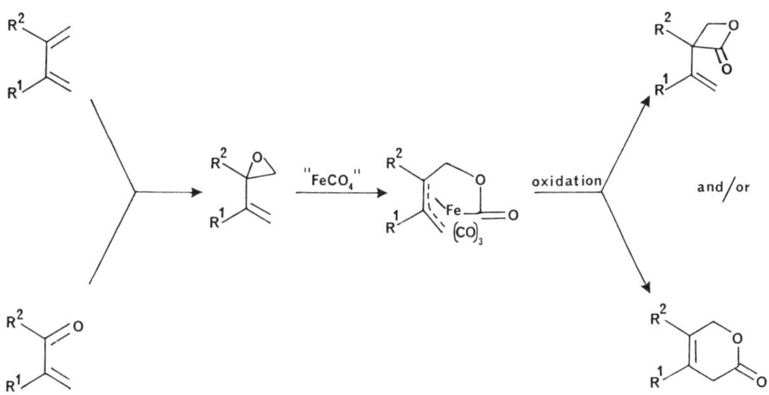

Before discussing our results, however, it is pertinent to comment briefly on the sources of reactive iron carbonyl species. Pentacarbonyliron, which is a toxic and somewhat volatile liquid, can be used to afford coordinatively unsaturated $Fe(CO)_x$ under either irradiation or thermal conditions. However, the use of the easily handled nonacarbonyldi-iron is much better. We find this reacts excellently with vinyl epoxides in hydrocarbon solvents under ultra sonification conditions. Alternatively, Cotton[3] has shown that $Fe_2(CO)_9$ dissociates in tetrahydrofuran to give a $Fe(CO)_4$.THF species; this too we find reacts nicely with vinyl epoxides. $Fe_3(CO)_{12}$, although it is very easily handled, turns out to be a poor source of coordinatively unsaturated iron.

Reactive Iron Carbonyl Sources

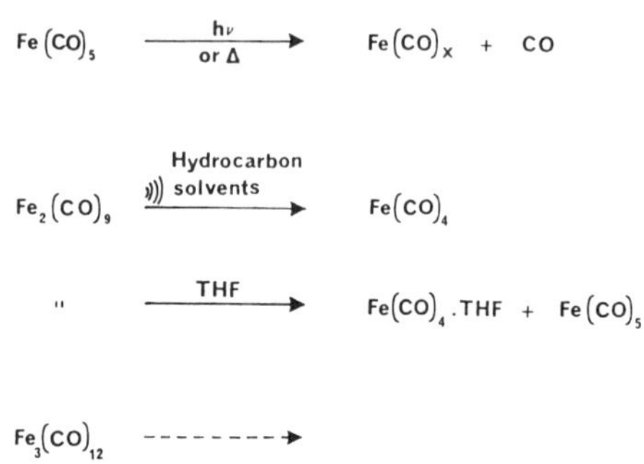

A large number of π-allyltricarbonyliron complexes have been prepared by our group, although for this paper only a limited and representative number of examples will be discussed together with their application for natural product synthesis.

Cyclopentanone, upon reaction with vinylmagnesium bromide followed by I_2 -assisted elimination of water and epoxidation by buffered peracetic acid gives a vinyl epoxide in good overall yield. This reacts with $Fe(CO)_5$ to give two easily separated tricarbonyliron lactone complexes. These we designate as syn and anti isomers to represent the relative placements of the iron tricarbonyl moiety and the bridging oxygen atoms.

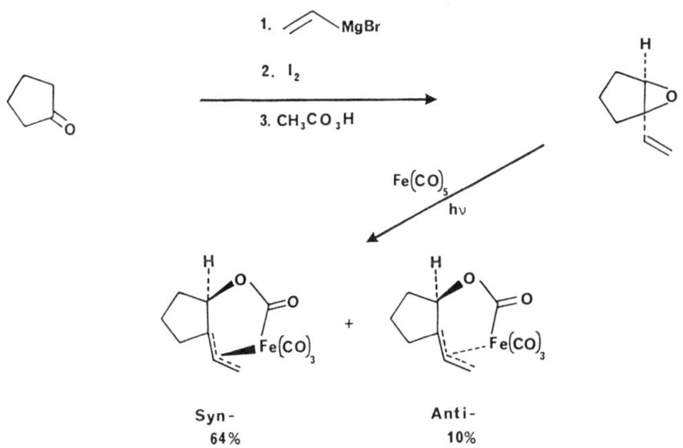

A very important factor is that we find that these two complexes behave differently upon oxidation with ceric ammonium nitrate. From a synthetic point of view this observation could be advantageous. The syn complex for example, upon oxidation with Ce^{IV} in acetonitrile at room temperature, gave both the cis-fused β-lactone and the δ-lactone in 51% and 29% yields respectively. On the other hand the anti isomer only gave the δ-lactone. The corresponding β-lactone is apparently precluded by ring strain.

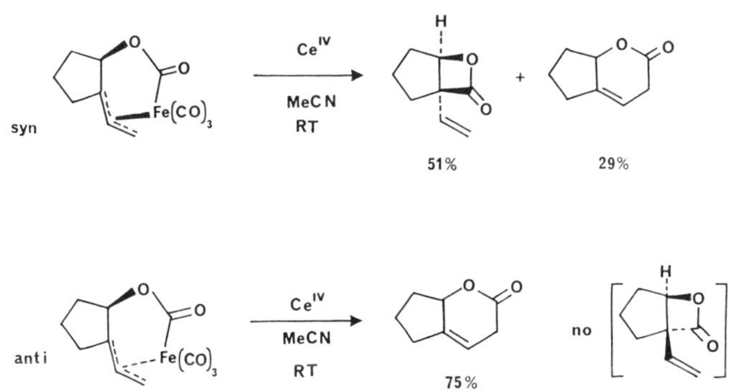

In fact, unless ring strain or other features are present, the major product of oxidation of tricarbonyliron lactone complexes is always the small ring β-lactone. In a further example, formyl cyclopentene was converted to a vinyl epoxide by treatment with dimethylsulphonium methylide. This was converted to an iron complex which upon oxidation with Ce^{IV} gave the β-lactone as the only product in essentially quantitative yield.

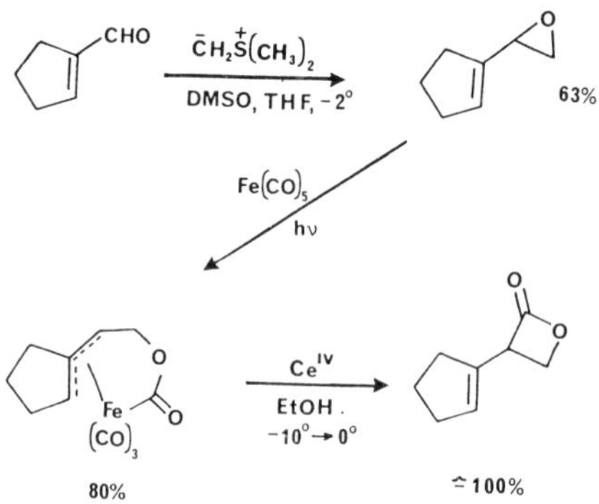

In a similar way the pentanyl substituted butadiene derivative gave the regiospecific vinyl epoxide on oxidation with m-chloroperbenzoic acid. This afforded two isomeric iron complexes with $Fe(CO)_5$ which on oxidation gave β-lactonic products. Once again the stereochemistry of the products reflects the initial geometry of the starting iron complex.

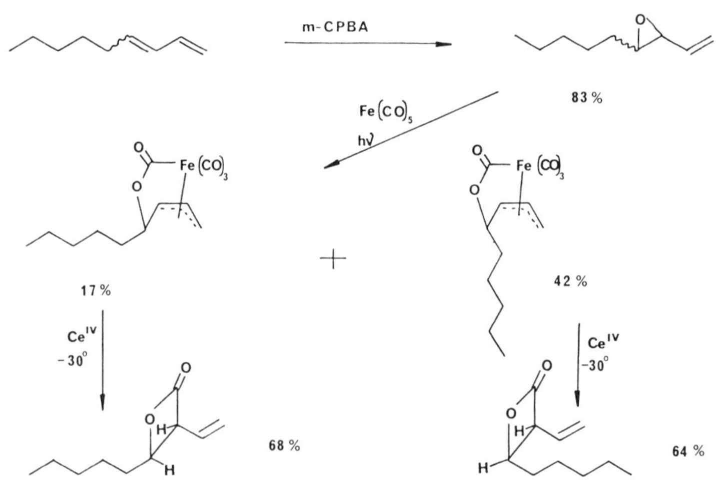

This new methodology appears, therefore, to be a novel way of preparing β-lactones from dienes or enals, but looks worthless as a method for preparing the natural δ-lactones. For this reason we sought a modification which would permit direct conversion of the complexes to δ-lactones. The best of these appeared to be that of exhaustive carbonylation.[4] Thus we find that the pentanyl substituted tricarbonyliron lactone complex, prepared previously, gave the natural product massoia lactone, in one step, by reaction with carbon monoxide at 60 atmos at 190°C in a sealed system using benzene as the solvent.

Massoia lactone

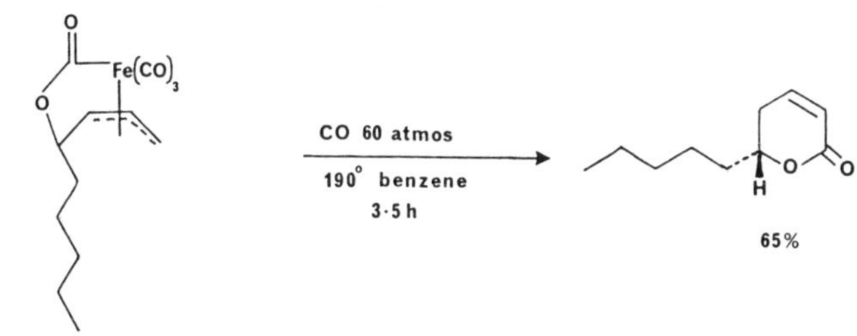

This was a pleasing result which we subsequently were able to translate into the synthesis of three other natural products: the carpenter bee pheromone; the oriental hornet, Vespa orientalis, pheromone; and the antibiotic malyngolide.

Carpenter Bee Sex Pheromone

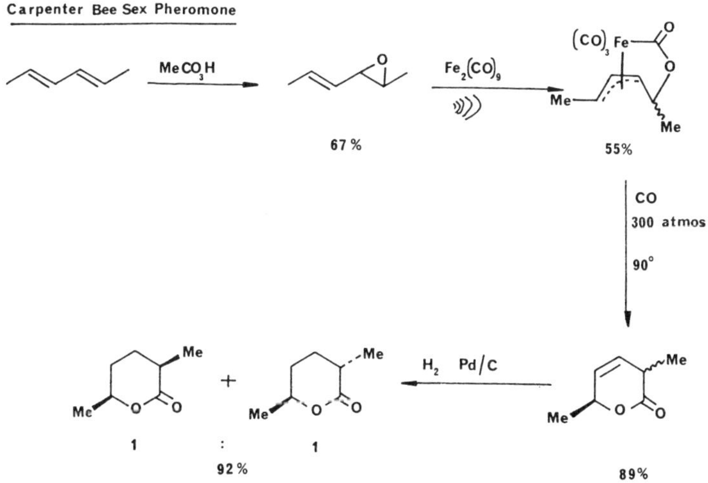

Vespa orientalis Pheromone

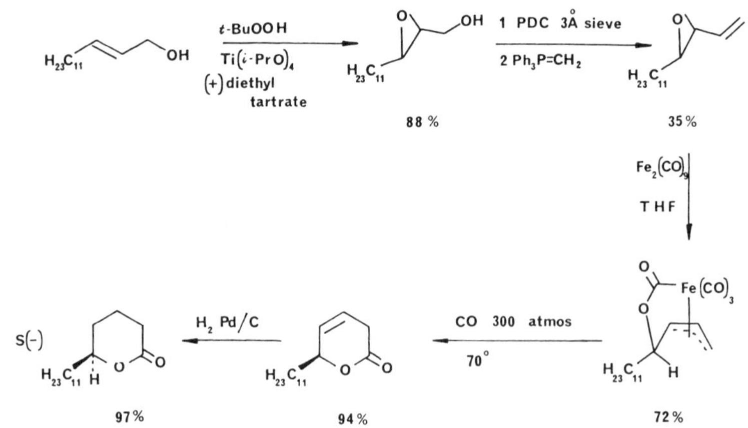

Malyngolide

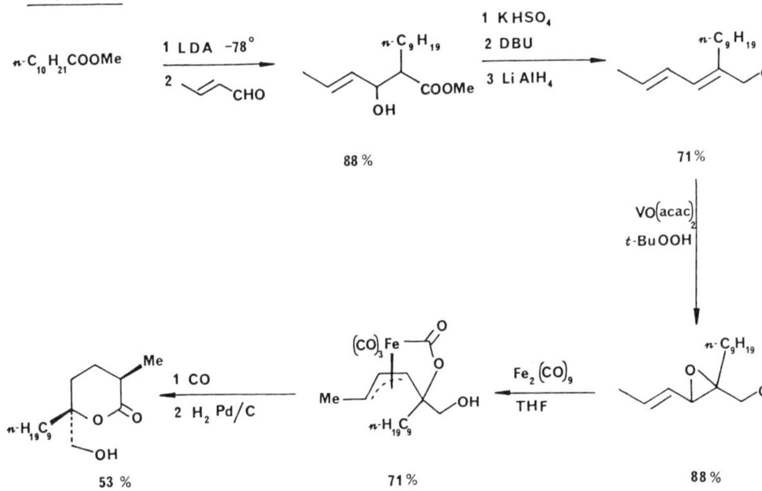

We were thus able to use the metal carbonyl reactions to prepare, in a reasonably selective fashion, both β- and δ-lactones. However, owing to the observation that the smaller ring β-lactones were the predominant product under the oxidation conditions, we speculated that the reaction would be more useful if it could be applied to β-lactam synthesis.

Several pharmaceutically important β-lactams have been investigated recently which have stimulated considerable synthetic interest. These include the carbopenems such as thienamycin, and the monocyclic β-lactams such as the nocardicins and the monobactams of which sulfazacin is an example.

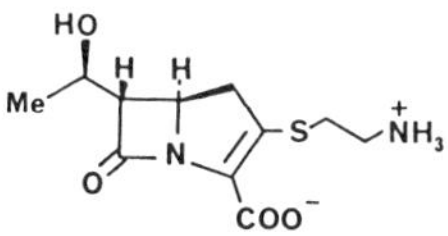

Thienamycin

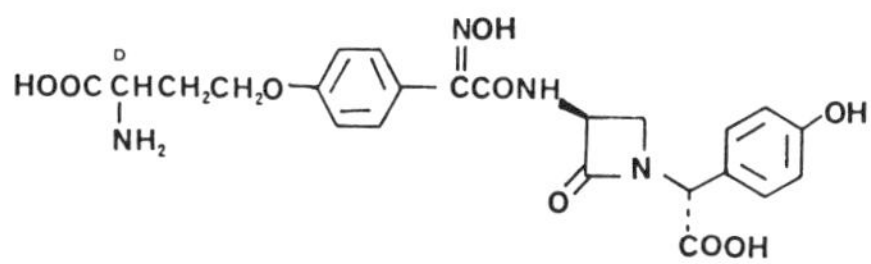

Nocardicin A

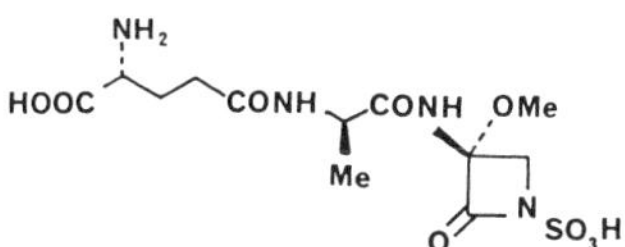

Sulfazacin

In order to test these ideas we find that π-allyl tricarbonyliron lactone complexes react with amines in the presence of various Lewis acids, by an S_N2'-like process, to give the corresponding lactam complexes.[5] In other words, there is no need to synthesise these lactam complexes from vinyl aziridines which are often notoriously difficult to prepare.

In a model study the cyclopentanyl tricarbonyliron complexes prepared earlier could be reacted with benzylamine in the presence of zinc chloride/tetramethylethylenediamine complex to give the corresponding lactam complex in 68% yield. Again these lactam complexes we found to be a nicely crystalline and air stable species which is obviously important if these are to be conveniently used in synthesis. The lactam complex upon oxidation with Ce^{IV} gave the β-lactam as the only observable product in excellent yield. The benzyl group could be removed using sodium in liquid ammonia to provide the parent β-lactam. While this debenzylation procedure has been used in other systems, its use in β-lactam chemistry is much less common.

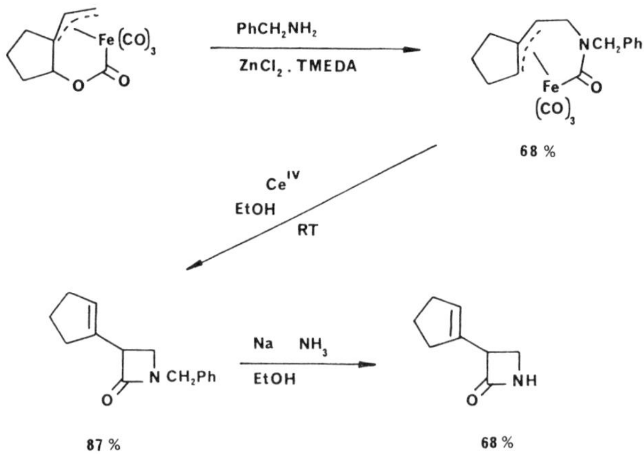

The overall sequence to an alkenyl-substitued β-lactam described above should be contrasted with the normal method of preparation involving the use of a diene with chlorosulphonyl isocyanate where the opposite regioisomers would be obtained.

In an attempt to apply the new methodology to more pharmaceutically interesting β-lactams, such as hydroxyethyl analogues of the monobactams, further reactions were studied. For example, the tricarbonyliron lactone complex derived from isoprene monoepoxide was treated with benzylamine, this time in the presence of diethylaluminium chloride as catalyst. The resulting amine complex was oxidised to the β-lactam in 89% yield. This could be deprotected and sulphonated or, alternatively, the isopropenyl group could be cleaved by ozonolysis and reduced to the hydroxyethyl derivative. In turn this could be deprotected and sulphonated to produce the corresponding hydroxyethyl analogue.

Encouraged by these results we wished to apply the method to more ambitious target molecules. In the first of these, the π-allyltricarbonyliron lactone complex from isoprene monoepoxide was reacted with (S)-dibenzyl-p-hydroxyphenylglycine and a Lewis acid to give a separable mixture of diastereoisomers in 48% and 47% yields respectively with complete retention of the introduced chirality.

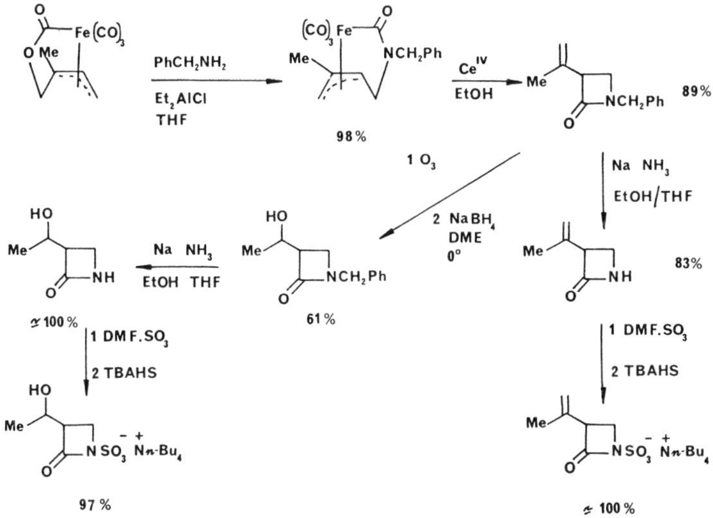

(–) 3-Aminonocardicinic Acid

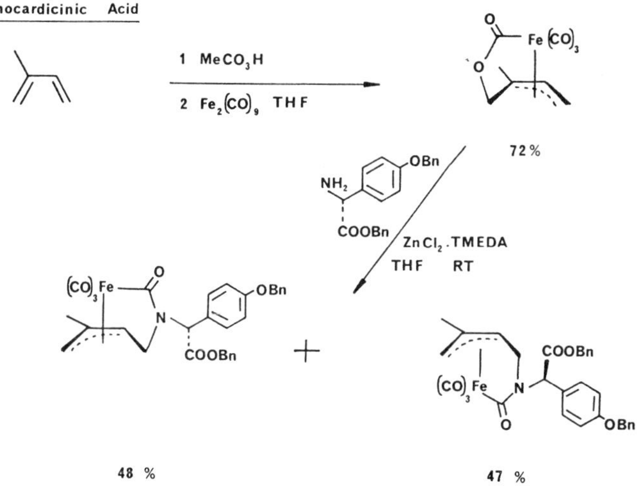

These upon oxidation afforded the isopropenyl β-lactams which could be separately ozonised and reduced with K-selectride to give different diastereo-isomeric ratios of hydroxyethyl nocardicin derivatives. Removal of the benzyl groups gave the deprotected nocardicin analogues.

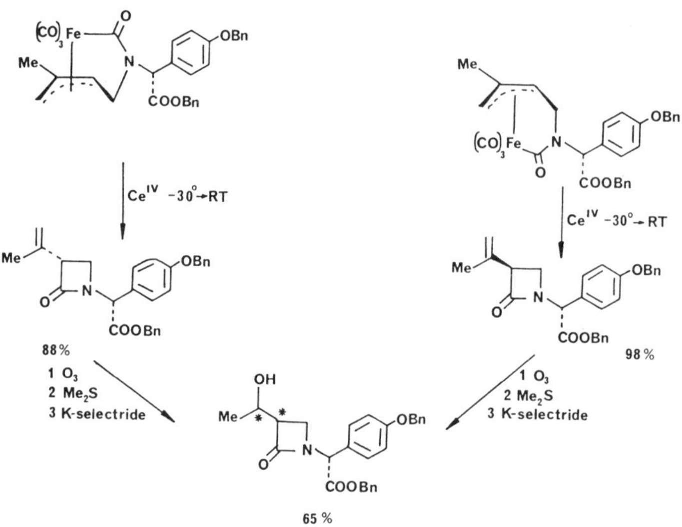

Mesylation of the protected hydroxyethyl derivative followed by elimination gave α-methylene derivative which was cleaved by ozonolysis to the 3-oxo derivative. As this had previously been converted to 3-aminonocardicinic acid[6] and hence the nocardicins, our route therefore now constitutes a formal total synthesis of these species.

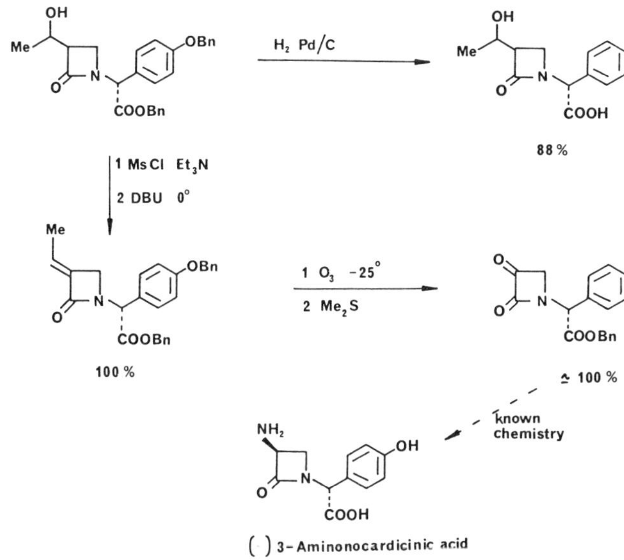
OH
Me
OBn
COOBn
H₂ Pd/C
COOH
88 %
1 MsCl Et₃N
2 DBU 0°
100 %
1 O₃ −25°
2 Me₂S
≏ 100 %
known chemistry
NH₂
( ) 3-Aminonocardicinic acid

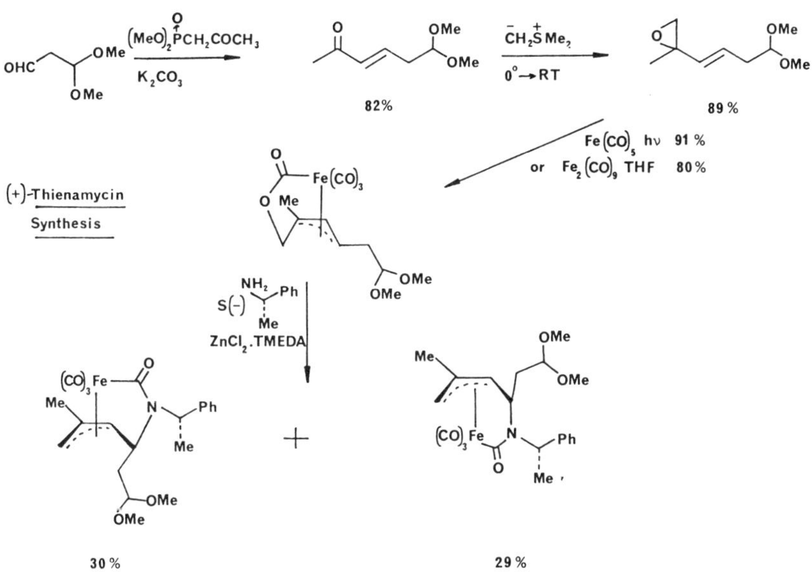
OHC
OMe
(MeO)₂PCH₂COCH₃
K₂CO₃
82%
CH₂SMe₂
0°→RT
89 %
Fe(CO)₅ hν 91 %
or Fe₂(CO)₉ THF 80 %
Fe(CO)₃
(+)-Thienamycin Synthesis
S(−)
ZnCl₂.TMEDA
Ph
30 %
29 %

Lastly, 3,3-dimethoxypropanal was reacted with the anion from dimethyl-2-oxoprop-1-yl phosphonate to give an enone which was readily cyclopropanated to produce a substituted vinyl epoxide. This was converted to the π-allyltricarbonyliron lactone complex then reacted with a chiral amine, S(-)α-phenylethylamine, to produce a separable diastereoisomeric mixture of tricarbonyliron lactam complexes.

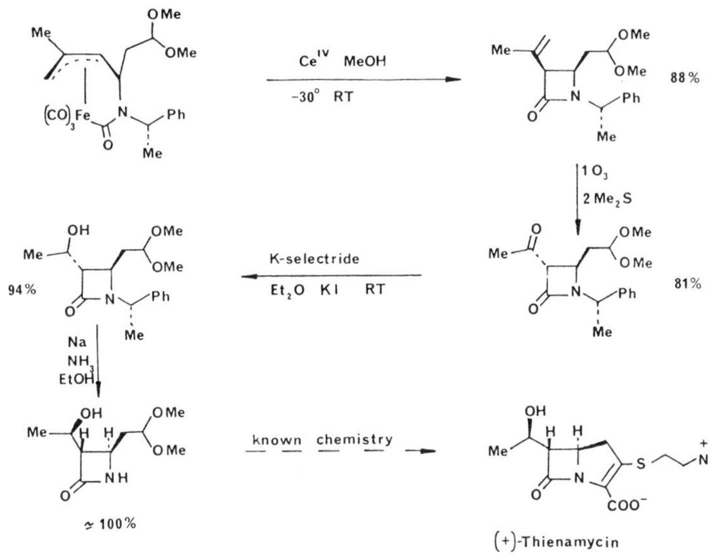

Oxidation of one of these complexes with ceric ammonium nitrate at -30°C produced the cis-β-lactam in 88% yield. Ozonolysis of the isopropenyl side chain followed by epimerisation on silica gel gave the 3-acetyl derivative exclusively. Reduction of the carbonyl group with K-selectride and deprotection gave a β-lactam which had previously been converted to thienamycin.[7] Thus again the route consitutes a formal total synthesis.

We hope we have demonstrated that π-allyltricarbonyliron complexes can be useful precursors for the synthesis of a wide range of lactones and lactams including several natural products.

We thank the Science and Engineering Research Council and Beecham Pharmaceuticals for financial support, and Johnson Matthey PLC for a CASE award (to DMH).

---

1. R.F. Heck and C.R. Boss, J. Am. Chem. Soc., 1964, 86, 2580, H.D. Murdock, Helv. Chim. Acta., 1964, 47, 936, R. Aumann, K. Fröhlick and H Ring, Angew. Chem. Internat., Edn., 1974, 13, 275, K-N Chen, R.M. Moriarty, B.G. De Boer, M.R. Churchill and H.J.S. Yeh, J. Am. Chem. Soc., 1975, 97, 5602, Y. Becker, A. Eisenstadt and Y. Shvo, Tetrahedron, 1974, 30, 839, G.D. Annis, S.V. Ley, J. Chem. Soc., Chem. Commun., 1977, 581.

2. G.D. Annis, C.R. Self and R. Sivaramakrishnan, J. Chem. Soc., Perkin I. 1981, 270.

3. Eg. F.A. Cotton and P. Lahuerta, Inorg. Chem., 1975, 15, 116 and references therein.

4. R. Aumann, H. Ring, C. Krüger and R. Goddard, Chem. Ber., 1979, 112, 3644.

5. G.D. Annis, E.M. Hebblethwaite and S.V. Ley, J. Chem. Soc., Chem. Commun., 1980, 297, G.D. Annis, E.M. Hebblethwaite, S.T. Hodgson, D.M. Hollinshead and S.V. Ley, J. Chem. Soc., Perkin Trans I, 1983, 2851.

6. K. Chimba, M. Mori, and Y. Ban, J. Chem. Soc., Chem. Commun., 1980, 770.

7. T. Kametani, S-P. Huang, T. Nagahara, and M. Ihara, Heterocycles 1981, 16, 65.

The "Chemist's Enzymes" as Templates for the Synthesis of Biologically Active Molecules

By B. M. Trost

DEPARTMENT OF CHEMISTRY, UNIVERSITY OF WISCONSIN, 1101 UNIVERSITY AVENUE, MADISON, WI. 53706, U.S.A.

Introduction

The synthesis of complex organic molecules requires the ability to manipulate selectively one particular site among a plethora of many in a stereocontrolled fashion. Nature has solved this most demanding task by the evolution of enzymes, exquisite and complex molecules that are the pinnacle of catalysts. Their selectivity has focused a great deal of attention on better use of enzymes and whole cells, through genetic engineering, to solve the problems of man. The extraordinary selectivity of enzymes also represents a limitation that restricts their applications to but a tiny fraction of the complex molecules that are required by man.

While a great deal of the current effort in catalysis focuses on commodity chemicals, the design of man-made "enzymes" for complex synthesis is beginning to unfold a new volume in the history of synthesis. In this context, the term man-made "enzymes" is not used to denote proteins of unnatural sequences that are catalysts, but rather refers to any catalysts that approach enzymes in their selectivity. For the chemist, a transition metal and its attendant ligands form an "active site" which can be varied electronically and sterically in order to tune its reactivity and, consequently, its selectivity. Such a catalyst has the promise of becoming the "chemist's enzymes".

Templates for Allyl Alkylation

Palladium complexes convert relatively inert functionalities such as allyl carboxylates, sulfones, amines, and nitro compounds into reactive alkylating agents.[1] To illustrate, the polyfunctional sulfone **1** can undergo clean allylic alkylation to give **2** in which the palladium selectively ionizes the sulfone substituent, or typical olefination of a carbonyl group to give **3**.[2] In neither case is a protecting group required. Subsequent

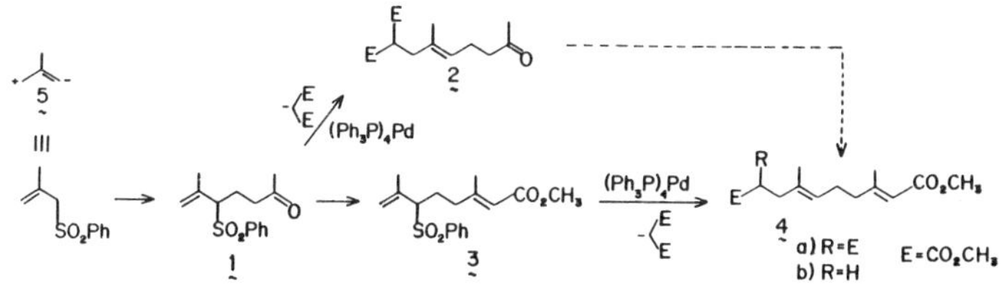

palladium catalyzed alkylation of 3 to 4, again without protecting groups, provides 4a which is easily decarbomethoxylated to the dimethyl ester of the sex pheromone of the Monarch butterfly 4b. Since 1 derives from methallylphenyl sulfone in which the carbon bearing the sulfone behaves as a nucleophilic center and, in the palladium reaction, the allyl sulfone unit serves as an electrophilic center, allyl sulfones can now be considered 1,3-dipoles of the type represented in 5. In the absence of a nucleophile, eliminations to dienes occur smoothly.[2] The ambident behavior of sulfones serves effectively to utilize this dual reactivity in the synthesis of Vitamin A analogs as shown in eq 1.

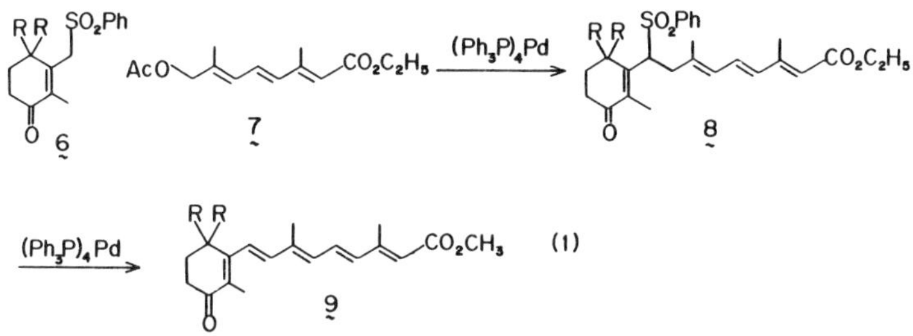

(1)

After serving to generate a carbon nucleophile in 6 for a palladium catalyzed allylic alkylation with allyl acetate 7, the resultant allyl sulfone 8 now can serve as a leaving group to create the Vitamin A analog 9.

A major reshuffling of the normal rules of reactivity occurs with these metal templates in terms of stereochemistry. While synonymous with displacement has been inversion of configuration, palladium catalyzed displacements proceed with net retention of configuration![3] To illustrate, the vinyl epoxide 10 requires a displacement by the internal nitrogen with retention to give the isoquinuclidine 11, a precursor to the iboga alkaloid,

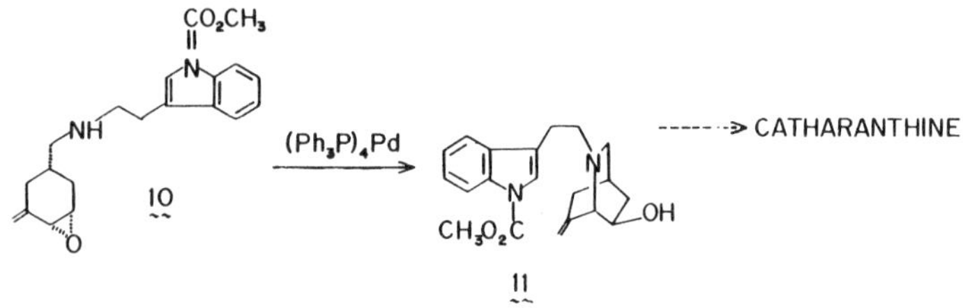

catharanthine - an important building block to the clinically important anti-tumor agents vinblastine and vincristine.[4] Indeed, exposing 10 to a palladium catalyst smoothly isomerizes it to the requisite isoquinuclidine.

Manipulating the nature of the metal template profoundly influences the nature of the alkylation reaction. Thus, alkylation of 12 with a palladium template leads preferentially to 14. By simply switching to a molybdenum template, the

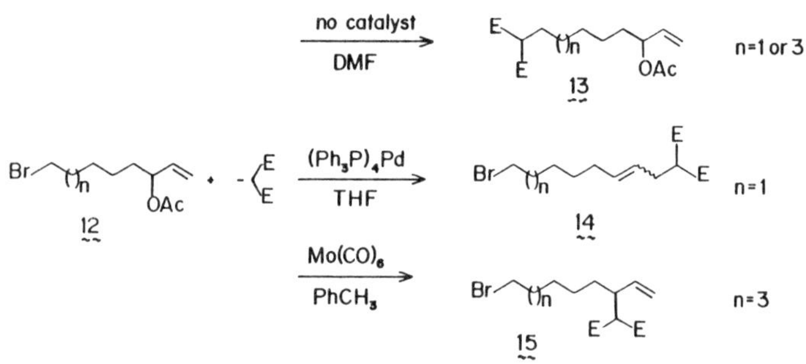

regioselectivity switches.[5] Fine tuning the molybdenum catalyst permits fine tuning of the regioselectivity as shown in eq 2.[6] By

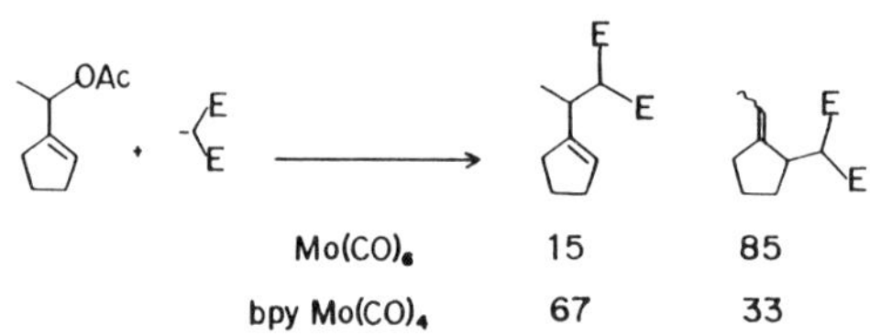

(2)

substituting σ-donor ligands such as 2,2'-bipyridyl (bpy) for the strong π-acceptor ligands CO reorients the reaction from alkylation at the endocyclic to the exocyclic carbon.

Increasing the steric demands of the metal template can have a further effect on regioselectivity. Thus, for cinnamyl acetate, a

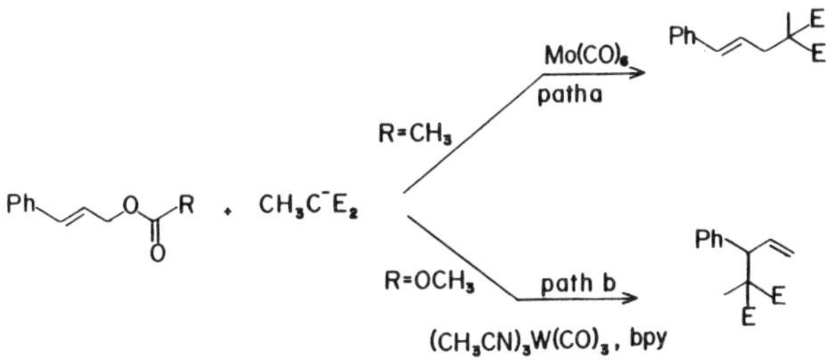

(3)

molybdenum catalyst directs reaction to the less substituted carbon; for cinnamylmethyl carbonate, a sterically more demanding tungsten catalyst steers the nucleophile to the more substituted carbon.[7] In essence, this regioselectivity can be understood by examining the initial products of alkylation I and II, in which

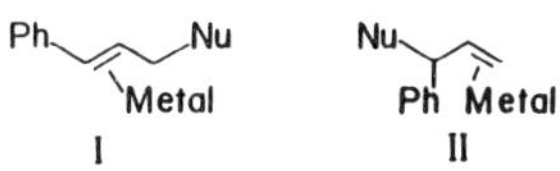

the organic product remains bound to the metal. To the extent that the steric demands of the template dictate the regioselectivity, coordination to the sterically less crowded monosubstituted olefin as in II will be preferred. The effect of tungsten catalysts on regioselectivity is general for a range of aromatic systems including heteroaromatic systems as shown in eq 4 and 5.

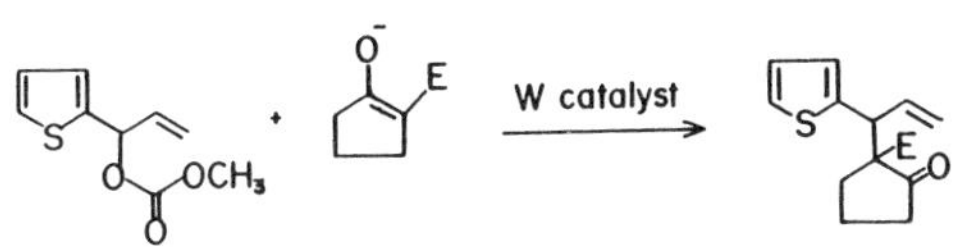

(4)

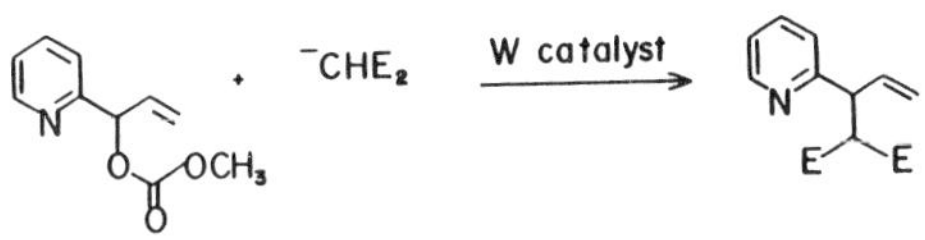

(5)

A dramatic example of how selective reactions can be made by appropriately choosing the catalyst is demonstrated by reaction of 16 with the tungsten catalyst. This catalyst shows a dramatic

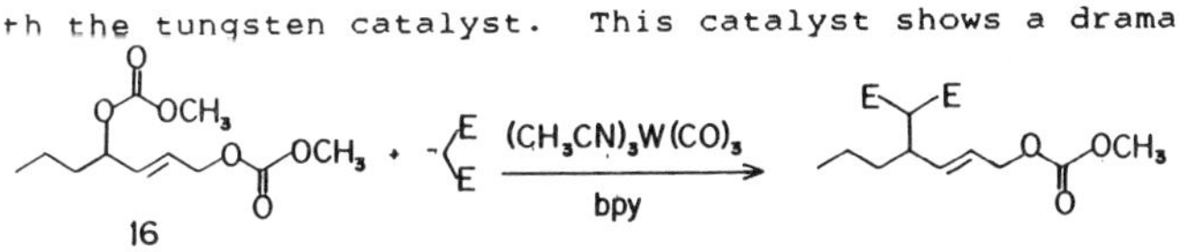

16

dependence of the rate of reaction on the degree of substitution of the carbon bearing the allyl leaving group in the order primary < secondary < tertiary. Indeed, the differences are so large that in 16, where a primary *versus* secondary carbonate competition exists, only ionization of the carbonate at the secondary carbon initiated by the metal initiates alkylation.

With this degree of control of selectivity in organic reactions possible, attention is now focused on applying these principles.

A Cyclopentane Annulation

Cycloaddition reactions represent the most important method for construction of six membered rings. A similar sequence for constructing five membered rings, compounds of great importance in the realm of bioactive natural products, would enhance our flexibility and efficiency in synthesizing such molecules. By combining allylic activation of allylic carboxylates with the silylophilic properties of oxygen nucleophiles, a general reaction of the type represented in eq 6 emerged.[8] The structural

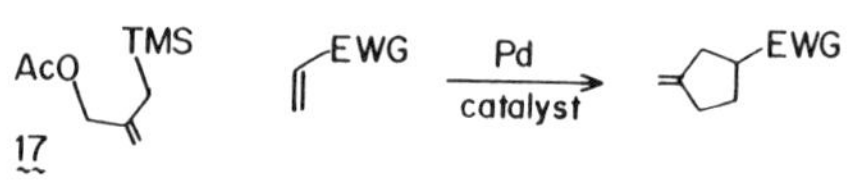

(6)

requirement of the acceptor is that it bear at least one electron withdrawing group (EWG). The structural flexibility with respect to the silyl acetate 17 remains to be established. We explored electron donating groups such as methyl[9] as in 18 (eq 7),

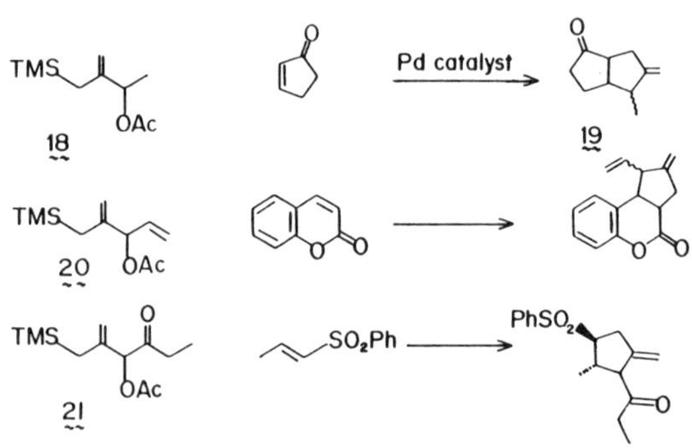

conjugating but relatively neutral substituents as in 20 (eq 8)[10] and electron withdrawing groups as in 21 (eq 9).[11] In each and every case, reaction proceeds well and with high regioselectivity.

Surprisingly, the regioselectivity is independent of the nature of the substituent.

How can such an approach be used to develop strategy in synthesis of natural products? In embarking upon answering such a question, initial efforts focused on relatively simple natural products. Let's consider the terpene albene (22).[12] The key

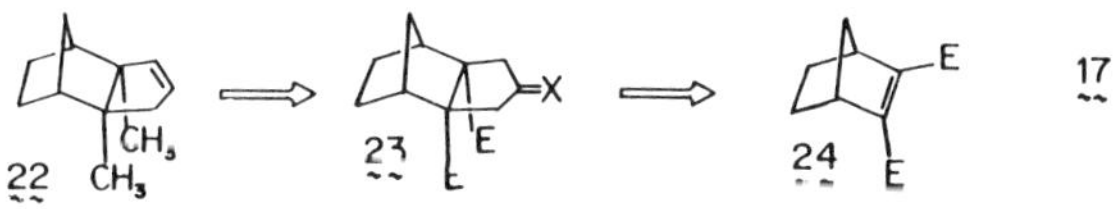

question is the *exo* stereochemistry of the cyclopentene fused onto the bicyclo[2.2.1]heptane nucleus. If one envisions the olefin to derive from a ketone (as in 23, X=O) and that the methyl groups may derive from a carbonyl substituent such as a carbomethoxy, a reasonable precursor of albene is 23 X=O. Since an exocyclic methylene group is a synthon for a carbonyl group, 23 X=CH_2 becomes a rational precursor. It is then obvious that a cycloaddition of 17 to an acceptor such as 24 would be an excellent beginning. Such a cycloaddition strategy beautifully resolves the stereochemical problem since an addition to a norbornene should be heavily favored from the *exo* direction. Indeed, such a sequence succeeds admirably and provides a simple five step synthesis in an overall 21% yield (see Scheme 1). The efficiency stems from the total absence of protecting group chemistry.

SCHEME 1.

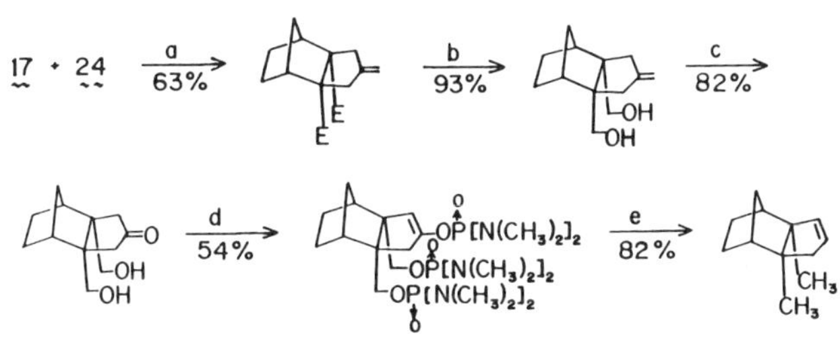

(a) $(iC_3H_7O)_3P$, $Pd(OAc)_2$, THF reflux. (b) LAH, ether, 0°. (c) O_3, CH_2Cl_2, CH_3OH, -78° then $(CH_3)S$. (d) $KN(TMS)_2$, DME, HMPA, $[(CH_3)_2N]_2P(O)Cl$, 0°. (e) Li, $C_2H_5NH_2$, THF, $t\text{-}C_4H_9OH$.

A more challenging target is 2 β-hydroxyjatrophone (25), an anti-tumor agent of limited supply. Based upon concepts to be developed later with respect to large ring synthesis, we focused on lactone 26 as a precursor of the cyclopentanoid portion of 25.

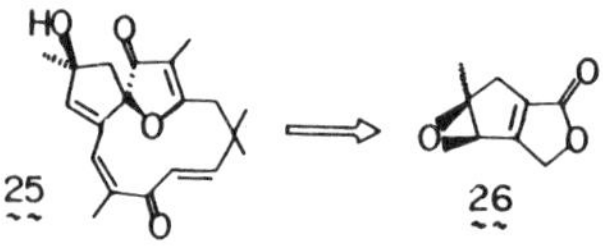

Envisioning a Diels-Alder adduct with cyclopentadiene as a protecting group for the α,β conjugated double bond of 27 and that

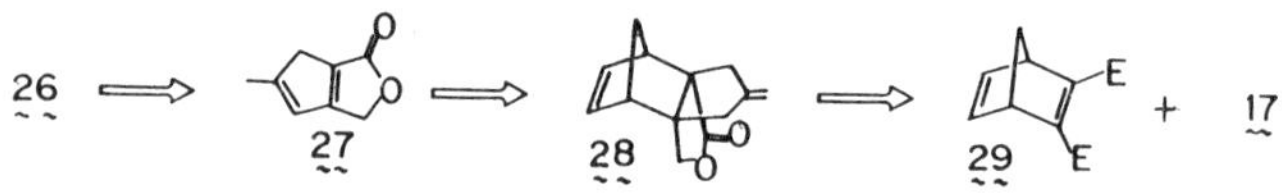

an exocyclic methylene group can be coaxed endocyclic, the tetracyclic compound 28 becomes the logical precursor. It becomes obvious that a cycloaddition of 17 to 29, a close relative to that used in the albene synthesis, could provide a ready entry. Indeed, as shown in Scheme 2, this sequence follows directly as outlined in the retrosynthetic analysis to constitute a six step synthesis in 56% overall yield![13] Undoubtedly, the efficiency of this sequence derives, in no small measure, from the absence of protecting groups.

SCHEME 2. A Synthesis of 2β-Hydroxyjatrophone Fragment

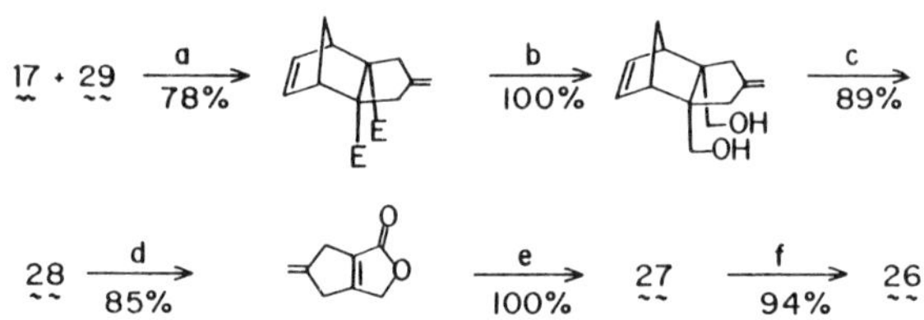

(a) $(Ph_3P)_4$, dppe, THF reflux. (b) LAH, ether, 0°. (c) $(C_5H_5N)_2CrO_3$, CH_2Cl_2, celite, 0°. (d) FVT, 470° @ 0.4 torr. (e) DBU, CH_2Cl_2, rt. (f) MCPBA, CH_2Cl_2, rt.

Loganin (30), an important biosynthetic intermediate, can also be examined from the perspective of cycloaddition methodology. In

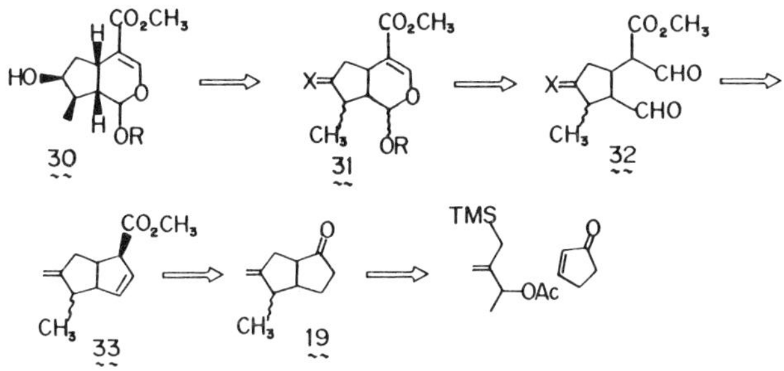

particular, the ability of a ketone to be a precursor of an alcohol and the ability to use the carbonyl group to adjust the stereochemistry of the methyl group permits 31 (X=O) and, correspondingly, 31 (X=CH_2) to be a likely precursor. The dihydropyran ring of 31 can be more explicitly recognized as a tautomeric form of a dialdehyde as in 32. Since oxidative cleavage of a cycloalkene conveniently unmasks the sensitive dialdehyde, cyclopentene 33 becomes the target. The β,γ-unsaturated ester can derive from a simple ketone which factors the problem of loganin synthesis to 19, the adduct obtained previously (see eq 7). Eq. 10 illustrates the conceptual

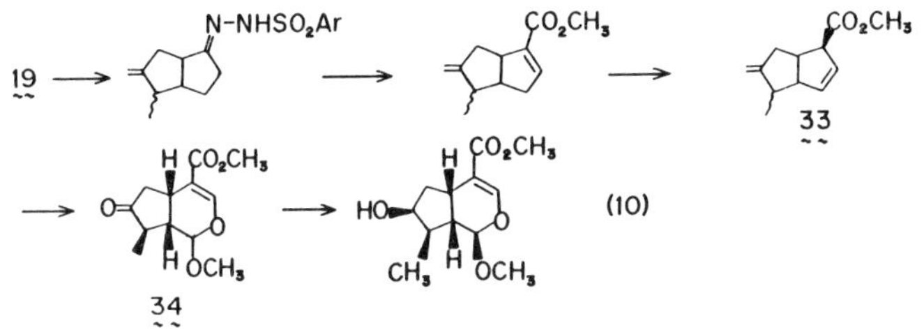

simplicity of this approach.[10] The conversion of the ketone 19 to the β,γ-unsaturated ester 33 uses the deconjugation of the α,β-unsaturated ester, which, in turn, derives from a Shapiro reaction. Ozonolysis unmasks the requisite functionality in the form of the ketone 34 which has previously been converted to loganin. By using latent functionality for the highly reactive groups of loganin, high synthetic efficiency evolves.

As a final target for cycloaddition methodology, let's consider brefeldin A (35), a cyclopentanoid macrolide. In

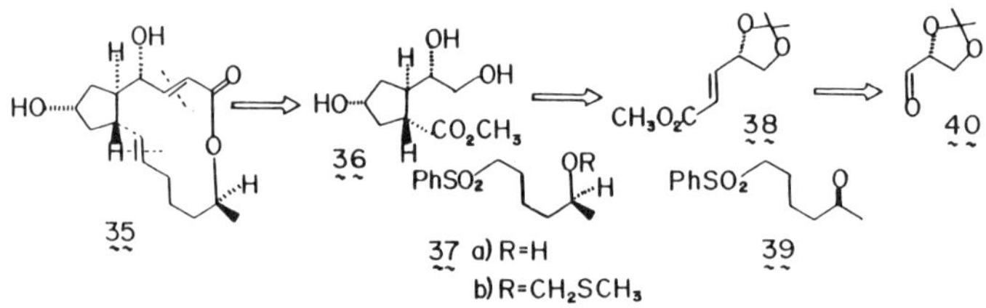

considering this problem, the question of control of absolute and relative stereochemistry as well as creation of the correct molecular framework must be addressed. Using an alkylative elimination and a reduction of a β-hydroxysulfone as olefination methods, dissection at the double bonds of 35 produces 36 and 37 as the logical precursors. Four of the five chiral centers of 35 end up in 36 with the fifth finding itself in 37. Such a convergent strategy demands that both 36 and 37 be obtained enantiomerically pure. For the latter, we turned our attention to our collaborators in chemical synthesis, the microbes, to reduce the ketone 39 with absolute stereochemical control. Unfortunately, most common yeasts choke on this substrate. Professor Charles Sih has an extensive collection of microbes among which we found that the yeast C. guillermonde nicely reduces 39 to enantiomerically pure 37 of the correct absolute configuration.[14,15]

The cyclopentane unit 36 represents the major challenge. Recognizing that the C(7) hydroxyl group can derive by stereocontrolled reduction of a ketone and an exocyclic olefin is a synthon of a carbonyl group, we now recognize that a stereocontrolled cycloaddition to 38 would generate three of the four asymmetric centers of 36, the fourth to be developed by the reduction of a ketone. Since 38 naturally derives by classical olefination of the glyceraldehyde derivative 40 and since the latter arises by the cleavage of the bis(acetonide) of mannitol, 38 and consequently 36 would be available with both the correct relative and absolute stereochemistry.

Scheme 3 outlines the sequence to 36. Most gratifying, the cycloaddition proceeds with good diastereoselectivity. The major adduct has the stereochemistry depicted in 41 and required for the synthesis of brefeldin A. It arises by attack on the acceptor as shown in 42.

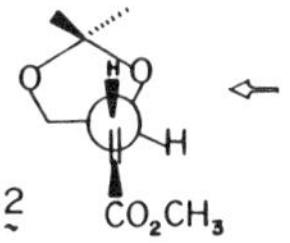

42

SCHEME 3. Synthesis of Cyclopentanoid Unit of Brefeldin A

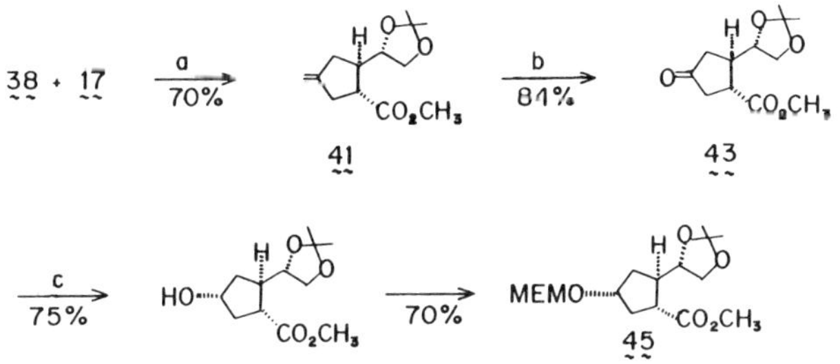

(a) $(iC_3H_7O)_3P$, $Pd(OAc)_2$, $PhCH_3$. (b) O_3, CH_3OH, -78° then $(CH_3)_2S$. (c) 2,6-$(tC_4H_9)_2$-4-$(CH_3)C_6H_2OAl(iC_4H_9)_2$, $PhCH_3$, -78°. (d) MEM-Cl, DMAP, $(iC_3H_7)_2NC_2H_5$, CH_2Cl_2, 0°.

With three of the four centers of 36 established, creation of the fourth and final asymmetric center requires selective reduction of 43. All common reducing agents produced as the major product the epimer at C(7). In fact, most recently, this epimer has been isolated as a co-metabolite with brefeldin A from Curvularia lunata.[17] While we could have adopted this new metabolite as our objective, we felt that doing so would be failing to address the problem. Since it appeared that the sterically more accessible side of 43 was *cis* to the ester, we envisioned temporarily coordinating a bulky metal to this ester with the hope that enhanced shielding of the α-face as in 44 would

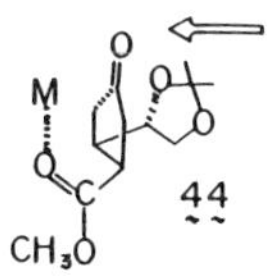

44

direct the nucleophile to the β (desired) face. Use of Yamamoto's reagent (prepared *in situ* from 2,6-di-t-butyl-4-methylphenol and diisobutylaluminum hydride) nicely fulfilled our hopes and completed all the centers of 36 and consequently brefeldin A.

Introduction of the two chains used somewhat non-conventional olefination methods, the key step of each being outlined in eq 11 and 12 respectively. The closure of the seco acid has been

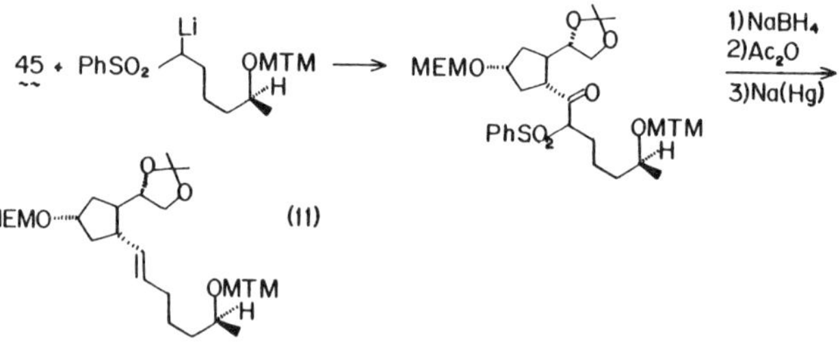

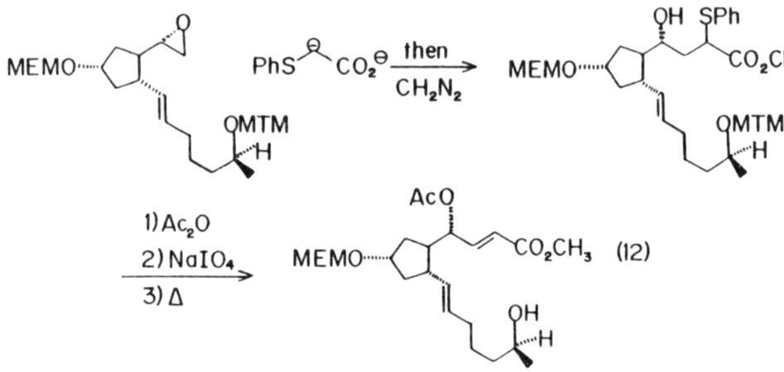

accomplished by earlier workers.[17] Thus, the first enantiocontrolled total synthesis which avoids a resolution evolves from the cycloaddition methodology.

Macrocyclization

The widespread occurrence of large ring natural products, many of which arise biosynthetically from acyclic precursors, indicates the facility by which nature can accomplish macrocyclization. Can the "chemist's enzymes" do as well?[18] A hopeful sign evolved out of a study of the cyclization of 46. This cyclization can

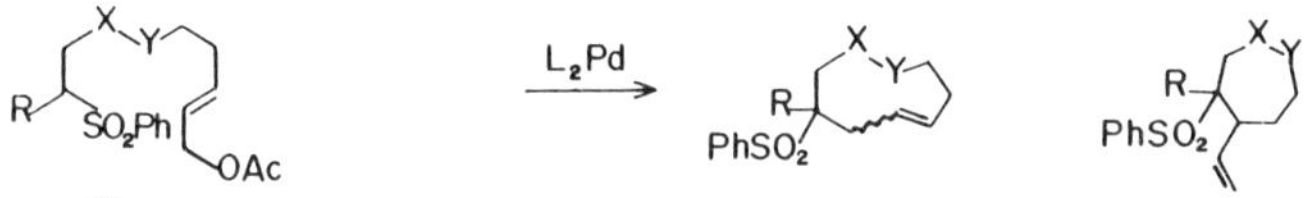

46

	R	X,Y	L_2		
(a)	CO_2CH_3		Ph_3P	only	----
(b)	CO_2CH_3	CH_2CH_2	Ph_3P	----	only
(c)	SO_2Ph	CH_2CH_2	$Ph_2PCH_2CH_2Ph_2$	only	----

generate either a seven or nine membered ring. Conventional wisdom dictates that seven membered ring formation is much more favorable than nine. Nevertheless, the palladium template directs the cyclization of 46a to give only the nine membered ring lactone.[19] Part of the reason undoubtedly derives from the removal of some transannular interactions by inserting an ester linkage in the chain. Such transannular interactions are likely a major source of the problem of medium and large ring formation. Indeed, reinserting this interaction, as in the all carbon chain 46b, totally reorients the reaction to give only the seven membered ring product. Nevertheless, increasing the steric demands of the nucleophile as in 46c and decreasing the steric demands of the template by substituting 1,2-bis(diphenylphosphino)ethane for triphenylphosphine again redirects the cyclization back to the nine membered ring. Thus, palladium templates combined with appropriate choice of nucleophile and intervening chain allow any ring size to form efficiently.

A second aspect of macrocyclizations by standard chemical methods has been the need for high dilution, a somewhat impractical approach for large scale reactions. Taking the cue from enzymes, a new approach was envisioned to obviate high dilution. In this strategy, a substrate is used that does not have a reactive nucleophilic or electrophilic site and the catalyst is embedded on an insoluble polymer support. As shown in eq 13, such a substrate must be absorbed onto the polymer and find

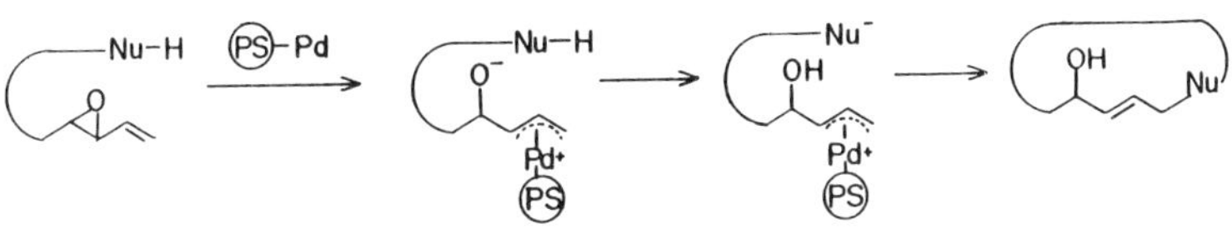

(13)

an "active site" before being activated. The combinatorial effect of having few active sites and fewer of them occupied at any given time gives an equivalent of pseudo high dilution. Indeed, this concept permits formation of both carbocyclic (eq 14)[20] and

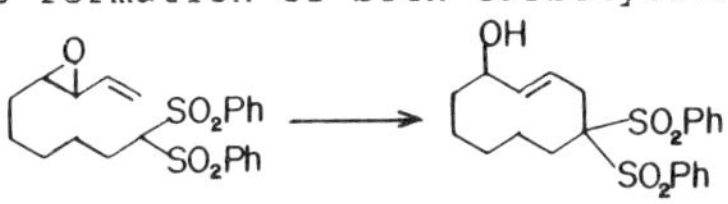

(14)

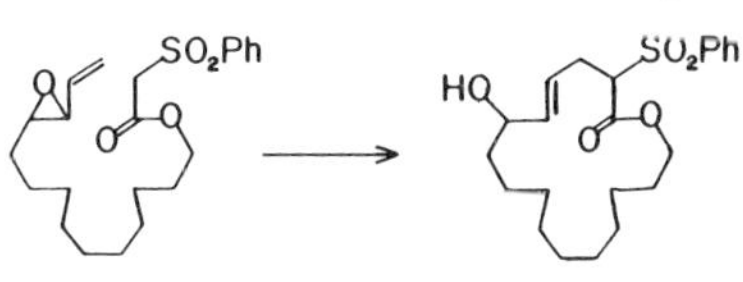

(15)

heterocyclic (eq 15)[21] rings and ring sizes that vary from 10 to as large as 27 at substrate concentrations of 0.1–0.5 M.

Let's consider two targets and the strategy that evolves when focusing on allylic alkylation as the critical ring forming reaction. Aspochalasin B (47) has a highly oxygenated macrocyclic

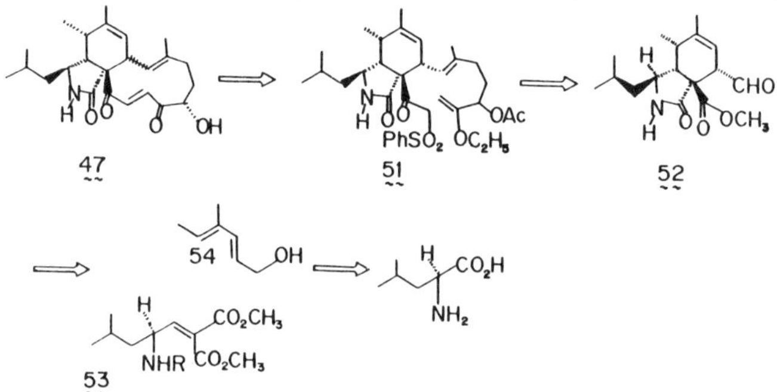

ring pattern as in 48 that is characteristic of many natural products. As shown in eq 16, insertion of a potential leaving

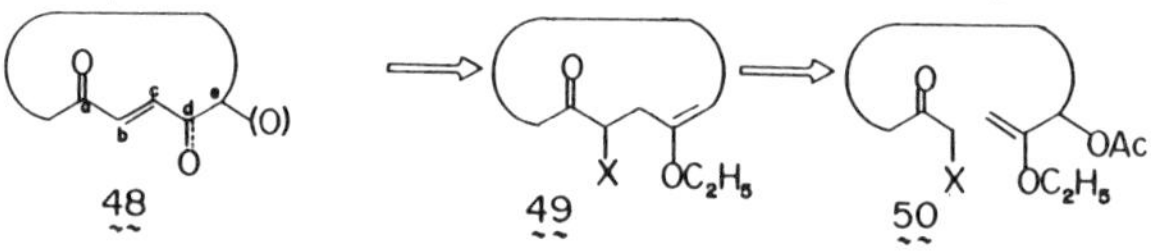

(16)

group and an enol ether permits unmasking of the sensitive olefin functionality late and allows formation of either the series that possesses oxygen at position e in 48 (by hydroxylation of 49) or lacks this oxygen (by hydration of 49). It is now apparent that 49 can derive from the macrocyclization of 50. Applying this reasoning to 47 simplifies it to 51 which, by using rather conventional methodology, can be envisioned to derive from 52. It is almost synonymous to realize the problem reduces to a stereocontrolled synthesis of a cyclohexene and to think of a Diels-Alder strategy. Such a strategy requires the alkylidenemalonate 53 and the methylsorbyl alcohol 54 as the Diels-Alder Partners. The former clearly derives from the amino acid leucine. Not only does this approach resolve the problem of the carbon skeleton, but the relative and absolute stereochemistry are also controlled. The absolute stereochemistry of leucine then blossoms into the stereochemistry of the entire molecule. Thus, we and the biological system employ the same raw material!

Conversion of leucine to 53 proceeds straightforwardly. Diels-Alder condensation gives a single regio- and stereoisomer, 55![22,23] This most unusual result reveals much about the

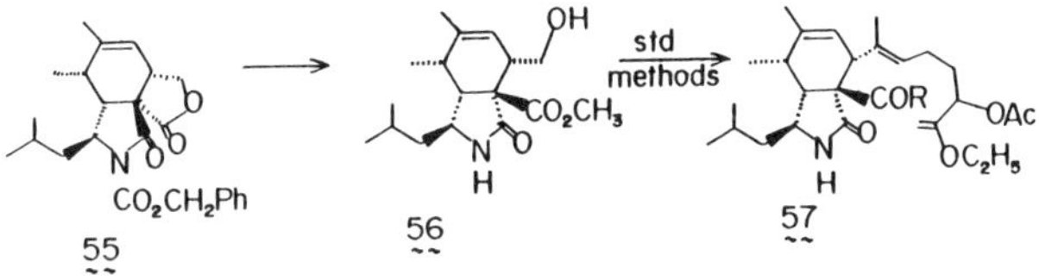

Diels-Alder reaction, but equally importantly, provides a source of the critical intermediate 56 which possesses the fully elaborated cyclohexene portion. Standard chain extension procedures convert 56 to 57 (R=OCH_3) and, currently, we focus on the final stage before cyclization. Nevertheless, a good model for the cyclization stems from the synthesis of antibiotic A26771B (58).

SCHEME 4. Retrosynthetic Analysis and Synthesis of Antibiotic A26771B

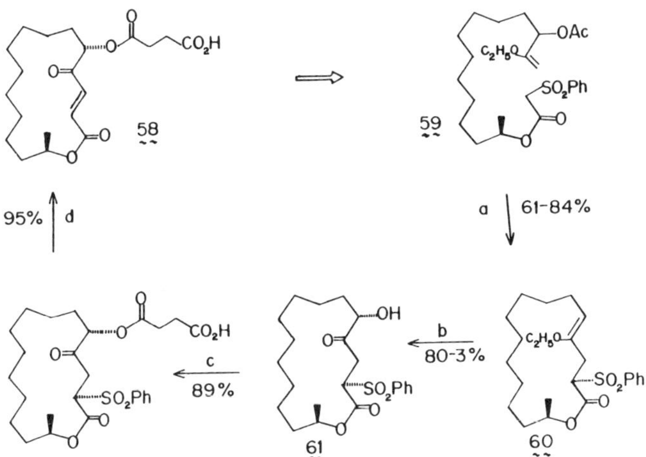

(a) $(Ph_3P)_4Pd$, BSA, dppp, $PhCH_3$,Δ . (b) cat. OsO_4, 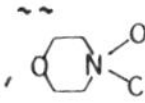 , H_2O, $(CH_3)_3CO$ or OsO_4, C_5H_5N. (c) 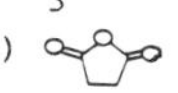 DMAP, CH_2Cl_2. (d) $(C_4H_9)_4NF$, THF, 0°.

The seco precursor 59 is available in 5 steps and 45% overall yield from 10-undecenal.[24] Cyclization as outlined in Scheme 4 gives a good yield of the ring system 60 as a single isomer. A bonus resulted from the use of 60 as an intermediate in that hydroxylation of the enol ether gave predominantly one of the two hydroxyl epimers as depicted in 61. Acylation and elimination of the elements of benzenesulfinic acid complete the sequence. Thus, the sulfone group plays a myriad of roles - (1) facilitating the formation of the requisite nucleophile for cyclization, (2) directing the stereochemistry of the hydroxylation, and (3) serving as a leaving group to introduce an olefin. Most importantly, the palladium based strategy for formation of a highly oxygenated macrocycle succeeds admirably (59% for the 4 steps in going from 59 -> 58) and bodes well for the aspochalasin B project.

A more challenging macrocycle in terms of the number of asymmetric centers is represented by the erythronolides A (62a) and B (62b) which are the aglycones of the clinically important

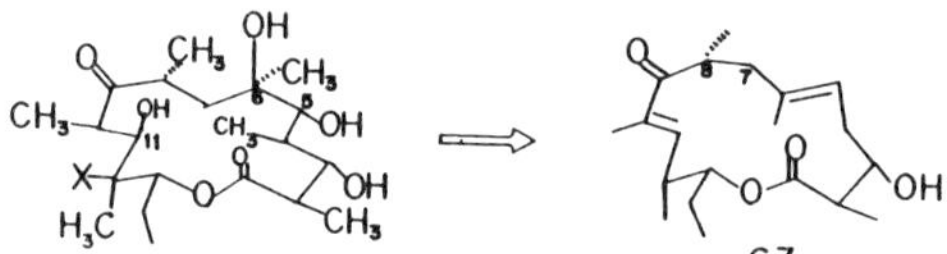

62 a) X=OH b) X=H

63

antibiotics, the erythromycins. Based upon earlier work, the functionality and stereochemistry of C(10) and C(11) of 62 can be created from an olefin. Combining this structural change with that of replacing the hydroxyl groups at C(5) and C(6) by a trans double bond leads to 63. Examination of the conformation of 63 suggests that a cis hydroxylation of the unconjugated double bond should recreate the functionality of the target. With the concept of metal catalyzed allylic alkylation in mind, formation of the macrocycle would involve creating the C(7)-C(8) bond from a seco derivative such as 64. A next logical disconnect is the ester

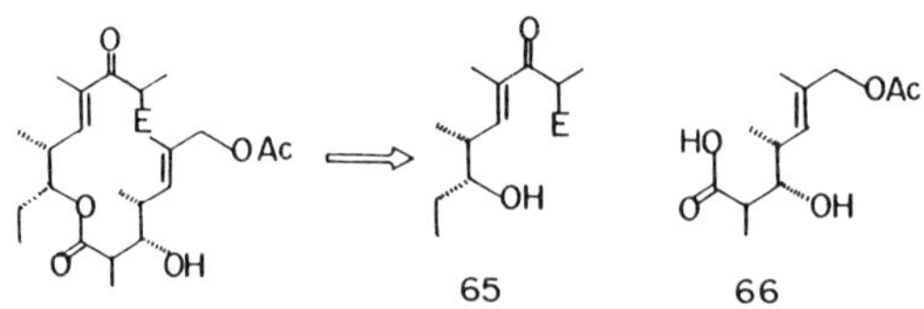

64 65 66

bond which creates the left (i.e., 65) and right (i.e., 66) fragments of erythronolide B. In such a convergent strategy, the joining of 65 and 66 demands that these fragments be of a single relative and absolute stereochemistry to avoid mixtures. Thus, the ten chiral centers of 62 reduce to two molecules - one containing only two and one containing only three chiral centers.

It is outside the scope of this report to detail the stereocontrolled syntheses of these fragments, but two different strategies have succeeded.[25] In the first, 67 is envisioned to be

the precursor of 65 and 68 that of 66. Inspection of 67 and 68 reveals that they are simply mirror images - *i.e.*, one enantiomer will translate into the alcohol half of erythronolide and the other enantiomer into the ester half. Thus, the classical problem of resolution in which only half of the product is useful is avoided, but a resolution of the alcohol that provides both enantiomers easily is required. Preparation of the O-methylmandelate ester of 67 and 68 nicely resolved this problem. HPLC completely separates the two diastereomeric mandelate esters and these, in turn, were converted into the optically pure alcohols. Furthermore, standard methodology converted 67 into 65

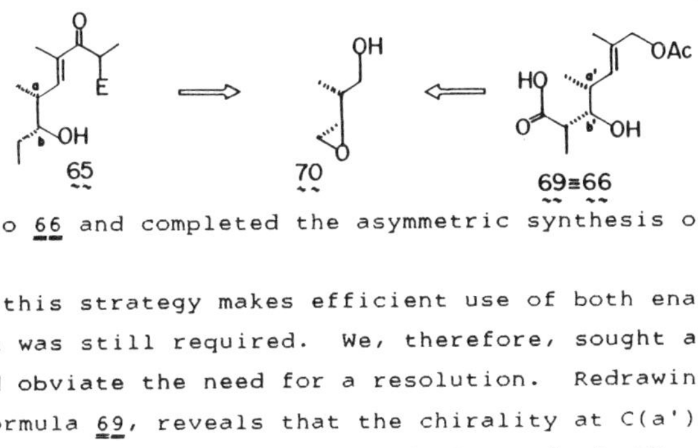

and 68 into 66 and completed the asymmetric synthesis of the two halves.

While this strategy makes efficient use of both enantiomers, a resolution was still required. We, therefore, sought a strategy that would obviate the need for a resolution. Redrawing 66, as is done in formula 69, reveals that the chirality at C(a') and C(b') of 69 is the same as C(a) and C(b) of 65. Indeed, 70 could serve as a common precursor for both halves provided that 70 could be obtained optically pure from readily available starting materials. Fortunately, such proved to be the case. Scheme 5 outlines the synthesis of 70 from the diethyl ester of natural D-(+)-tartaric acid in a straightforward manner and in an extraordinarily efficient 65% overall yield. Furthermore, 70 has now been converted into optically pure 65 and 66, setting the stage for the completion of our synthesis.

SCHEME 5. Asymmetric Synthesis of 2S,3R-1,2-Epoxy-4-hydroxy-3-methylbutane (70)

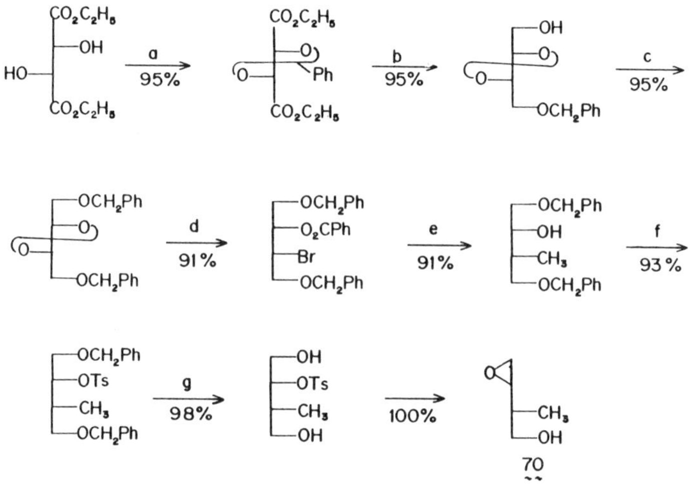

(a) PhCHO, $HC(OC_2H_5)_3$, TsOH. (b) LAH, ether, reflux. (c) NaH, $PhCH_2Br$, THF. (d) NBS, CCl_4, reflux. (e) $(CH_3)_2CuLi$, ether, 0°. (f) TsCl, C_5H_5N. (g) 10% Pd/ C, CH_3OH, HOAc. (h) NaOH, CH_3OH

In anticipation of that aspect, the critical cyclization and several subsequent transformations of the macrocycle were explored. As summarized in eq 17, not only does the palladium

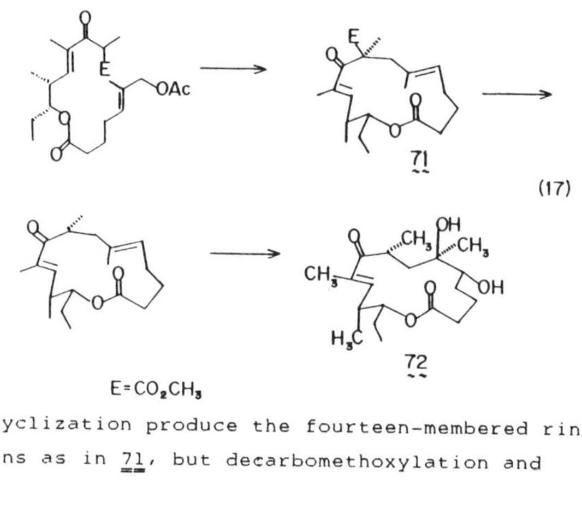

(17)

catalyzed cyclization produce the fourteen-membered ring of the erythromycins as in 71, but decarbomethoxylation and *cis*-hydroxylation give the simple analog **72**, confirming the validity of our approach.

Conclusions

The worlds of biology and chemistry continue to merge. Powerful structural tools reveal more clearly the chemical basis of life. The molecules, both simple and complex, reveal a path to enhance the quality of life. The molecules themselves and analogs, unnatural biomimics and inhibitors, agonists and antagonists, represent a continuing challenge for the synthetic chemist. Differentiating the subtle differences in chemical reactivity necessary to construct such complex molecules requires extraordinary levels of control that, to a large extent, was avoided in the past with a great loss in synthetic efficiency. The utilization of these compounds to benefit man requires the development of efficient syntheses. In part, this need will be provided through genetic engineering, but only a small fraction can succumb to such methods. The chemical equivalent of the life process, the development of the "chemist's enzymes," will be the major route. The development of such catalysts for complex synthesis is still embryonic.

Acknowledgement

I wish to thank the National Science Foundation and the General Medical Sciences Institute of the National Institutes of Health for their generous support of our programs. My collaborators on these projects, individually acknowledged in the references, are sincerely thanked for their steadfastness, creativity, and skill for transforming these ideas into reality.

References and Notes

1. For a review see Trost, B.M.; Verhoeven, T.R. *Compr. Organomet. Chem*. **1982**, *8*, 779. Also see Trost, B.M. *Pure Appl. Chem*. **1981**, *53*, 2357; *Accounts Chem. Res*. **1980**, *13*, 385.
2. Trost, B.M.; Schmuff, N.R.; Miller, M.J. *J. Am. Chem. Soc*. **1980**, *102*, 5979.
3. Trost, B.M.; Verhoeven, T.R. *J. Am. Chem. Soc*. **1980**, *102*, 4730.
4. Romero, A.G. unpublished work.
5. Trost, B.M.; Lautens, M. *J. Am. Chem. Soc*. **1983**, *105*, 3343.
6. Trost, B.M.; Lautens, M. *J. Am. Chem. Soc*. **1982**, *104*, 5543.
7. Hung, M.-H., unpublished work.

8. Trost, B.M.; Chan, D.M.T. J. Am. Chem. Soc. 1983, 105, 2315.
9. Trost, B.M.; Chan, D.M.T. J. Am. Chem. Soc. 1981, 103, 5972.
10. Nanninga, T.N. unpublished work.
11. Satoh, T. unpublished work.
12. Trost, B.M.; Renaut, P. J. Am. Chem. Soc. 1982, 104, 6668.
13. Balkovec, J. unpublished work in these laboratories.
14. We thank Professor C. Sih for collaborating in this project.
15. Lynch, J.; Renaut, P.; Steinman, D. unpublished work.
16. Gorst-Allman, C.P.; Steyn, P.S.; Rabie, C.J. J.C.S. Perkin I, 1982, 2387.
17. For a complete listing of earlier work see LeDrian, C.; Greene, A.E. J. Am. Chem. Soc. 1982, 104, 5473.
18. Trost, B.M.; Verhoeven, T.R. J. Am. Chem. Soc. 1980, 102, 4743.
19. Martina, D.; Verhoeven, T.R., unpublished work.
20. Trost, B.M.; Warner, R.W. J. Am. Chem. Soc. 1982, 104, 6112.
21. Warner, R.W., unpublished work.
22. Brickner, S.J.; Okawara, H., unpublished work.
23. After completion of our work, a similar approach was reported, see Schmidlin, T.; Burckhardt, P.E.; Waespe-Sarcevic, N.; Tamm, C. Helv. Chim. Acta, 1983, 66, 450.
24. Trost, B.M.; Brickner, S.J. J. Am. Chem. Soc. 1983, 105, 568.
25. Belletire, J.L.; Lubineau, A.; Brown, F.J.; Kerwin, J. unpublished work.

Manipulation of the Arachidonic Acid Cascade

By S. Moncada and J. R. Vane*

WELLCOME RESEARCH LABORATORIES, LANGLEY COURT, BECKENHAM, KENT BR3 3BS, U.K.

Introduction

The complex field of prostaglandins (PGs) provides a good example of rapid progress in scientific knowledge (with a little help from serendipity) providing strong potential for new therapy. The term "prostaglandin" was introduced by von Euler in 1935[1] to describe the material present in human semen and accessory genital glands which had vasodepressor and smooth-muscle stimulating activity. The activity was due to several different acidic lipids, and Bergström and Sjövall[2] isolated in pure form prostaglandin E_1 (PGE_1) and prostaglandin $F_{1\alpha}$ ($PGF_{1\alpha}$) and later several other stable prostaglandins. In point of fact, prostaglandins are made by all tissues except red blood cells and they are released very easily by perturbation of cell membranes induced mechanically or by chemicals.

In 1964, two groups[3,4] demonstrated that prostaglandins are biosynthesized from polyunsaturated fatty acids. Arachidonic acid (AA), the main precursor of prostaglandins in man, is the most common fatty acid present in cellular phospholipids and is obtained by desaturation and chain elongation from dietary linolenic acid (C18:2ω6) or directly from the diet.

Many of the earlier studies concentrated on prostaglandins E_2 and $F_{2\alpha}$ as these possessed potent and diverse biological activities and were available in pure form. Since 1973, other important derivatives of arachidonic acid have been discovered, including the prostaglandin endoperoxides (PGG_2 and PGH_2), thromboxane A_2 (TXA_2), prostacyclin (PGI_2) and the leukotrienes (previously known as slow reacting substances). Several of these eicosanoids are chemically unstable but all have potent biological activities.

Metabolism of Arachidonic Acid

Arachidonic acid is metabolized by cyclo-oxygenase, which is ubiquitous in mammalian tissues, to form the endoperoxide PGG_2 which then can lead to a number of products including TXA_2, prostacyclin and the stable PGs, E_2, D_2 and $F_{2\alpha}$ (see Figure 1). Much evidence has now accumulated to indicate that PGE_2 plays a major role in the development of inflammatory erythema, oedema and pain. More recent evidence suggests that prostacyclin also contributes to these responses.

In 1971, Vane, using cell-free homogenates of lung tissue, showed that aspirin, sodium salicylate or indomethacin caused a dose-dependent reduction in prostaglandin synthesis and proposed that this was the biochemical basis for the actions of aspirin-like drugs.[5]

In 1974, Hamberg and Samuelsson isolated two cyclic endoperoxides formed as intermediates in the transformation of arachidonic acid to prostaglandins, thus proving that prostaglandin synthesis involved two stages, cyclo-oxygenation followed by isomerization or reductive cleavage of the endoperoxides.[6] Non-steroid anti-inflammatory drugs inhibit the cyclo-oxygenase step by preventing the abstraction of hydrogen from C-13 and, therefore, blocking peroxidation at C-11 or C-15.[7] This action is highly specific, for similar abstraction and peroxidation reactions by lipoxygenases at other points in the fatty acid molecule are not inhibited, so they do not prevent the formation of leukotrienes. Furthermore, cyclo-oxygenase inhibitors do not prevent the generation of prostaglandins from cyclic endoperoxides.

Arachidonic acid is also metabolized by lipoxygenases to other arachidonate hydroperoxides (HPETEs) and the 5-lipoxygenase products lead to the leukotrienes (see Figure 1). There is increasing evidence that leukotrienes are involved in inflammatory responses. They have been proposed as important mediators of allergic broncho-constriction, leukocyte activation and vascular permeability (for reviews see 8, 9). The bronchoconstrictor mediator SRS-A is now known to be a mixture of leukotrienes. There is evidence to suggest that lipoxygenase activity is involved both in the generation of anaphylactic mediators and in the sensitization of smooth muscle to their contractile actions.[10,11] Thus, inhibitors not only of the cyclo-oxygenase enzyme (such as aspirin) which leads to the prostaglandins and thromboxanes but also of the lipoxygenase enzyme (which leads to the leukotrienes) should have a better anti-inflammatory profile than aspirin. Although selective inhibitors of lipoxygenase have been described there are no reports of their activity in vivo. The evidence that lipoxygenase inhibition leads to important pharmacological effects relies, at present, on compounds which also inhibit cyclo-oxygenase. A number of pyrazoline analogues of phenidone are dual inhibitors of both cyclo-oxygenase and lipoxygenase activity. Initially, phenidone and BW755C (3-amino-1-[m-(trifluoromethyl)-phenyl]-pyrazoline) were shown to inhibit lung and platelet enzymes.[12,13] BW755C has now been shown to prevent the synthesis of prostaglandins, hydroxy acids and leukotrienes in a variety of leukocyte preparations from different species including man (for review see 14).

Unlike indomethacin and the salicylates, BW755C does not enhance leukocyte migration at low doses and is equi-active in reducing oedema, prostaglandin synthesis and leukocyte accumulation. Furthermore, BW755C reverses the indomethacin-induced increase in leukocyte migration in acute experimental inflammation. BW755C reduces both polymorphonuclear and mononuclear leukocyte numbers in inflammatory exudates.[14] These observations support the hypothesis that release of chemotactic lipoxygenase products contribute to leukocyte accumulation. Another pathological state in which a reduction of leukocyte accumulation and activation could be beneficial is myocardial infarction. BW755C limits the area of infarction following coronary occlusion in dogs and this effect has been attributed to a reduction in leukocyte infiltration of the ischaemic tissues.[15,16] Because aspirin-like drugs have no therapeutic value in anaphylaxis, a selective lipoxygenase inhibitor is an obvious target for new drug development. In more general inflammatory responses, however, where prostaglandins are important mediators, dual inhibitors of cyclo-oxygenase and lipoxygenase are indicated.

Dual inhibitors should have similar properties to the anti-inflammatory steroids, which although inactive against cyclo-oxygenase in vitro[17] do inhibit phospholipase A_2, the enzyme responsible for cleaving arachidonic acid from its phospholipid stores in the cell membrane (for review see 18). This leads, presumably, to a reduction in all the products of arachidonic acid, including the leukotrienes. This action has been explained

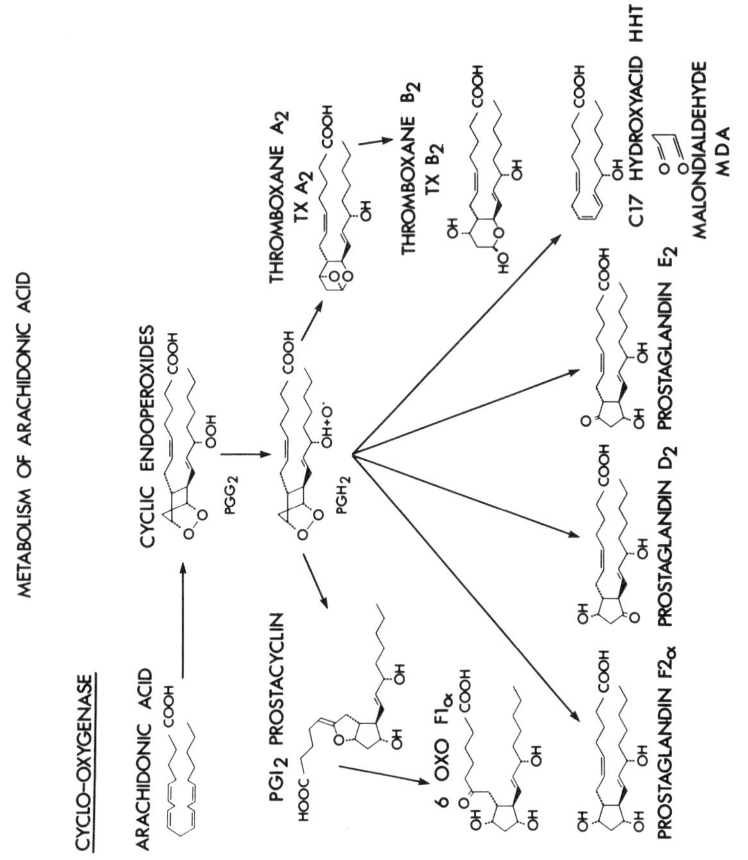
METABOLISM OF ARACHIDONIC ACID
CYCLO-OXYGENASE
ARACHIDONIC ACID
COOH
CYCLIC ENDOPEROXIDES
OOH
PGG2
OH+O·
PGH2
THROMBOXANE A2
TX A2
THROMBOXANE B2
TX B2
PGI2 PROSTACYCLIN
HOOC
6 OXO F1α
PROSTAGLANDIN F2α
PROSTAGLANDIN D2
PROSTAGLANDIN E2
C17 HYDROXYACID HHT
MALONDIALDEHYDE
MDA

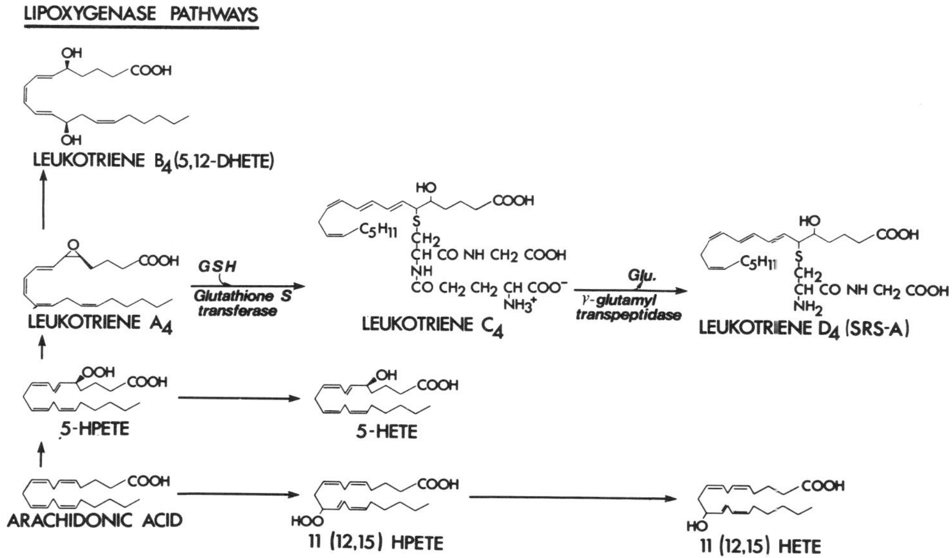

Figure 1

Pathways of arachidonic acid metabolism. Upper panel, conversion of arachidonic acid *via* the cyclo-oxygenase pathway to prostaglandins of the 2-series. Lower panel, conversion of arachidonic acid *via* the lipoxygenase pathway to leukotrienes.

by the recent discovery that corticosteroids induce the intra-cellular synthesis and release of an endogenous phospholipase inhibitor - a protein named macrocortin[19] or lipomodulin[20], which directly inhibits phospholipase A_2. Elucidation of its structure and synthesis of peptide analogues will surely be another important lead in the treatment of inflammatory diseases, including asthma.

The Prostacyclin/Thromboxane A_2 Balance and its Manipulation.

Prostacyclin is the main product of arachidonic acid in all vascular tissues so far tested including those of man. It is a strong hypotensive agent through vasodilation of all vascular beds studied, including the pulmonary and cerebral circulation (for review see 21).

The ability of the large vessel to synthesize prostacyclin is greatest at the intimal surface and progressively decreases towards the adventitia.[22] Production of prostacyclin by cultured cells from vessel walls also shows that endothelial cells are the most active producers of prostacyclin.[23,24]

In blood at 37°, the activity of prostacyclin (as measured by bioassay on vascular smooth muscle) has a half-life of 3 min[25,26] although there have been various reports of a stabilizing effect of plasma.[27] Alkaline pH increases the chemical stability of prostacyclin,[28,29] so that at pH 10.5 at 10°C, it has a half-life of 100 h. It is stabilized as a pharmaceutical preparation (epoprostenol) by freeze drying.

Prostacyclin is the most potent endogenous inhibitor of platelet aggregation yet discovered. This effect is short-lasting in vivo, disappearing within 30 minutes of stopping an intravenous administration. Prostacyclin disperses platelet clumps in vitro in various animal models of thrombus formation, and in the circulation of man (for review see 21).

Prostacyclin inhibits platelet aggregation by stimulating adenylate cyclase, leading to an increase in cAMP levels in the platelets.[30,31] In this respect prostacyclin is much more potent than either PGE_1 or PGD_2 and its effect is longer-lasting. Prostacyclin enhances Ca^{++} sequestration in platelet membranes.[32] Moreover, inhibitory effects on platelet phospholipase[33,34] and platelet cyclo-oxygenase[35] have been described. All these effects are related to its ability to increase cAMP in platelets. By inhibiting several steps in the activation of the arachidonic acid metabolic cascade, prostacyclin exerts an overall control of platelet aggregability.

Another substance with the same fatty acid precursor (arachidonic acid) is thromboxane A_2 (TXA_2) which is made by platelets, leads to their aggregation and potently constricts vascular and airway smooth muscle (for review see 36). TXA_2, which breaks down to TXB_2, has a chemical half-life at body pH and temperature of 30 sec, although the presence of albumin appears to stabilize the molecule. The reduction in cAMP levels brought about by many stimulators of platelet aggregation can be inhibited by thromboxane synthetase inhibitors, thus suggesting that TXA_2 formation itself can decrease platelet cAMP and lead to aggregation.

Prostaglandin endoperoxides, being precursors of substances with opposing biological properties have an important pivotal role in the regulation of haemostasis. On the one hand, TXA_2 produced predominantly by the platelets is a strong vasoconstrictor and induces platelet aggregation. On the other hand, prostacyclin produced by the vessel wall is a potent vasodilator and the most potent naturally occurring inhibitor of platelet aggregation known. Each substance likewise has opposing effects on cAMP concentrations in platelets and gives a balanced control mechanism that will, therefore, affect thrombus and haemostatic plug formation. We have developed the hypothesis that it is a balance between formation of prostacyclin by the vessel wall and of TXA_2 by the platelets which controls haemostasis.[37] Prostacyclin increases cAMP levels in cells other than platelets[38] raising the possibility that in these cells a balance with the thromboxane system exerts a similar homeostatic control of cell behaviour to that observed in platelets. Thus, the prostacyclin/TXA_2 system may have wider biological significance in cell regulation.

Prostacyclin and TXA_2 in disease. The main metabolites of arachidonic acid *via* the pulmonary cyclo-oxygenase pathway are now recognized as TXA_2 and prostacyclin. TXA_2 is a bronchoconstrictor, whereas prostacyclin, although not a very potent bronchodilator in humans, is effective in antagonizing bronchoconstriction induced by other agents and is also a vasodilator. It is possible that the balance between these compounds is important in the modulation of pulmonary function and that some pathological conditions impair this balance.

A number of diseases have been associated with an imbalance in the prostacyclin/TXA_2 system. Platelets from patients with arterial thrombosis, deep venous thrombosis, or recurrent venous thrombosis produce more PG endoperoxides and TXA_2 than normal and have a shortened survival time.[39] Platelets from rabbits made atherosclerotic by a high fat diet[40] or from patients who have survived myocardial infarction[41] are abnormally sensitive to aggregating agents and produce more TXA_2 than controls. Elevated TXB_2 levels have been demonstrated in the blood of patients with Prinzmetal's angina[42] and vasotonic angina.[43] Studies of thromboxane levels in coronary sinus blood of patients with unstable angina indicate that local TXA_2 release is associated with recent episodes of angina but it is difficult to determine whether this release is cause or effect.[44]

Platelets from rats made diabetic release more TXA_2 than normal, whereas their blood vessels show a reduced production of prostacyclin; these effects are reversed by chronic insulin treatment.[45] Prostacyclin production by blood vessels from patients with diabetes is depressed[46,47] and circulating levels of 6-oxo-$PGF_{1\alpha}$ are reduced in diabetic patients with proliferative retinopathy[48] although a direct association between reduced prostacyclin production and diabetic retinopathy has not been established.

Thrombocytopenic purpura (TTP), like diabetes, is associated with formation of microvascular thrombo-emboli, and a deficiency in prostacyclin production may be responsible for the increased platelet consumption which occurs in TTP.[49-51] This deficiency is postulated to be secondary to a lack of a "plasma factor" which normally stimulates prostacyclin production.[24] Prostacyclin production is significantly lower in umbilical and placental vessels from pre-eclamptic patients than in those from normally pregnant women.[52]

An increased prostacyclin production, resulting from an accumulation of a "plasma factor" which stimulates prostacyclin synthesis, has been suggested to explain the haemostatic defect in uraemic patients.[53] Patients with Bartter's syndrome excrete in the urine about four times as much 6-oxo-$PGF_{1\alpha}$ as control patients.[54] This has led to the suggestion that overproduction of prostacyclin mediates both the hyper-reninaemia and hyporesponsiveness to pressor agents observed in these patients.[54] Enhanced prostacyclin production by blood vessels of spontaneously hypertensive rats has been demonstrated,[55] although clinical studies have shown a diminished excretion of 6-oxo-$PGF_{1\alpha}$ in the urine of patients with essential hypertension.[56]

TXA_2 produced during ligation of the coronary artery of the dog produces arrhythmias.[57] Vasoconstriction induced by TXA_2 in the gastric mucosa of the dog contributes to gastric ulceration.[58] During transplant rejection there is an increased level of TXB_2 excreted in the urine preceding the acute crisis[59] and also, high levels of TXB_2 have been found in the venous blood of patients suffering from endotoxin shock.[60]

In general, it seems that in diseases where there is a tendency for thrombosis to develop, TXA_2 production is elevated or prostacyclin production reduced, or both. The opposite is found in some diseases associated with an increased bleeding tendency.

Aspirin and Thromboxane A_2 Synthetase Inhibitors. A consequence of the discovery of prostacyclin has been the need to re-assess the use of aspirin as an antithrombotic agent. Because aspirin inhibits cyclo-oxygenase it can prevent not only production of TXA_2 in platelets but also production of prostacyclin in vascular tissue.

Two important considerations have emerged. First, inhibition of the vascular cyclo-oxygenase, unlike that of the platelet cyclo-oxygenase which is inhibited by aspirin for the entire lifespan of the platelet, may persist for a much shorter period because of the generation of new enzyme. Second, the platelet cyclo-oxygenase seems to be more sensitive in vitro and in vivo than the vessel wall cyclo-oxygenase to the inhibitory action of aspirin,[61] although reports to the contrary have been published.[62]

During the past few years, attempts have been made to titrate aspirin dosage down to find one that will achieve inhibition of TXA_2 formation in man without affecting prostacyclin production. The present data suggest that this will be difficult (for review see 38). However, discrimination between effects on enzymes in platelets and vessel walls may not be the main problem associated with any lack of effectiveness of aspirin as an anti-thrombotic agent. Platelet aggregation is a complex mechanism which can involve several pathways, the generation and actions of TXA_2 being only one of them (the other two identified at present are the ADP and thrombin pathways). Only those stimuli that induce platelet aggregation via the release of TXA_2 will be substantially affected by aspirin treatment.[63] Since initiating processes for intravascular thrombosis are poorly understood, it is difficult to predict the results of aspirin treatment. Platelet aggregation during disseminated intravascular coagulation, venous thrombosis or arterial thrombosis may each have a different triggering mechanism, and it is also possible that platelet aggregation on a fissure in an atherosclerotic plaque or during coronary vasospasm might depend on the activation of different pathways. Interestingly, a recent placebo-controlled trial[64] has demonstrated a halving of the mortality in a group of patients with unstable angina taking one aspirin a day for 12 weeks

A selective inhibitor of thromboxane synthetase might be a superior antithrombotic agent to aspirin by allowing prostacyclin formation by vessel walls or other cells either from their own endoperoxides or from those released from platelets.[38] Work on selective thromboxane synthetase inhibitors is beginning to appear, including the first publications on administration of one of these compounds to man.[65,66] In two experimental studies,[67,68] TXA_2 synthetase inhibitors have shown a superior antithrombotic action to that of aspirin.

Manipulation of Prostacyclin Production. Lipid peroxides, such as 15-hydroperoxy arachidonic acid (15-HPAA), are potent (IC_{50} 0.48 µg/ml) and selective inhibitors of prostacyclin generation by vessel wall microsomes or by fresh vascular tissue.[69-72] High concentrations of lipid peroxides have been demonstrated in advanced atherosclerotic lesions.[73] Lipid peroxidation induced by free radical formation is known to occur in vitamin E deficiency, the ageing process and perhaps also in hyperlipidaemia accompanying atherosclerosis.[74] Accumulation of lipid peroxides in atheromatous plaques could predispose to thrombus formation by inhibiting generation of prostacyclin by the vessel wall without affecting TXA_2 production by platelets. Moreover, platelet aggregation is induced by 15-HPAA and this aggregation is not inhibited by adenosine or PGE_1.[75] Human atheromatous plaques from some patients did not produce prostacyclin.[76] Prostacyclin generation by atherosclerotic arterial tissue was significantly lower than from normal arterial tissue but no difference was found between early and advanced atherosclerotic lesions.[77] This suggests that the early "fatty streak" may be a biochemically critical stage of the atherosclerotic process. In normal rabbits the production of prostacyclin by the luminal surface of the aorta is abolished by de-endothelization over a period of about 70 days. However, the recovery of prostacyclin formation did not occur in rabbits made moderately hypercholesterolaemic by diet.[78] The same authors have also suggested that the accumulation of cholesteryl esters in the areas of damage of the vessel wall[79] is related to the decrease in prostacyclin synthesis since addition of prostacyclin enhanced acid cholesteryl ester hydrolase activity in cultured smooth muscle,[80] thus suggesting a positive feedback between the prostacyclin system and lipid accumulation in the vessel wall. In the rat, application of 15-HPAA to the outside of mesenteric vessels in vivo increased the rate of thrombus formation in response to superfusion with ADP.[81] Smooth muscle cells obtained from atherosclerotic lesions and cultured in vitro consistently produce less prostacyclin than normal vascular smooth muscle cells.[82] In addition, in rats a diet deficient in vitamin E leads to an increase in peroxide levels in the aortae and to a decrease in prostacyclin production in vitro.[83] It may, therefore, be worthwhile exploring whether attempts to reduce lipid peroxide formation by inhibiting peroxidation influence the development of atherosclerosis and arterial thrombosis.

It is important to point out that in vitro human diploid fibroblasts[84] and bovine smooth muscle and endothelial cells[85] lose their ability to produce prostacyclin during ageing while the production of other arachidonate acid metabolites like PGE_2, $PGF_{2\alpha}$ and TXA_2 increases. In addition, aortic smooth muscle cells obtained from old rats produce less prostacyclin in culture than those obtained from young animals.[86] This is due to a specific decrease in the prostacyclin synthetase activity since the cyclo-oxygenase activity was similar in both groups. Similar results have been obtained in vitro with fresh swine arteries.[87] Whether these changes are due to a specific damage of the prostacyclin synthetase as a result of increased lipid peroxidation with age remains to be investigated.

Raised concentrations of low density lipoprotein (LDL) are one of the risk factors associated with ischaemic heart disease[88-90] whereas high density lipoprotein (HDL) is thought to protect against the disease.[89,90] Nordoy and co-workers[91] discovered that LDL reduces the release of a prostacyclin-like substance by human endothelial cells and others later showed that LDL inhibited, whereas HDL stimulated, prostacyclin synthesis.[92] A mixture of low LDL and high HDL also stimulated prostacyclin synthesis. Gryglewski and Szczeklik[93] have confirmed that LDL inhibits prostacyclin synthesis and have additionally shown that the LDL fraction of lipoproteins taken from healthy and hyperlipidaemic volunteers was more susceptible to peroxidation during the extraction procedure than the HDL fraction. Although their most recent results[94] do not support the idea that it is the lipid peroxide associated with LDL which inhibits prostacyclin synthesis, they still found inhibition of prostacyclin synthesis by the LDL taken from some patients with coronary heart disease.

Cell proliferation in vitro is inhibited by substances which stimulate cAMP formation.[95] Cell growth in tissue culture,[96] including vascular smooth muscle culture,[97] is inhibited by PGE_1. Possibly prostacyclin has a role in the regulation of cell growth in the vascular wall. If so, smooth muscle proliferation in atherosclerotic plaques might be a consequence of inhibition of prostacyclin generation by lipid peroxides.

Thrombin, calcium ionophore A23187 and trypsin all stimulate prostacyclin production by cultured endothelial cells.[98] It is interesting that all of these substances cause platelet aggregation and probably act by stimulating phospholipases which liberate free arachidonic acid from cellular phospholipids. Prostacyclin production by cultures of vascular endothelial cells is also enhanced by platelet-poor plasma[24,99] and by platelet-derived growth factor.[100] Control of prostacyclin production is possible, therefore, by a variety of agents acting locally at the level of the cell-vessel wall interaction but the precise mechanisms acting at any time have yet to be clarified.

Chemical stimulation of prostacyclin release from the vessel wall has been reported[101] and confirmed.[102] Bendrofluazide used to treat patients with essential hypertension caused significant increases in circulating levels of 6-oxo-$PGF_{1\alpha}$ suggesting a possible contribution of prostacyclin to the reduction in peripheral resistance seen with thiazide treatment.[103] Similarly, the vasodilator nitroglycerine in clinically attainable concentrations enhances prostacyclin production by cultured human endothelial cells.[104]

In contrast, prostacyclin formation by kidney microsomes[105] and rabbit isolated hearts[106] is reduced by nicotine. These observations, together with the fact that umbilical arteries from smoking mothers release significantly less prostacyclin than those from non-smoking control subjects,[107] indicate that chemically-induced inhibition of prostacyclin production is also a factor to be considered.

Manipulation of diet. Eicosapentaenoic acid (EPA; C20:5ω3) is a polyunsaturated fatty acid like arachidonic acid but has a higher degree of unsaturation. It gives rise to prostaglandins of the '3' series and when incubated with vascular tissue leads to the release of an anti-aggregating substance.[108] Synthetic Δ^{17} prostacyclin or PGI_3 is as potent an anti-aggregating agent as prostacyclin. In contrast, TXA_3 has a weaker pro-aggregating activity than TXA_2.[108] The fatty acid available for PG biosynthesis in Greenland Eskimos is mainly EPA, unlike that in Caucasians which is mainly arachidonic acid.[109] These differences might explain why Eskimos have a low incidence of myocardial infarction, low blood cholesterol levels, an increased tendency to bleed, together with reduced ex vivo platelet aggregability.[110] The plasma concentration of cholesterol, triglyceride, low and very low density lipoprotein (VLDL) is low in Eskimos, whereas that of high density lipoprotein is high.[111]

EPA inhibits platelet aggregation in platelet-rich plasma stimulated by ADP, collagen, AA, and a synthetic analogue of PGH_2.[108] EPA also inhibits aggregation in aspirin- and imidazole-treated platelets and inhibits thrombin-induced aggregation (for review see 38). It is clear, therefore, that both prostaglandin dependent and independent pathways of platelet aggregation are inhibited by EPA in vitro. In vivo, however, EPA is incorporated into platelet phospholipids, to some extent replacing arachidonic acid and exerting an antithrombotic effect by competing with released arachidonic acid for cyclo-oxygenase and lipoxygenase[112,113] or by being converted to the less pro-aggregatory PGH_3 and TXA_3.[108]

Studies in man in which the diet is supplemented with fish, fish oil or EPA have consistently demonstrated a reduction in platelet aggregability and in TXB_2 production during ex vivo platelet aggregation to collagen, as well as an increase in bleeding time.[114-117]

Fish oil feeding increased the incorporation of EPA into phospholipids as measured in heart and liver of cats and the platelets of dogs.[118,119] Brain infarct volume after experimentally-induced cerebral ischaemia and the neurological deficit were less in cats fed fish oil than in a corresponding control group.[118] In dogs fed fish oil, thrombosis and subsequent infarct size induced by electrical stimulation were reduced (3% compared to 25% in control group) with less than 30% ectopic beats after 19 hours compared with 80% in the control group at the same time.[119]

Overall then, the present evidence suggests that research into the fatty acid precursors in our diet may lead to new, simple, but potentially valuable agents.

Clinical Potential of Prostacyclin and its Analogues.

Prostacyclin is prepared as a stable freeze-dried preparation (Epoprostenol: "Flolan", Wellcome) for administration to man. Intravenous infusion of prostacyclin in healthy volunteers leads to a dose-related inhibition of platelet aggregation, dispersal of circulating platelet aggregates, arteriolar vasodilatation, increases in skin temperature, facial flushing and sometimes headache.

Extracorporeal circulation of blood brings it into contact with artificial surfaces which cannot generate prostacyclin. In the course of such procedures thrombocytopaenia and loss of platelet function occur and make an important contribution to the bleeding problems following charcoal haemoperfusion and prolonged cardiopulmonary bypass in man. Formation of microemboli during cardiopulmonary bypass may also contribute to cerebral complications which sometimes follow this procedure. Platelet damage and thrombocytopaenia were prevented by prostacyclin in various animal models of extracorporeal circulation (for reviews see 38,120).

In patients with fulminant hepatic failure undergoing charcoal haemoperfusion[121] prostacyclin infusion prevented the fall in platelet count and elevation of β-thromboglobulin seen in the control patients. Gimson and his colleagues[122] have now made almost 200 charcoal haemoperfusions on a daily basis using prostacyclin for platelet protection in the treatment of 76 patients with fulminant hepatic failure and have obtained an overall survival rate of 38%, compared with a survival rate of 15% in patients under standard intensive care measures. Survival rate was higher (65%) in those patients in whom treatment was started early.

Several double-blind clinical trials of prostacyclin in cardiopulmonary bypass have been published.[123-129] The treatment groups showed a preservation of platelet number and function, with a reduction in the blood loss in the first 18 hours after operation.

As first shown in dogs, prostacyclin can safely be used in man to replace heparin as the sole antithrombotic agent during haemodialysis. Prostacyclin can potentiate the effects of heparin [130] and itself has a small indirect anticoagulant effect. Platelets stimulated by low doses of aggregating agents accelerate clotting by providing a surface upon which coagulation factors can combine and react more efficiently.[131] Prostacyclin, by preventing platelet activation, inhibits the shortening of clotting time produced when either kaolin or collagen is incubated with platelet-rich plasma.[132,133] Platelets release antiheparin activity, which reduces the anticoagulant effect of heparin <u>in vitro</u>. Prostacyclin, by inhibiting this release and by preventing the development of procoagulant activity, can enhance the action of heparin by as much as one hundred per cent.[124,132]

Because heparin therapy in some patients is complicated by thrombocytopaenia and thrombo-embolic episodes,[134,135] and <u>in vitro</u> heparin can cause platelet aggregation and potentiate aggregation caused by other aggregating agents,[136] prostacyclin therapy will allow a substantial reduction or even elimination of heparin during haemodialysis, especially if anticoagulation is contra-indicated.

In peripheral vascular disease prostacyclin was of benefit in open trials in terms of relief of ischaemic pain and ulcer healing.[137-142] Placebo-controlled double-blind trials are now in progress and the results of the first to be reported are encouraging.[143] Dramatic differences were observed between placebo- and prostacyclin-treated patients at 5 days, 1 month and at the 6 month follow-up.

Prostacyclin also induced significant and long-lasting improvements in Raynaud's phenomenon and the striking reduction in the frequency, duration and severity of the disease seen in open trials[144] have been confirmed in a double-blind study.[145]

Gryglewski and his colleagues in Cracow were the first to demonstrate the beneficial effects of infusion of prostacyclin in ischaemic disease of the legs. They also showed effectiveness in retinal vein occlusion.[146] They have now described improvements following prostacyclin infusion in 10 patients with ischaemic stroke.[147]

Prostacyclin has been successfully used in a few cases of pulmonary hypertension.[148,149] In a group of patients with pulmonary hypertension secondary to mitral valve stenosis, prostacyclin caused a dose-dependent pulmonary vasodilatation with no observed side effects.[150] In both of the above studies prostacyclin was more effective than PGE_1. Single case studies have suggested that prostacyclin may be useful in the treatment of patent ductus arteriosus[151] and pre-eclamptic toxaemia.[152]

Yui and co-workers[153] described beneficial effects of intravenous infusion of prostacyclin in 9 patients with severe congestive heart failure refractory to digitalis and diuretics. Mean pulmonary and systemic pressures and vascular resistances were reduced, and heart rate, cardiac index and stroke index were all increased during the infusion, with facial flushing as the only side effect. Acute administration of prostacyclin has been shown to be beneficial in angina, having effects similar to those of the short acting nitrates.[142,154]

Mundy and co-workers[155] infused prostacyclin and protected transplanted kidneys from hyperimmune rejection in dogs. Leithner and colleagues[156] have now shown in 8 patients with chronic renal transplant rejection that intravenous infusion of prostacyclin at 5 ng/kg/min for 5 days resulted in less platelet consumption by the kidneys and an improvement in transplant function.

Clearly, there are many clinical conditions which may respond to prostacyclin treatment and its place (or that of stable analogues) in therapeutics will be defined in the next few years. Some of these conditions are pre-eclamptic toxaemia,[157] haemolytic uraemic syndrome,[158] thrombocytopaenic purpura,[159] peptic ulceration,[58] the thrombotic complications associated with transplant rejection,[155] the prevention of tumour metastasis[160] and the treatment of pulmonary embolism.[161]

It is difficult to ascribe the long-term benefits of prostacyclin in chronic obstructive diseases of the circulation either to the anti-platelet activity or to the direct vasodilator action, which are normally over within 20-40 minutes of stopping administration. However, in addition to its well-known vasodilator and anti-aggregating actions, prostacyclin shares with other prostaglandins a "cytoprotective activity" as yet not clearly defined, but which may be important in explaining certain therapeutic effects of prostacyclin. In some experimental models of myocardial infarction, prostacyclin reduces infarct size[162-164] and arrhythmias[165] and also decreases oxygen demand[164] and the release of cathepsin D and creatinine phosphokinase from infarcted areas.[166] In other studies on the effects of prostacyclin on lung injury in sheep, Demling and co-workers[167] found that prostacyclin protected the lungs against injury induced by endotoxin. A beneficial effect of prostacyclin has also been reported in endotoxin shock in the dog[168] and in the cat[169] where it improves splanchnic blood flow and reduces the formation and release of lysosomal hydrolases (cathepsin D). The effects of hypoxic damage in the cat isolated perfused liver are also substantially reduced by prostacyclin.[170] Monden and Fortner[171] have preserved for up to 48 hours and then successfully transplanted canine livers, using a combination of refrigeration, Sacks' solution and prostacyclin.

All these effects could be related to a result obtained recently in our laboratory. The addition of prostacyclin during the separation from blood and the subsequent washing of platelets substantially improves their immediate functionality in vitro, but in addition, whereas normal in vitro platelet survival time is about 4 to 8h, platelets prepared with the addition of prostacyclin remain functional for more than 72h.[172] This effect is not accompanied by a prolonged increase in cAMP level in platelets, thus separating it from the classical anti-aggregating effect.[173] Interestingly, there has been a study demonstrating a dissociation between anti-aggregating and cytoprotective effects of a prostacyclin analogue in a model of acute myocardial ischaemia.[174] All these results suggest that some of the therapeutic effects of prostacyclin might be related to this cytoprotective effect and point to even wider indications for prostacyclin in cell or tissue preservation in vivo and in vitro.

Although the instability of prostacyclin is an advantage in limiting its biological activity when used in extra-corporeal circulations the possible difficulties of using an unstable substance intravenously as a medicine, together with the desire to separate from each other the three main activities (vasodilator, anti-platelet and cytoprotective) of prostacyclin has led to a world-wide search for stable and more specific analogues. The prostacyclin analogue (5E) 6a-carbaprostaglandin I_2 (carbacyclin) has proved to be a potent inhibitor of platelet aggregation in human plasma.[175] Studies in animals have shown carbacyclin to be a close mimic of prostacyclin with respect to its haemodynamic and platelet actions. Carbacyclin is chemically stable at physiological temperatures and pH but metabolically unstable (its anti-aggregating and vasodilator action is short-lived once an infusion is terminated). The development of potent, stable prostacyclin analogues with biological profiles very close to that of the parent compound thus represents the first stage in drug design. The next step is the development of analogues with selective biological actions.

Summary

There are several enzymes in the cascade leading to the oxidation of fatty acids in the body which are susceptible to selective inhibition.

Aspirin-like drugs only affect cyclo-oxygenase but new substances are on the way which inhibit both lipoxygenase and cyclo-oxygenase, promising a better anti-inflammatory profile which may extend into myocardial infarction and asthma. Within the cyclo-oxygenase pathway there is also the opportunity of selective inhibition of thromboxane formation by newly synthesized substances. And then, the advent of prostacyclin has led to exciting concepts, especially in a kind of hormone replacement therapy. The development of orally active prostacyclin analogues with a long duration of action in humans will provide drugs for the control of platelet aggregation and thromboembolic disorders superior to those known at present and may well form the basis of future antithrombotic therapy.

Chemical stimulation of endogenous prostacyclin may prove to be possible and even the composition of the fats in cell membranes may be changed by dietary manipulation to reduce disease. The discovery of macrocortin, which inhibits the release of arachidonic acid so that all its products are affected, offers another therapeutic avenue to be explored.

Indeed, there is a lot of exciting work still to be done and in the next 20 years we should see a further substantial attack on disease processes.

Acknowledgment

We would like to thank Mrs. A. Higgs and Mrs. A. Skinner for their help in the preparation of this manuscript.

References

[1]U.S. von Euler, Klin Wochenschr. 1935, 14, 1182.

[2]S. Bergström and J. Sjövall, Acta Chem. Scand., 1960, 14, 1693.

[3]S. Bergström, H. Danielson and B. Samuelsson, Biochim. Biophys. Acta, 1964, 90, 207.

[4]D.A. Van Dorp, R.K. Beerthuis, D.H. Nugteren and H. Vonkeman, Biochim. Biophys. Acta, 1964, 90, 204.

[5]J.R. Vane, Nature, 1971, 231, 232.

[6]M. Hamberg and B. Samuelsson, Proc. Natl. Acad. Sci. USA., 1974, 71, 3400.

[7]M.I. Siegel, R.T. McConnell, N.A. Porter, J.L. Selph, J.F. Truax, R. Vinegar and P. Cuatrecasas, Biochem. Biophys. Res. Commun., 1980, 92, 688.

[8]M.A. Bray, Br. Med. Bull., 1983, 39, 249.

[9]P.J. Piper, Br. Med. Bull., 1983, 39, 255.

[10]R. Patterson, J.L. Pruzansky and K.E. Harris, J. Allergy Clin. Immunol., 1981, 61, 444.

[11]J.J. Adcock and L.G. Garland, Br. J. Pharmacol., 1980, 69, 167.

[12]G.J. Blackwell and R.J. Flower, Prostaglandins, 1978, 16, 417.

[13]G.A. Higgs, R.J. Flower and J.R. Vane, Biochem. Pharmacol., 1979, 28, 1959.

[14]G.A. Higgs, S. Moncada and J.R. Vane, "Anti-rheumatic Drugs", Praeger, London, 1983, p. 11.

[15]K.M. Mullane and S. Moncada, Prostaglandins, 1982, 24, 255.

[16]B.R. Lucchesi, S.R. Jolly, M.B. Baslie and G.D. Abrams, Fed. Proc., 1982, 41, 1737.

[17]R.J. Flower, R. Gryglewski, K. Herbaczynska-Cedro and J.R. Vane, Nature, 1972, 238, 104.

[18]R.J. Flower, TIPS, 1981, 2, 186.

[19]G.J. Blackwell, R. Carnuccio, M. Di Rosa, R.J. Flower, L. Parente and P. Persico, Nature, 1980, 287, 147.

[20]F. Hirata, E. Schiffman, K. Venkatasubramanian, D. Solomon and J. Axelrod, Proc. Natl. Acad. Sci. USA., 1980, 77, 2533.

[21]J.R. Vane, S. Bunting and S. Moncada, Int. Rev. Exper. Path., 1982, 23, 162.

[22]S. Moncada, A.G. Herman, E.A. Higgs and J.R. Vane, Thromb. Res., 1977, 11, 323.

[23]B.B. Weksler, A.J. Marcus and E.A. Jaffe, Proc. Natl. Acad. Sci, USA., 1977, 74, 3922.

[24]D.E. MacIntyre, J.D. Pearson and J.L. Gordon, Nature, 1978, 271, 549.

[25]G.J. Dusting, S. Moncada and J.R. Vane, Br. J. Pharmacol., 1977, 62, 414.

26G.J. Dusting, S. Moncada and J.R. Vane, Br. J. Pharmacol., 1978, 64, 315.

[27]M.J. Gimeno, L. Sterin-Borda, E.S. Borda, M.A. Lazzari and A.L. Gimeno, Prostaglandins, 1980, 19, 907.

[28]R.A. Johnson, D.R. Morton, J.H. Kinner, R.R. Gorman, J.C. McGuire, F.F. Sun, N. Whittaker, S. Bunting, J. Salmon, S. Moncada and J.R. Vane, Prostaglandins, 1976, 12, 915.

[29]M.J. Cho and M.A. Allen, Prostaglandins, 1978, 15, 943.

[30]R.R. Gorman, S. Bunting and O.V. Miller, Prostaglandins, 1977, 13, 377.

[31]J.E. Tateson, S. Moncada and J.R. Vane, Prostaglandins, 1977, 13, 389.

[32]R. Kaser-Glanzmann, M. Jakabova, J. George and E. Luscher, Biochim. Biophys. Acta, 1977, 466, 429.

[33]E.G. Lapetina, C.J. Schmitges, K. Chandrabose and P. Cuatrecasas, Biochem. Biophys. Res. Commun., 1977, 76, 828.

[34]M. Minkes, M. Stanford, M. Chi, G. Roth, A. Raz, P. Needleman and P. Majerus, J. Clin. Invest., 1977, 59, 449.

[35]C. Malmsten, E. Granström, and B. Samuelsson, Biochem. Biophys. Res. Commun., 1976, 68, 569.

[36]B.J.R. Whittle and S. Moncada, Br. Med. Bull., 1983, 39, 232.

[37]S. Moncada, R. Gryglewski, S. Bunting and J.R. Vane, Nature, 1976, 263, 663.

[38]S. Moncada, Br. J. Pharmacol., 1982, 76, 3.

[39]M. Lagarde and M. Dechavanne, Lancet, 1977, i, 88.

[40]T. Shimamoto, M. Kobayashi, T. Takahashi, Y. Takashima, M. Sakamoto and S. Morooka, Jap. Heart J., 1978, 19, 748.

[41]A. Szczeklik, R.J. Gryglewski, J. Musial, L. Grodzinska, M. Serwonska and E. Marcinkiewicz, Thromb. Diath. Haemorrh., 1978, 40, 66.

[42]R.I. Lewy, J.B. Smith, M.J. Silver, J. Saia, P. Walinsky and L. Wiener, Prostaglandins Med., 1979, 2, 243.

[43]R.M. Robertson, D. Robertson, J. Roberts, R.L. Maas, G.A. Fitzgerald, G.C. Friesinger and J.A. Oates, New Eng. J. Med., 1981, 304, 998.

[44].P.D. Hirsh, L.D. Hillis, W.D. Campbell, B.G. Firth and J.T. Willerson, New Eng. J. Med., 1981, 304, 685.

[45]H.E. Harrison, A.H. Reece and M. Johnson, Life Sci., 1978, 23, 351.

[46]M. Johnson, H.E. Harrison, A.T. Raftery and J.B. Elder, Lancet, 1979, i, 325.

[47]T.M.E. Davis, E. Brown, D.R. Finch, M.D. Mitchell and R.C. Turner, Br. Med. J., 1981, 282, 1259.

[48]C.T. Dollery, L.A. Friedman, C.N. Hensby, E. Kohner, P.J. Lewis, M. Porta and J. Webster, Lancet, 1979, ii, 1365.

[49]G. Remuzzi, R. Misiani, D. Marchesi, M. Livio, G. Mecca, G. de Gaetano and M.B. Donati, Lancet, 1978, ii, 871.

[50]G. Remuzzi, E.C. Rossi, R. Misiani, D. Marchesi, G. Mecca, G. de Gaetano and M.B. Donati, Seminars in Thrombosis and Haemostasis, 1980, 6, 391.

[51]D.M. Cocchetto, L. Cook, A.E. Cato and E. Niedel, Seminars in Thrombosis and Haemostasis, 1981, 7, 43.

[52]G. Remuzzi, D. Marchesi, G. Mecca, R. Misiani, E. Rossi, M.B. Donati and G. De Gaetano, Lancet, 1980, ii, 310.

[53]G. Remuzzi, A.E. Cavenaghi, G. Mecca, M.B. Donati and G. de Gaetano, Thromb. Res., 1977, 11, 919.

[54]H.G. Gullner, C. Cerletti, F.C. Bartter, J.B. Smith and J.R. Gill, Jr., Lancet, 1979, ii, 767.

[55]C.R. Pace-Asciak, M.C. Carrara, G. Rangaraj and K.C. Nicolaou, Prostaglandins, 1978, 15, 1005.

[56]J.H. Grose, M. Lebel and F.M. Gbeassor, Clin. Sci., 1980, 59, 121s.

[57]S.J. Coker, J.R. Parratt, I. McA. Ledingham and I.J. Zeitlin, Nature, 1981, 291, 323.

[58]B.J.R. Whittle, G.L. Kauffman and S. Moncada, Nature, 1981, 292, 472.

[59]M.L. Foegh, J.F. Winchester, M. Zmudka, G.B. Helfrich, C. Cooley, P.W. Ramwell and G.E. Schreiner, Lancet, 1981, ii, 431.

[60]H.D. Reines, P.V. Halushka, J.A. Cook, W.C. Wise and W. Rambo, Lancet, 1982, ii, 174.

[61]J.W. Burch, N.L. Baenziger, N. Stanford and P.W. Majerus, Proc. Natl. Acad. Sci. USA., 1978, 75, 5181.

[62]E.A. Jaffe and B.B. Weksler, J. Clin. Invest., 1979, 63, 532.

[63]M.A. Packham and J.F. Mustard, Blood, 1977, 50, 555.

[64]H.D. Lewis, Jr., J.W. Davis, D.G. Archibald, W.E. Steinke, T.C. Smitherman, J.E. Doherty III, H.W. Schnaper, M.M. LeWinter, E. Linares, J.M. Pouget, S.C. Sabharwal, E. Chesler and H. DeMots, New Eng. J. Med., 1983, 309, 396.

[65]H.M. Tyler, C.A.P.D. Saxton and M.J. Parry, Lancet, 1981, i, 629.

[66]J. Vermylen, L.O. Carreras, J.V. Schaeren, G. Defreyn, S.J. Machin and M. Verstraete, Lancet, 1981, i, 1073.

[67]J.W. Aiken, R.J. Shebuski, O.V. Miller and R.R. Gorman, J. Pharmacol. Exp. Ther., 1981, 219, 299.

[68]W.E. Yarger, D.D. Schocken, R.H. Harris, R. McRae, M.A. Boyd, C.H. Best, B. Kennedy, J. Gill and N.W. Schrader, J. Clin. Invest., 1980, 65, 400.

[69]S. Bunting, R. Gryglewski, S. Moncada and J.R. Vane, Prostaglandins, 1976, 12, 897.

[70]R.J. Gryglewski, S. Bunting, S. Moncada, R.J. Flower and J.R. Vane, Prostaglandins, 1976, 12, 685.

[71]S. Moncada, R.J. Gryglewski, S. Bunting and J.R. Vane, Prostaglandins, 1976, 12, 715.

[72] J.A. Salmon, D.R. Smith, R.J. Flower, S. Moncada and J.R. Vane, Biochim. Biophys. Acta, 1978, 523, 250.

[73] J. Glavind, S. Hartmann, J. Clemmesen, K.E. Jessen and H. Dam, Acta Pathol. Microbiol. Scand., 1952, 30, 1.

[74] T.F. Slater, "Free Radical Mechanisms in Tissue Injury", Pion Ltd., London, 1972,

[75] H.S. Mickel and J. Horbar, Lipids, 1974, 9, 68.

[76] V. D'Angelo, S. Villa, M. Mysliwiec, M.B. Donati and G. de Gaetano, Thromb. Diath. Haemorrh., 1978, 39, 535.

[77] H. Sinzinger, W. Feigl and K. Silberbauer, Lancet, 1979, ii, 469.

[78] A. Eldor, D.J. Falcone, D.P. Hajjar, C.R. Minick and B.B. Weksler, Am. J. Pathol., 1982, 107, 186.

[79] D.J. Falcone, D.P. Hajjar and C.R. Minick, Am. J. Pathol., 1980, 99, 81.

[80] D.P. Hajjar, B.B. Weksler, D.J. Falcone and C.R. Minick, Fed. Proc., 1981, 40, 351 (Abstract).

[81] R.H. Bourgain, F. Six and R. Andries, Artery, 1980, 8, 96.

[82] J. Larrue, M. Rigaud, D. Daret, J. Demond, J. Durand and H. Bricaud, Nature, 1980, 285, 480.

[83] M. Okuma, H. Takayama and H. Uchino, Prostaglandins, 1980, 19, 527.

[84] S.I. Murota, Y. Mitsui and M. Kawamura, Biochim. Biophys. Acta, 1979, 574, 351.

[85] A. Ager, J.L. Gordon, S. Moncada, J.D. Pearson, J.A. Salmon and M.A. Trevethick, J. Cell Physiol., 1982, 110, 9.

[86] W.C. Chang, S.I. Murota, J. Nakao and H. Orimo, Biochim. Biophys. Acta, 1980, 620, 159.

[87] R.S. Kent, B.B. Kitchell, D.G. Shand and A.R. Whorton, Prostaglandins, 1981, 21, 483.

[88] J.H. Medalie, H.A. Kahn, H.N. Neufeld, E. Riss and U. Gouldbourt, J. Chron. Dis., 1973, 26, 63.

[89] D. Streja, G. Steiner and P.O. Kwiterovich, Ann. Intern. Med., 1978, 89, 871.

[90] W.B. Kannel, W.P. Castelli and T. Gordon, Ann. Intern. Med., 1979, 90, 85.

[91] A. Nordoy, B. Svensson, D. Wiebe and J.C. Hoak, Circ. Res., 1978, 43, 527.

[92] J. Beitz and W. Forster, Biochim. Biophys Acta, 1980, 620, 352.

[93] R.J. Gryglewski and A. Szczeklik, "Clinical Pharmacology of Prostacyclin", Raven Press, New York, 1981, p. 89.

[94] A. Szczeklik, R.J. Gryglewski, B. Domagala, A. Zmuda, J. Hartwich, E. Wozny, M. Grzywacz, J. Madej and T. Gryglewska, Prostaglandins, 1981, 22, 795.

[95] I.H. Pastan, G.S. Johnson and W.B. Anderson, Ann. Rev. Biochem., 1975, 44, 491.

[96]G.S. Johnson, I. Pastan, C.V. Peery, J. Otten and M.C. Willingham, "Prostaglandins in Cellular Biology and the Inflammatory Process", Plenum Press, New York, 1972, p. 195.

[97]J.J. Huttner, E.T. Gwebu, R.V. Panganamala, G.E. Milo, D.G. Cornwell, H.M. Sharma and J.C. Geer, Science, 1977, 197, 289.

[98]B.B. Weksler, C.W. Ley and E.A. Jaffe, J. Clin. Invest., 1978, 62, 923.

[99]G. Defreyn, M. Vergara Dayden, S.J. Machin and J. Vermylen, Thromb. Res., 1980, 19, 695.

[100]S.R. Coughlin, M.A. Moskowitz, B.R. Zeiter, H.N. Antoniades and L. Levine, Nature, 1980, 288, 600.

[101]J. Vermylen, D.A.F. Chamone and M. Verstraete, Lancet, 1979, i, 518.

[102]L.O. Carreras, D.A.F. Chamone, P. Kerckx and J. Vermylen, Thromb. Res., 1980, 19, 663.

[103]J. Webster, C.T. Dollery and C.N. Hensby, Clin. Sci., 1980, 59, 125s.

[104]R.I. Levin, E.A. Jaffe, B.B. Weksler and K. Tack-Goldman, J. Clin. Invest., 1981, 67, 762.

[105]P. Alster and A. Wennmalm, Clin. Physiol., 1981, 1, 105.

[106]A. Wennmalm, Br. J. Pharmacol., 1980, 69, 545.

[107]C.H. Dadak, C. Leithner, H. Sinzinger and K. Silberbauer, Lancet, 1981, i, 94.

[108]R.J. Gryglewski, J.A. Salmon, F.B. Ubatuba, B.C. Weatherley, S. Moncada and J.R. Vane, Prostaglandins, 1979, 18, 453.

[109]J. Dyerberg, H.O. Bang, E. Stoffersen, S. Moncada and J.R. Vane, Lancet, 1978, ii, 117.

[110]J. Dyerberg and H.O. Bang, Lancet, 1979, ii, 433.

[111]H.O. Bang and J. Dyerberg, Acta Med. Scand., 1972, 192, 85.

[112]B.R. Culp, B.G. Titus and W.E.M. Lands, Prostaglandins Med., 1979, 3, 269.

[113]P. Needleman, A. Raz, M.S. Minkes, J.A. Ferrendelli and H. Sprecher, Proc. Natl. Acad Sci USA., 1979, 76, 944.

[114]W. Seiss, P. Roth, B. Scherer, I. Kurzmann, B. Bohlig and P.C. Weber, Lancet, 1980, i, 441.

[115]T.A.B. Sanders, D.J. Naismith, A.P. Haines and M. Vickers, Lancet, 1980, i, 1189 .

[116]J.H. Brox, J.E. Kille, S. Gunnes and A. Nordoy, Thromb. Haem., 1981, 46, 604.

[117]M. Thorngren and A. Gustafson, Lancet, 1981, ii, 1190.

[118]R.L. Black, B. Culp, D. Madison, O.S. Randall and W.E.M. Lands, Prostaglandins Med., 1979, 3, 257.

[119]B.R. Culp, W.E.M. Lands, B.R. Luccesi, B. Pitt and J. Romson, Prostaglandins, 1980, 20, 1021.

[120] J.R. Vane, J. Endocrinol., 1982, 95, 3P.

[121] A.E.S. Gimson, R.D. Hughes, P.J. Mellon, H.F. Woods, P.G. Langley, J. Canalese, R. Williams and M.J. Weston, Lancet, 1980, i, 173.

[122] A.E.S. Gimson, S. Braude, P.J. Mellon, J. Canalese and R. Williams, Lancet, 1982, ii, 681.

[123] J.G. Bennett, D.B. Longmore and J. O'Grady, "Clinical Pharmacology of Prostacyclin", Raven Press, New York, 1981, p. 201.

[124] S. Bunting, J. O'Grady, J-N. Fabiani, E. Terrier, S. Moncada and J.R. Vane, "Clinical Pharmacology of Prostacyclin", Raven Press, New York, 1981, p. 181.

[125] J. Chelly, A. Tricot, A. Garcia, J-C. Boucherie, J-N Fabiani, J. Passalecq and Ch. Dubost, "Clinical Pharmacology of Prostacyclin", Raven Press, New York, 1981, p. 209.

[126] K. Radegran, N. Egberg and C. Papaconstantinou, Scand. J. Thoracic Cardiovasc Surg., 1981, 15, 263.

[127] I.D. Walker, J.F. Davidson, A. Faichney, D. Wheatley and K. Davidson, "Clinical Pharmacology of Prostacyclin", Raven Press, New York, 1981, p. 195.

[128] D.B. Longmore, J.G. Bennett, P.M. Hoyle, M.A. Smith, A. Gregory, T. Osivand and W.A. Jones, Lancet, 1981, i, 800.

[129] C.R. Noback, J.H. Tinker, M.P. Kaye, G.R. Holcomb and J.R. Pluth, Circulation, 1980, 62 Suppl.3, 1242.

[130] S. Bunting, S. Moncada, J.R. Vane, H.F. Woods and M.J. Weston, "Prostacyclin", Raven Press, New York, 1979, p. 361.

[131] A.J. Marcus, J. Lipid Res., 1978, 19, 793.

[132] S. Bunting and S. Moncada, Br. J. Pharmacol., 1980, 69, 268P.

[133] S. Bunting, P.M. Simmons and S. Moncada, Thromb. Res., 1981, 21, 89.

[134] P.L. Cimo, J.L. Moake, R.S. Weinger, Y. Ben-Menachem and K.G. Khalil, Am. J. Hematol., 1979, 6, 125.

[135] C.V. Hussey, V.M. Bernhard, M.R. McLean and J.E. Fobian, Ann. Clin. Lab. Sci., 1979, 9, 487.

[136] E.W. Salzman, R.D. Rosenberg, M.H. Smith, J.N. Lindon and L. Favreau, J. Clin. Invest., 1980, 65, 64.

[137] D. Negus, " Hormones and Vascular Disease", Pitman Medical, London, 1981, p. 181.

[138] A.G. Olsson, Lancet, 1980, ii, 1076.

[139] B.J.H. Pardy, J.D. Lewis and H.H.G. Eastcott, Surgery, 1980, 88, 826.

[140] O. Soreide, L. Segadahl, A. Trippestad and H. Engedal, Scand. J. Thoracic Cardiovasc. Surg., 1982, 16, 71.

[141] A. Szczeklik, R. Nizankowski, S. Skawinska, J. Szczeklik, P. Gluszko and R.J. Gryglewski, Lancet, 1979, i, 1111.

[142]A. Szczeklik and R. Gryglewski, "Clinical Pharmacology of Prostacyclin", Raven Press, New York, 1981, p. 159.

[143]J.J.F. Belch, A. McKay, B. McArdle, P. Lieberman, J.G. Pollock, G.D.O. Lowe, C.D. Forbes and C.R.M. Prentice, Lancet, 1983, i, 315.

[144]P.M. Dowd, M.F.R. Martin, E.D. Cooke, S.A. Bowcock, R. Jones, P.A. Dieppe and J.D.T. Kirby, Br. J. Dermatol., 1982, 106, 81.

[145]J.J.F. Belch, P. Newman, J.K. Drury, F. McKenzie, H. Capell, P. Lieberman, C.D. Forbes and C.R.M. Prentice, Lancet, 1983, i, 313.

[146]H. Zygulska-Mach, E. Kostka-Trabka, A. Niton and R.J. Gryglewski, Lancet, 1980, ii, 1075.

[147]R.J. Gryglewski, S. Nowak, E. Kostka-Trabka, K. Bieron, A. Dembinska-Kiec, B. Blaszczyk, J. Kusmiderski, E. Markowska and S. Szmatola, Pharm. Res. Commun., 1982, 14, 879.

[148]W.D. Watkins, M.B. Peterson, R.K. Crone, D.C. Shannon and L. Levine, Lancet, 1980, i, 1083.

[149]L.J. Rubin, B.M. Groves, J.T. Reeves, M. Frosolono, F. Handel and A.E. Cato, Circulation, 1982, 66, 334.

[150]J. Szczeklik, A. Szczeklik and R. Nizankowski, Lancet, 1980, ii, 1076.

[151]J.E. Lock, P.M. Olley, F. Coceani, P.R. Swyer and R.D. Rowe, Lancet, 1979, i, 1343.

[152]J. Fidler, M.J. Bennett, M. de Swiet, C. Ellis and P.J. Lewis, Lancet, 1980, ii, 31.

[153]Y. Yui, H. Nakajima, C. Kawai and T. Murakami, Am. J. Cardiol., 1982, 50, 320.

[154]G. Bergman, R. Daly, L. Atkinson, M. Rothman, P.J. Richardson, G. Jackson and D.E. Jewitt, Lancet, 1981, i, 569.

[155]A.R. Mundy, M. Bewick, S. Moncada and J.R. Vane, Prostaglandins, 1980, 19, 595.

[156]C. Leithner, H. Sinzinger and M. Schwarz, Prostaglandins, 1981, 22, 783.

[157]J. Fidler, C. Ellis, M.J. Bennett, M. de Swiet and P.J. Lewis, "Clinical Pharmacology of Prostacyclin", Raven Press, New York, 1981, p. 141.

[158]J. Webster, L.K. Borysiewicz, A.J. Rees and P.J. Lewis, "Clinical Pharmacology of Prostacyclin", Raven Press, New York, 1981, p. 77.

[159]G.A. Fitzgerald, L.J. Roberts II, D. Maas, A.R. Brash and J.A. Oates, "Clinical Pharmacology of Prostacyclin", Raven Press, New York, 1981, p. 81.

[160]K.V. Honn, B. Cicone and A. Skoff, Science, 1981, 212, 1270.

[161]T. Utsunomiya, M.M. Krausz, C.R. Valeri, D. Shepro and H.B. Hechtman, Surgery, 1980, 88, 25.

[162]B.F. Jugdutt, G.M. Hutchins, B.H. Bulkley and L.C. Becker, Clin. Res., 1979, 27, 177A.

[163]M.L. Ogletree, A.M. Lefer, J.B. Smith and K.C. Nicolaou, Eur. J. Pharmacol., 1979, 56, 95.

[164] L.G.T. Ribeiro, T.A. Brandon, D.G. Hopkins, L.A. Reduto, A.A. Taylor and R.R. Miller, Am. J. Cardiol., 1981, 47, 835.

[165] V.A. Starnes, R.K. Primm, R.L. Woolsey, J.A. Oates and J.W. Hammon, J. Cardiovasc. Pharmacol., 1982, 4, 765.

[166] R. Ohlendorf, E. Perzborn and K. Schrör, Thromb. Res., 1980, 19, 447.

[167] R.H. Demling, M. Smith, R. Gunther, M. Gee and J. Flynn, Surgery, 1981, 89, 257.

[168] J.R. Fletcher and P.W. Ramwell, Circ. Shock., 1980, 7, 299.

[169] A.M. Lefer, J. Tabas and E.F. Smith III, Pharmacol., 1980, 21, 206.

[170] H. Araki and A.M. Lefer, Am. J. Physiol., 1980, 238, H176.

[171] M. Monden and J.G. Fortner, Ann. Surg., 1982, 196, 38.

[172] S. Moncada, M. Radomski and J.R. Vargas, Br. J. Pharmacol., 1982, 75, 165P.

[173] G.J. Blackwell, M. Radomski, J.R. Vargas and S. Moncada, Biochim Biophys Acta, 1982, 718, 60.

[174] K. Schrör, R. Ohlendorf and H. Darius, J. Pharmacol. Exp. Ther., 1981, 219, 243.

[175] B.J.R. Whittle, S. Moncada and J.R. Vane, "Medicinal Chemistry Advances", Pergamon Press, London, 1981, p. 141.

Chemical Studies on Slow Reacting Substances/Leukotrienes

By E. J. Corey and A. Marfat*

DEPARTMENT OF CHEMISTRY, HARVARD UNIVERSITY, CAMBRIDGE, MA. 02138, U.S.A. and DEPARTMENT OF MEDICINAL CHEMISTRY, PFIZER CENTRAL RESEARCH, GROTON, CT. 06340, U.S.A.

Introduction

This article summarizes research at Harvard over the past five years on a new class of biologically active mammalian natural products. The "slow reacting substances", which are also known by the corresponding acronym "SRS", are a family of chemical substances produced in the body as part of the immunochemical response, that is, the sequence of events by which viruses, bacteria, and also certain indigenous cells such as tumor cells or damaged tissue are recognized and dealt with. The name derives from a characteristic ability to cause slow and prolonged contraction of a test smooth muscle strip maintained under tension in a suitable physiological bath. This response contrasts with the rapid and short lasting contraction produced by substances such as histamine. The SRSs are currently also referred to as "leukotrienes," a term recently introduced by Prof. Bengt Samuelsson of the Karolinska Institutet, Stockholm. Although the knowledge of the roles of the leukotrienes in normal or abnormal physiological states is now only at a rudimentary level, it is already clear that these substances are significantly involved in immunoregulation and in a variety of diseases including asthma and various allergic conditions.

The first evidence for the existence of the SRSs came from immunological experiments with various antigens. It has been known for several decades that whereas the injection of a foreign nontoxic protein such as serum albumin into experimental animals usually produces no ill effects, a second injection of the same substance days or even weeks later can cause a very acute illness and in the extreme, death within minutes. The injection in these cases induces an allergic state in which the experimental animal has become hypersensitive to the initially injected "antigen" or "allergen." The same sequence of operations is involved in "immunization", however, there is a very different response: the production of antibodies which are specific for the antigen and which serve to neutralize additional antigen without harmful affect. By contrast, in hypersensitivity the combination of antigen with antibody triggers a poorly controlled response and an excessive release of various mediators which serve as the immediate cause of pathological change. Among these mediators are the slow reacting substances. The extreme form of hypersensitivity is called anaphylaxis.

The slow reacting substances were first recognized as biological factors in 1938 by Feldberg and Kellaway in England[1] who found that the perfusion of guinea pig lung with various antigenic materials resulted in the generation of agents which caused contraction of guinea pig ileum. They introduced the acronym SRS to designate the factor(s) responsible for the observed physiological response of this test muscle. Later it was shown by the same group that the synthesis of these compounds can be induced by challenge with conventional allergens.[2] Some time later Brocklehurst at Eli Lilly Laboratories in England[3] demonstrated that during an anaphylactic event materials are generated which show biological activity similar to the Feldberg and Kellaway SRS. Because he could not ascertain whether the anaphylactic material and the previously described SRS were the same or different substances, he modified the acronym to SRSA to indicate the former. It is now known that these substances are indeed the same and a distinction between SRS and SRSA is unnecessary.

Work carried out by a number of investigators in the 1960s, most noteworthy Austen's group[4] at Harvard Medical School, demonstrated that the slow reacting substances behave as potent mediators of airway constriction and bronchial spasm in human asthma and of allergic hypersensitivity. The action of this SRS is not blocked by antihistamines, a fact consistent with the ineffectiveness of antihistamines in the treatment of asthma and other forms of hypersensitivity. It is of special interest that the symptoms of asthma have been produced by an aerosol administration of SRSs into the lungs of experimental animals and that SRSs have been detected in fluid from the lungs of asthmatic humans.[4c]

For several years it had been known that these slow reacting substances are produced from arachidonic acid as a biochemical predecessor. They are therefore members of the same large biochemical family as the prostaglandins and thromboxanes. I have introduced the term "eicosanoid" to denote this general class of compounds which is derived from the 20-carbon precursor arachidonic acid.

Evidence has accumulated over the years that various cells of the body, which are either components of circulating blood or initially circulating cells that have become localized in tissue, are involved in the immune response hypersensitivity. These cells are listed along with certain salient characteristics in Table 1.

Table 1. Cells involved in hypersensitivity

1. **Mast Cell:** widely dispersed in connective tissue; immunological "lookout" or "sentinel" function at sites of potential invasion by harmful agents. Receptors at cell surface bind ca. 5×10^5 IgE molecules. Activation causes release of granules/discharge of various mediators and factors. Immediate hypersensitivity (IH*).

2. **Polymorphonuclear leukocytes (PMNL's):** (white blood cells): ~95% neutrophils, ~1% basophils, ~1% eosinophils. IH*

3. **Lymphocytes:** Bursa produced (B cells) and Thymus produced (T cells); circulating; Ag specific; delayed hypersensitivity; attract or inhibit migration of macrophages.

4. **Macrophages:** develop in marrow, transported to tissues via blood; nonspecific destroyer cells; pro- and antiinflammatory effects.

The mast cell is widely distributed and occurs in all kinds of connective tissue being especially abundant in the respiratory system, gastrointestinal tissue, skin and generally those parts of the body that are exposed to external agents. It has been described as a kind of immunological watchdog or sentinel cell because of the ability to detect foreign material. A mechanism for this detection has emerged from extensive biological and biochemical studies (see Table 2). Each mast cell has at its surface about a half a million receptors which are specific for one kind of immunoglobin, the immunoglobin E variety (IgE). These receptors bind IgE and detect foreign substances (allergens) having an affinity for the IgE. Such allergens activate mast cells causing degranulation and secretion of a number of biological agents. Mast cell degranulation has been implicated in immediate hypersensitivity.[4c]

Table 2. Mast cell activation and mediator release

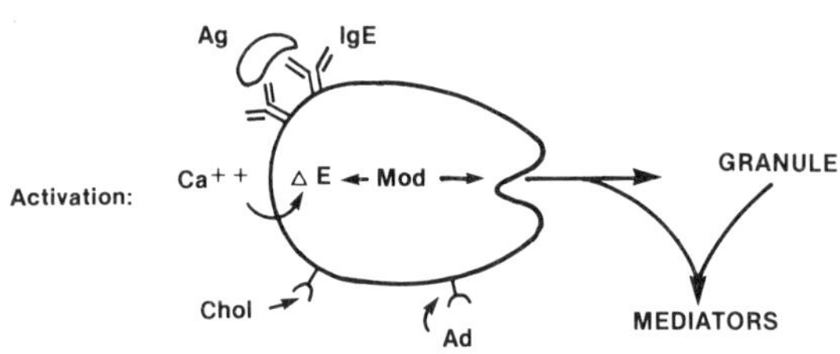

Mediators: histamine, SRS, platelet activating factor, chemotactic factors (peptide, lipid), heparin, enzymes.

SRS can be isolated from cultured mastocytoma cells.

Cell Interactions: Attract macrophages, neutrophils, eosinophils.

Modulation: Release inhibited by increase in c-AMP; α-adrenergic, cholinergic activity decrease c-AMP, β-agonists increase c-AMP. (Theophylline blocks PDE, increases c-AMP.)

A great deal is known about the morphology and function of the mast cell, though much remains to be discovered. Each mast cell possesses roughly half a million surface receptors that are specific for immunoglobin E which binds to these receptors through its "fixed region." The "variable region" of the IgE structure projects outward from the surface of the cell and is available for specific binding to a suitable antigen. Whenever two IgE molecules attached to adjacent receptors are bridged by a single molecule of antigen, a signal is sent to the cell which changes the character of the cell membrane in such a way as to allow influx of calcium ions into the mast cell. This influx of calcium triggers a series of biochemical processes the end result of which is degranulation. The processes of mast cell activation are also regulated *within* the cell in a rather complicated way. The modulation depends, for example, on the level of cyclic AMP within the cell. This is in turn dependent upon the presence of other modulators such as prostaglandins and on adrenergic and cholinergic agents acting on receptors at the cell surface. In general the higher the concentration of cyclic AMP, the less the tendency for the cell to be activated for the granule release action.

The biological mediators released in degranulation include histamine, slow reacting substances, the so-called platelet activating factor (a phospholipid), heparin, various chemotactic factors which cause the attraction to the mast cell of other types of cells such as the macrophages, and various enzymes, especially proteolytic enzymes which have a capacity to break down tissue. SRS compounds can actually be isolated from cultured mast tumor cells (mastocytoma cells). This provides part of the evidence that SRS is produced by the mast cell. Electron microscopy has demonstrated that the release of chemotactic factors can signal macrophages to approach the mast cell surface and accept granules (and mediators) in a direct transfer.

In current therapy theophylline is a widely used drug for the treatment of asthma. It evidently works by blocking the enzyme phosphodiesterase thereby preserving ADP, the predecessor of cyclic AMP, and increasing the level of cyclic AMP. That increase tends to diminish the release reaction and the hypersensitivity response in asthmatic individuals. Steroids such as cortisone are also used in medicine. They block arachidonic acid release and SRS synthesis. Thus mast cells occupy a position on the perimeter of the body's defense system and they trigger important cellular events of the immune response. Macrophages (Table 3) are produced in the marrow, transported to tissues by way of the blood, and subsequently lodge and reside in tissue for several weeks. They take up their stations and they serve as nonspecific destroyer cells once a signal from the other parts of the immune system, the trigger mechanisms, is received. Because they can produce massive non-selective digestion of the material around them, macrophages can be a major source of inflammation. Macrophages also possess IgE receptors. Further, they are capable of ingesting and modifying antigens (by an obscure process) so as to increase their immunogenicity, thereby amplifying the immune response. Macrophages secrete SRS and also powerfully destructive agents, for example, superoxide ion (O_2^-), hydrogen peroxide, collagenase, galactase, and other chemicals which are capable of breaking down biological matter. Thus they are protectors against viruses, tumor cells, and other biological threats.

Table 3. Macrophages

Macrophages (MP's): possess IgE receptors; ingest antigens (Ag's) and process them to increase immunogenicity; attach to the surface of mast cells effecting granule transfer.

MP's produce SRS *in vitro* from arachidonic acid under stimulation by zymosan; also produce PGE_2, PGI_2, TXA_2, PAF.

MP's secrete superoxide, H_2O_2, collagenase, elastase (and other proteases), lipases, phosphatases, complement components, endogenous pyrogens.

MP's inhibit replication of microbes, tumor cells and viruses (interferon).

The neutrophils which also function as "killer cells" in many ways are similar to the macrophages. They produce a comparable assortment of powerful cell- and tissue-disintegrating substances: proteases, hydrogen perioxide, and the like. They, too, are attracted by chemotactic factors secreted by the trigger cells.

Slow reacting substances are now known to be produced in a variety of tissues and cells. These include mast cells and mast tumor cells (mastocytoma cells), rat basophilic leukemia cells, macrophages, human and guinea pig lung tissue (after sensitization), and cells in the inflamed rat peritoneal cavity. Four different SRSs have now been obtained from these sources.

Structure and Synthesis of SRSs

The chemical nature of the SRSs was completely unknown for many years. However, as mentioned above, it has become clear that these materials are formed biosynthetically from arachidonic acid, the same C_{20}-tetraunsaturated fatty acid which gives rise to prostaglandins (PGs) such as $PGF_{2\alpha}$ and PGE_2.[5] SRS biosynthesis can be stimulated by the calcium ionophore A23187 (which transports Ca^{++} across membranes)[6] and by addition of various thiols (such as cysteine)[7] to certain SRS synthesizing cells. Of considerable importance to the clarification of the structure of SRSs were the findings that these substances could be refined by reversed phase high performance liquid chromatography (RP-HPLC),[8] that they exhibited strong ultraviolet absorption (centered at *ca.* 280 nm),[8] and that they are deactivated by soybean lipoxygenase, an enzyme specific for fatty acids containing the structural subunit: (*cis, cis*)-CH = CH-CH_2-CH = CH-.[9]

Although our group at Harvard had long been involved in chemical and synthetic studies on the prostaglandins and had followed developments in the research on SRSs, chance was to play a key role in our activities in this area. Professor Bengt Samuelsson, a former collaborator and also a former postdoctoral associate of mine (1961-1962), happened to be visiting in the Boston area in March of 1977. Over lunch in a restaurant in Harvard Square he described some interesting studies which were in progress in his lab involving the metabolism of arachidonic acid in polymorphonuclear leukocytes (PMNs), principally the neutrophil type. In work done collaboratively with a postdoctoral fellow, Pierre Borgeat, he found that arachidonic acid is converted by PMNs in a very low yield to a new substance which could be isolated in microgram amounts. It was demonstrated that this metabolite contained two hydroxyl groups in addition to the oxygens originally present in arachidonic acid and that these hydroxyl groups were at carbons 5 and 12. Further, it was shown that the absolute chirality of the stereocenters was 5-(*S*), 12-(*R*). On the basis of these data and also ultraviolet absorption, he proposed the structure in Figure 1 (middle, center), with the geometry of the three conjugated bonds being only tentative since the data did not allow a final decision on this point.

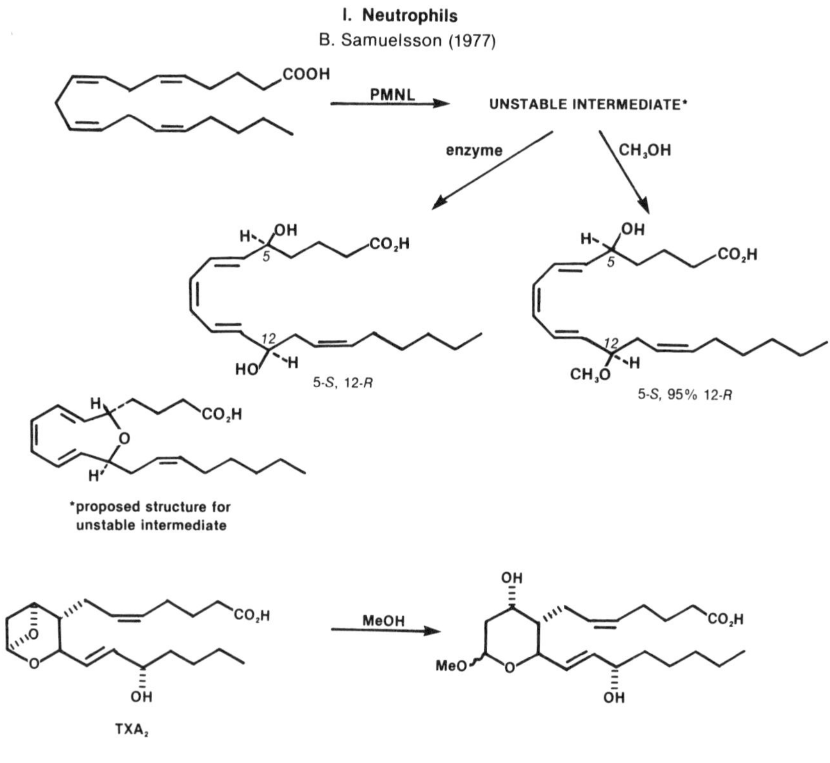

Figure 1. Eicosanoid production in polymorphonuclear leukocytes (PMNL)

This metabolite was found not to have contractile activity in measurements involving a conventional smooth muscle test preparation, guinea pig ileum. It was therefore conjectured that this substance might be a transformation product of a much less stable, but biologically active and more important molecule. This reasonable hypothesis finds a parallel in Samuelsson's work on metabolites of arachidonic acids in platelets, the thromboxanes A_2 and B_2: Arachidonic acid is converted in platelets into an extremely unstable substance, thromboxane A_2, which has a half life of only *ca.* thirty-three seconds, but which is a powerful inducer of platelet aggregation. The major metabolite of thromboxane A_2 can be trapped by addition of methanol and converted to thromboxane B_2 methyl ether.

A parallel methanol trapping experiment in the neutrophil system led to isolation of a new product, which was formed at the expense of the C_{20}-5, 12-dihydroxy acid that would normally be obtained. It was found to be a monomethyl ether having one free OH and one COOH group. By microchemical degradation and mass spectroscopy, it was shown that the free hydroxyl function was attached at C-5 and the methoxy group at C-12. Thin layer chromatographic analysis indicated that the methanol trapping product was homogeneous and it seemed from degradation studies to have the R-configuration at C-12. On the basis of all these data, Samuelsson deduced the formulas shown in Figure 1 for the monomethyl ether and the unstable intermediate. It was considered that the reactive oxa-cyclononatriene intermediate combines with methanol by a non-enzymic S_N2 process in the trapping experiment.

For various reasons this interpretation seemed to us unlikely. Why should the oxa-cyclononatriene structure react with methanol only at C-12 and not at C-5? Why should it not react by an S_N1 process to form mixtures of isomeric methoxy compounds? Is not the oxa-cyclononatriene structure so strained as to be an unlikely biosynthetic product? We had encountered similar structures having one more ring member, cyclodeca-1,3,5-trienes, years before in our synthetic studies on costunolides[10] (Figure 2). These were very unstable even in aprotic solution at 20°C and it was clear that the oxa-cyclononatriene structure would be vastly more strained and less stable. Stimulated by these doubts and considering the usual lipoxygenase mode of oxygenation of arachidonic acid, I proposed the alternative structure for the PMN-derived predecessor of the 5,12-diol and the pathway for biosynthesis which are shown in Figure 3. This scheme could explain enzymic 5,12-diol formation, but it required that the methanol quenching reaction produce at least a mixture of two epimeric C-12 methoxy compounds (an S_N1 mechanism would clearly be expected) and probably other products as well (e.g., two C-6 methoxy products). These considerations suggested the desirability of a re-examination of the methanolysis product to ascertain whether more than one isomer could be detected.

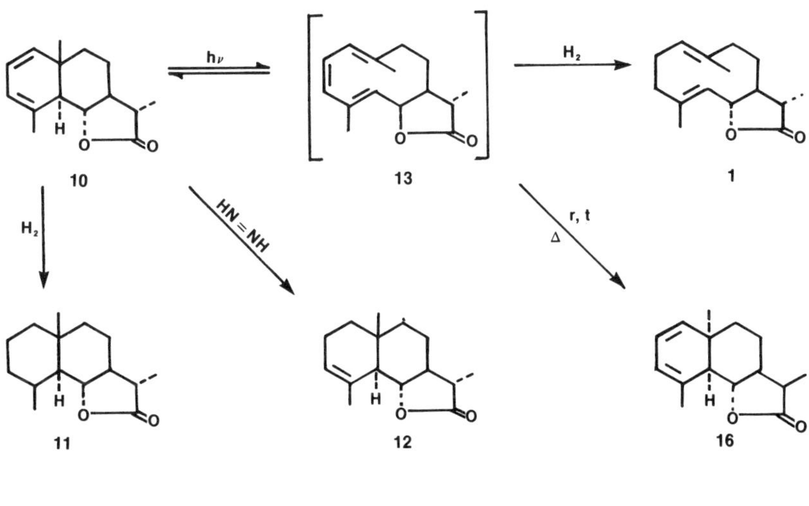

E.J. Corey *et al., J. Am. Chem. Soc.*, **87**, 5736 (1965).

Figure 2. Generation of an unstable cyclodecatriene

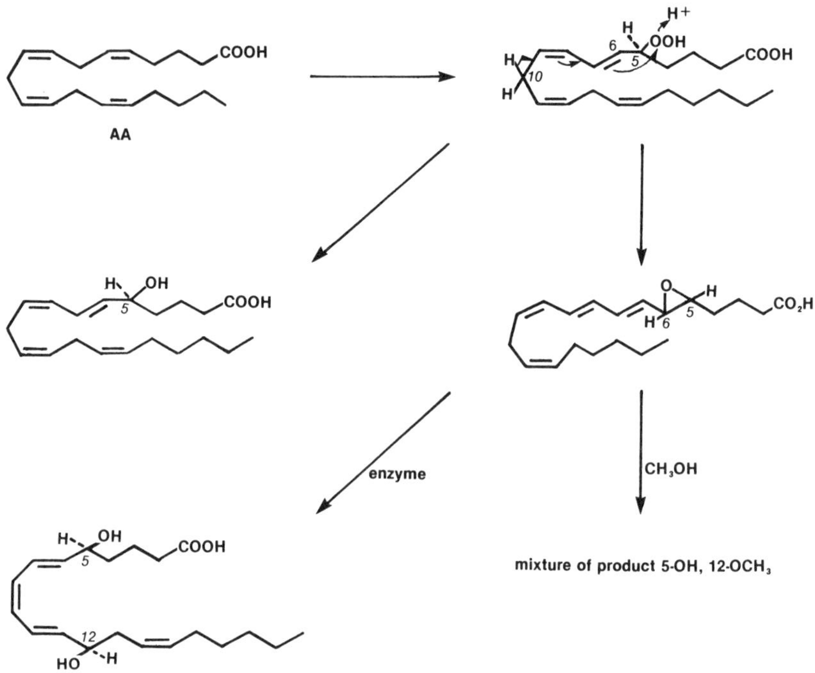

See, P. Borgeat and B. Samuelsson, *Proc. Nat. Acad. U.S.*, **76**, 3213 (1979).

Figure 3. Alternate proposal for PMNL intermediate (March 1977)

The proposal outlined in Figure 3 is mechanistically reasonable. In fact, some years ago we encountered the allylic hydroperoxide to epoxide conversion with studies in the triterpene series (Figure 4).[11] The triene-conjugated oxirane structure for the metastable intermediate which is shown in Figure 3 was expected to be quite readily hydrolyzed in water at pH7 (even faster at lower pH) but much more stable at higher pH and therefore an experimental pH-lifetime profile of the intermediate from the neutrophil system was indicated. The olefinic stereochemistry shown in Figure 3 for the oxirane intermediate seemed the most likely on mechanistic grounds. The oxirane unit in principle could be either *trans* substituted or *cis* substituted with the former being more likely on purely energetic grounds. It seemed appropriate to test the oxirane proposal by synthesis and this was subsequently undertaken at Harvard despite the discrepancy between Samuelsson's experimental observation of a single methanolysis product from interception of the unstable neutrophil intermediate and the more complex results expected on the basis of the oxirane structure.

Our first synthesis of the oxirane structure postulated for the metastable neutrophil product is outlined in Figure 5. It was designed to produce a (racemic) mixture of *cis* and *trans* substituted oxiranes, but to be *unambiguous* with regard to the geometry of the four double bonds. It was our intention to separate the *cis* and *trans* oxiranes resulting from the synthesis, to assign structures from pmr data and to compare each pure isomer with the Samuelsson intermediate. The synthesis proceeded satisfactorily, as has already been described both in published form[12] and in a lecture presented at the Fourth International Meeting on Prostaglandins in Washington D.C. on May 28, 1979. The *cis*- and *trans*-5,6-oxides of methyl eicosa-7,9-*trans*, 11,14-*cis*-pentaenoate were isolated and it was demonstrated that these could be saponified in aqueous base to the corresponding carboxylate ions (from which the starting methyl ester could be obtained with methyl sulfate). Unfortunately, the mixture of *cis* and *trans* epoxy esters could not be preparatively separated by thin layer chromatography. A mixture of the two was therefore sent to Professor Samuelsson to enable comparison with his metastable intermediate.

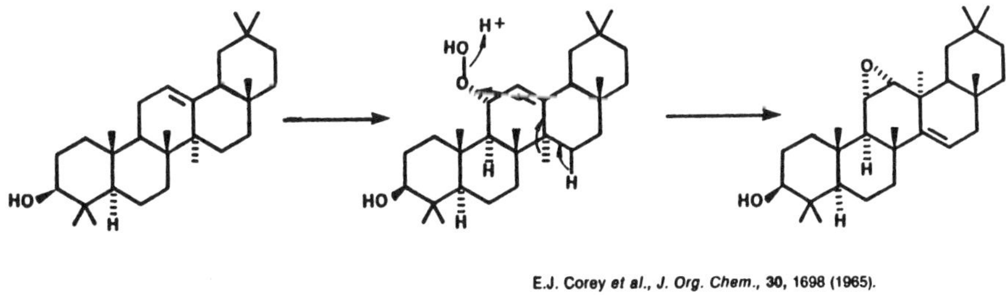

Figure 4. Allylic hydroperoxide to epoxide transformation

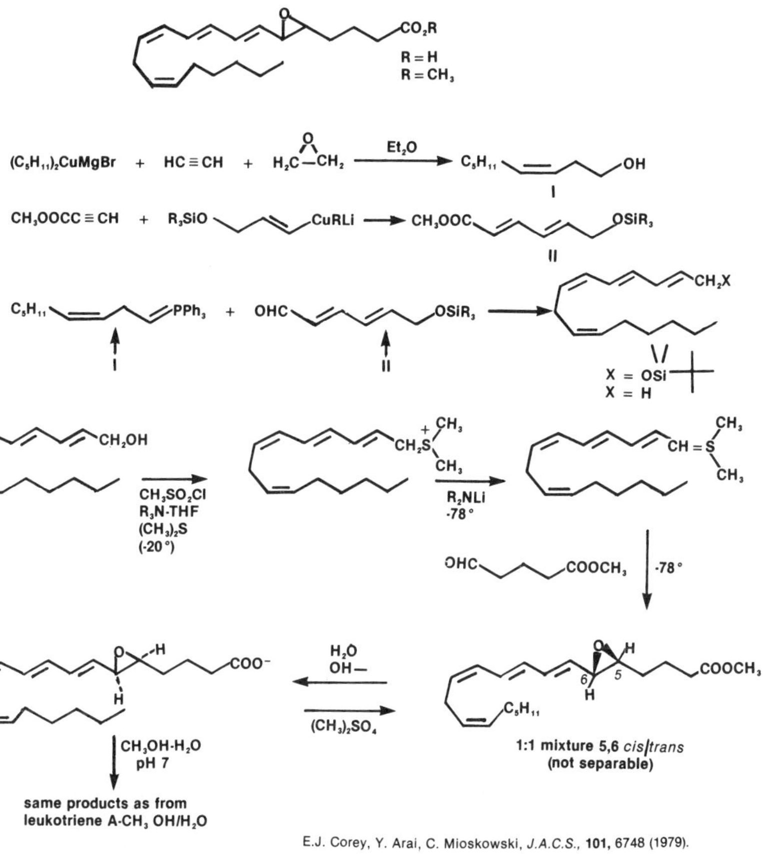

Figure 5. Synthesis of (±)-5,6-oxido-7,9-trans-11,14-cis-eicosapentaenoic acid

In the meantime Borgeat and Samuelsson had completed the reinvestigation of the metastable intermediate from neutrophils using HPLC for analysis and had found that mixtures of several methoxy or dihydroxy compounds are indeed formed in quenching experiments using, respectively, methanol or aqueous acid quencher. In view of these results and others, the 5,6-oxirane formulation was accepted for the intermediate, and they reported the interpretation outlined in Figure 6.[13] Comparative quenching experiments with the biosynthesized intermediate and our synthetic oxide showed essentially indistinguishable behavior; both were quite stable in alkaline solution (pH 10) at 24° as expected.

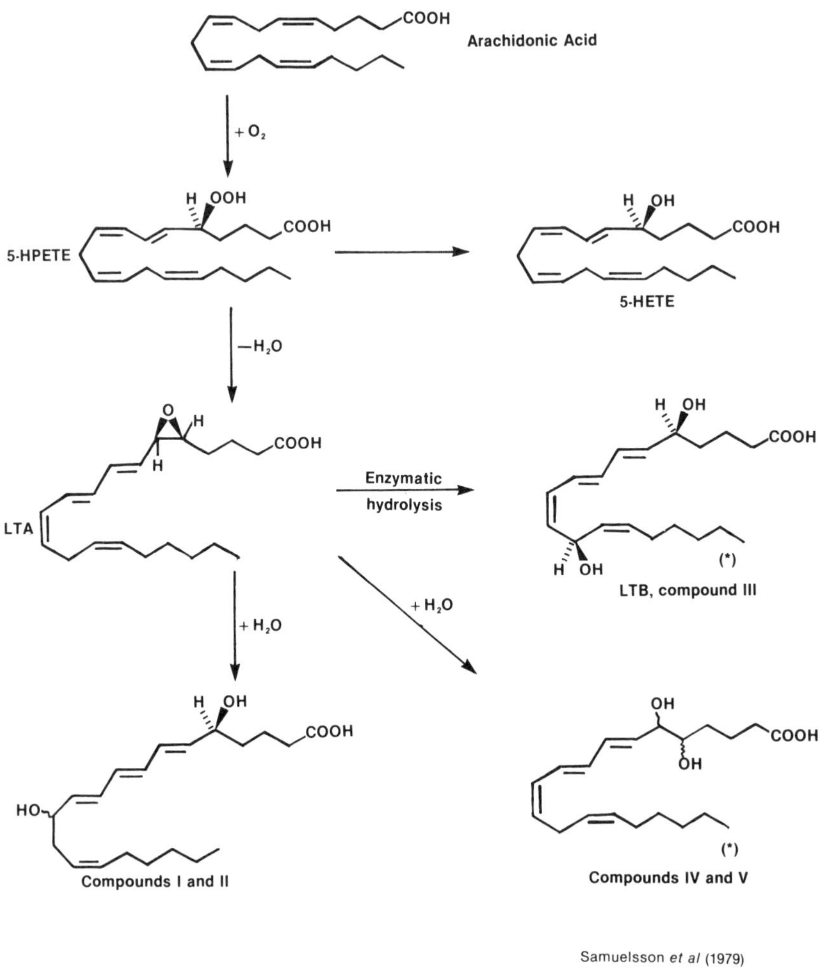

Figure 6. PMNL-neutrophil pathway

Then a really dramatic development occurred in Stockholm. Professor Samuelsson and his colleagues had the inspired idea that perhaps this oxirane structure was an intermediate in the formation of the slow reacting substances. They speculated that one could understand the role of the sulfhydryl compounds such as cysteine to increase production of the slow reacting substances as a reaction of the oxirane intermediate, leukotriene

A, as it came to be called, with a thiol group in a nucleophilic ring opening to give a sulfurcarbon bond, the resulting product being a conjugate of an amino acid with a fatty acid moiety. Based on this consideration they carried out an incubation of tritiated arachidonic acid with mastocytoma cells in the presence of[14] C-labeled cysteine and the ionophore A23187. They were able to separate from the mixture by HPLC a very low yield of material which contained both[14] C and tritium, which had the same ultraviolet absorption and reactivity toward soybean lipoxygenase as SRSs. Samuelsson and coworkers therefore proposed the scheme shown in Figure 7 for the formation of the SRS from mastocytoma cells and this material as leukotriene C.[14] Because of the incorporation of radiocarbon label from cysteine, they suggested that a cysteinyl substituent was present in leukotriene C; this was placed at C-6 in view of the ultraviolet data (UV_{max}280 nm) and the reactivity toward soybean lipoxygenase to form a conjugated tetraene.[9,14]

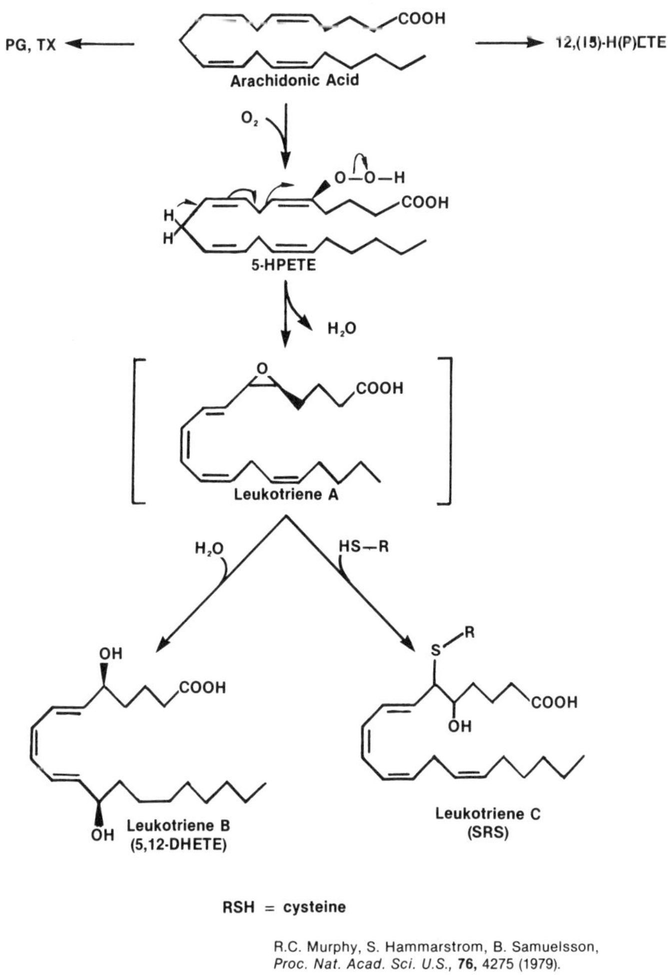

Figure 7. Structure of SRS — Samuelsson proposal

Professor Samuelsson had informed us of his interesting experiment in April of 1979 and since we had synthetic (±)-leukotriene A in hand we embarked on the synthesis of the cysteine and other thiol conjugates so that synthetic and native compounds could be compared. In this way we hoped to establish the SRS structure in all detail (including stereochemistry) and to make adequate amounts of these hitherto rare and elusive agents available for study.

By the use of the readily available and simple model compound shown in Figure 8 we were able to establish suitable reaction conditions for effecting the desired S_N2 displacement at the oxirane unit. In addition, other conditions were found for directing oxirane ring opening via an S_N1 pathway, which was found to afford a different structural isomer, as indicated in Figure 8. The structures of the S_N2 and S_N1 products were rigorously established by physical measurements, especially pmr spectroscopy.

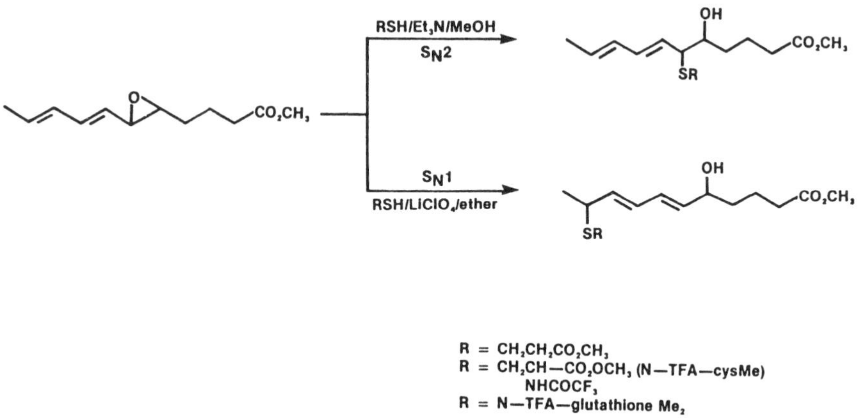

Figure 8. Position selectivity in displacement of RSH

Using the S_N2 reaction conditions which were developed with the model compound (Figure 8), the synthetic racemic mixture of 5,6-*cis* and 5,6-*trans*-epoxides of methyl eicosa-7,9-*trans*-11,14-*cis*-pentaenoate was transformed into a mixture of 4 diastereomeric 6-*L*-cysteinyl thiol conjugates as shown in Figure 9. The displacement was conducted using N-trifluoroacetyl-*L*-cysteine methyl ester in methanol containing excess triethylamine at 23° and the resulting material was deprotected by aqueous base (all under an argon atmosphere, since intermediates and products of these reactions were found to be extremely sensitive to air and free radical initiators). Reversed phase HPLC analysis revealed equal amounts of four diastereomeric compounds as expected on the basis of the starting material being a mixture of (±)-*cis* and (±)-*trans* epoxides. When these synthetic compounds were compared with the two SRSs isolated from mastocytoma cells in Samuelsson's laboratory, it was clear that they were different. In fact, the differences between the synthetic compounds and the native leukotriens (C-1 and C-2) were so large by RP-HPLC that it seemed likely that a different (and more polar) cysteine-containing substituent was attached to the C_{20} chain in leukotriene C-1 and C-2. It transpired from comparisons performed in Stockholm that one of the 4 isomeric glutathione conjugates was *identical with naturally derived leukotriene C-1* (LTC-1) as shown by UV absorption, RP-HPLC behavior, bioassay using guinea pig ileum and reaction with lipoxygenase from soybean (see Figure 10).[15,16]

At the same time the comparison of synthetic and native LTC-1 was being made, we were completing a second synthetic route to leukotriene A. This one was designed to provide optically active *trans*-5,6-oxide with the 5-(*S*)-chirality. The scheme of synthesis[17] which is outlined in Figures 11 and 12 is very effective and allows the direct production of chiral LTA.[18] Reaction of this synthetic LTA with glutathione in methanol containing triethylamine gave LTC-1 directly as the sole coupling product; it was identical with naturally derived LTC-1. This synthetic LTA has also been converted to the cysteinyl conjugate (LTE) and the cysteinyl glycine conjugate (LTD). Both are biologically active and both have been shown to be identical with naturally occurring SRSs.[19,20] These interconversions are summarizd in Figure 13. Further, it was shown that LTA undergoes

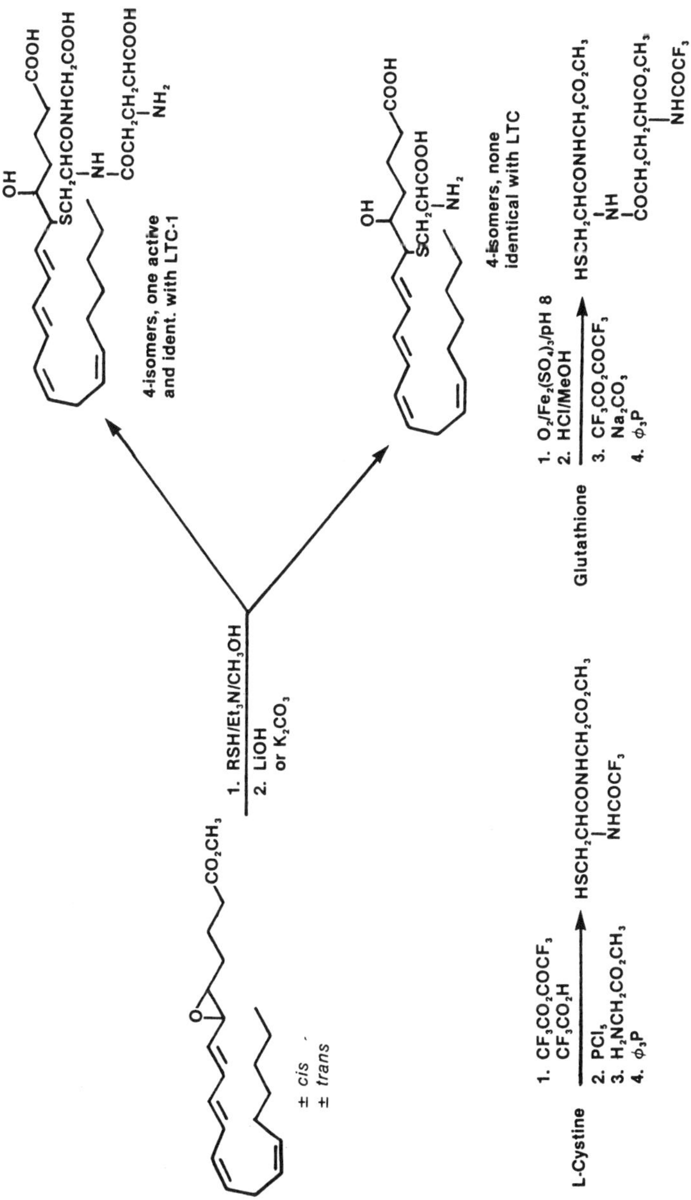

Figure 9. Synthesis of (±)-C_{20}-6-cys and (±)-C_{20}-6-glutathione

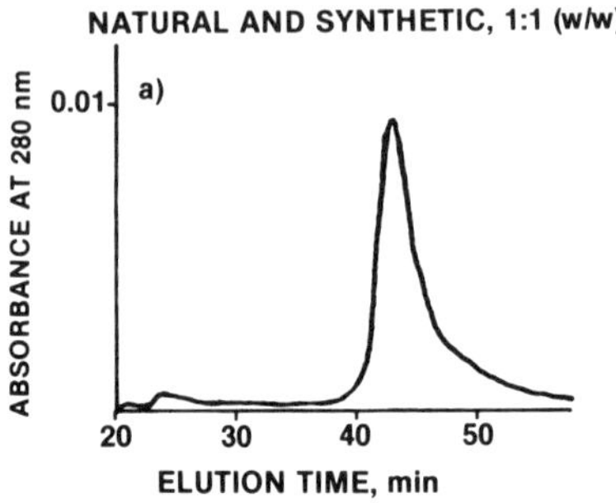

a) Cochromatography of synthetic and natural leukotriene C-1 on reverse phase high pressure liquid chromatography (Nucleosil C_{18}, 5 μ; methanol/ water 65:35 (v/v) + 0.01% acetic acid, 1 ml/min).

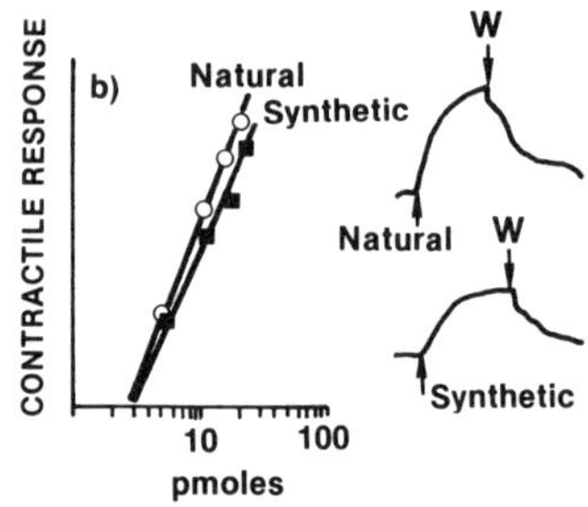

b) Bioassay on guinea pig ileum of synthetic and natural leukotriene C-1.

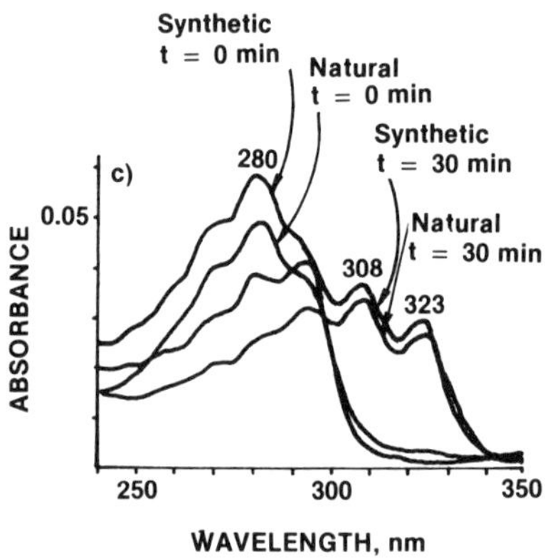

c) Ultraviolet spectra of synthetic and natural leukotriene C-1 before (t = 0 min) and after treatment with soybean lipoxygenase (t = 30 min). Spectra were recorded in Tyrode's buffer. (Hammarstrom *et al*[16-17]).

Biochem. Biophys. Res. Commun., **92**, 946 (1980).

Figure 10.

enzymic conversion to LTB, the 5,12-dihydroxyeicosanoid which Samuelsson originally isolated from polymorphonuclear leukocytes (PMNs). The correct structure for this 5,12-diol, now termed leukotriene B,[22] which has recently been established in our laboratory,[23] is shown in Figure 13. Leukotriene A has also been isolated from human polymorphonuclear leukocytes and compared with synthetic LTA.[24]

Leukotrienes C, D, and E can all be detected in the lung fluids of human asthmatic patients.[4c] Leukotriene D, which is more active than leukotriene C (by a factor of 6 on guinea pig ileum and of *ca.* 100 on guinea pig pulmonary parenchema), occurs in larger amounts and, in fact, accounts for most of the activity. Leukotriene D is *ca.* 6000 times as active as histamine on a molar basis toward the small airways of guinea pig lung. Leukotrienes C, D, and E produce their constrictive effects mainly on the small airways and from all indications are major mediators of asthma. Leukotriene C is transformed successively in tissue or serum into leukotrienes D and E. Synthesis of the last two actually predated isolation of the native LTs.

Leukotriene B, initially thought to be without biological activity, turns out to be powerfully bronchoconstricting. It is also a potent chemotactic agent for neutrophils and macrophages (at *ca.* 1 ng/ml). It is likely that this substance too plays an important role in allergic and inflammatory states. Two syntheses of

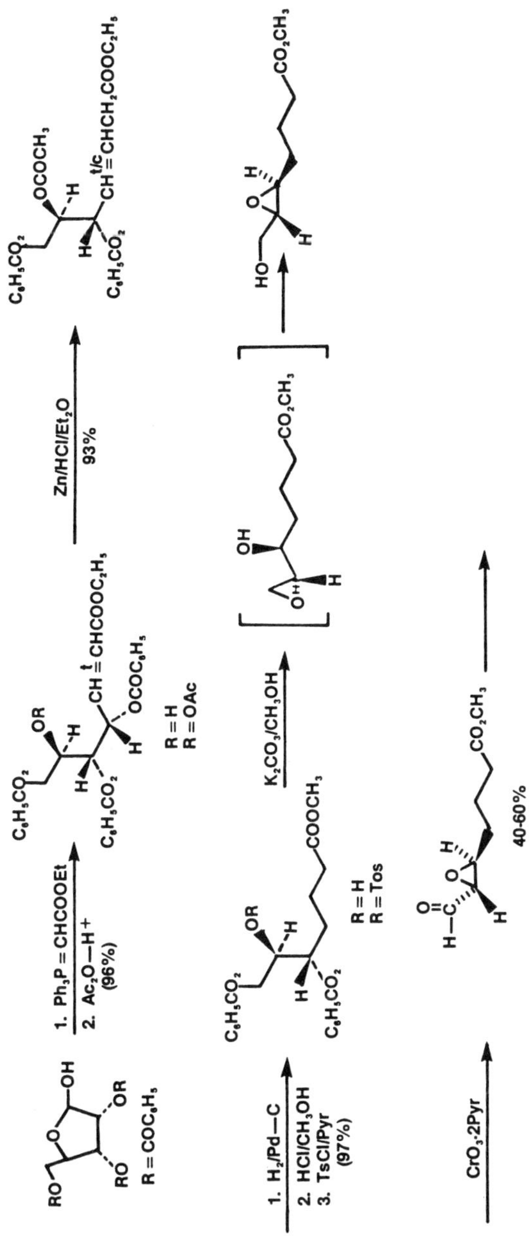

J. Am. Chem. Soc., **102**, 1436 (1980).

Figure 11. Stereospecific synthesis of leukotrienes A and C-1

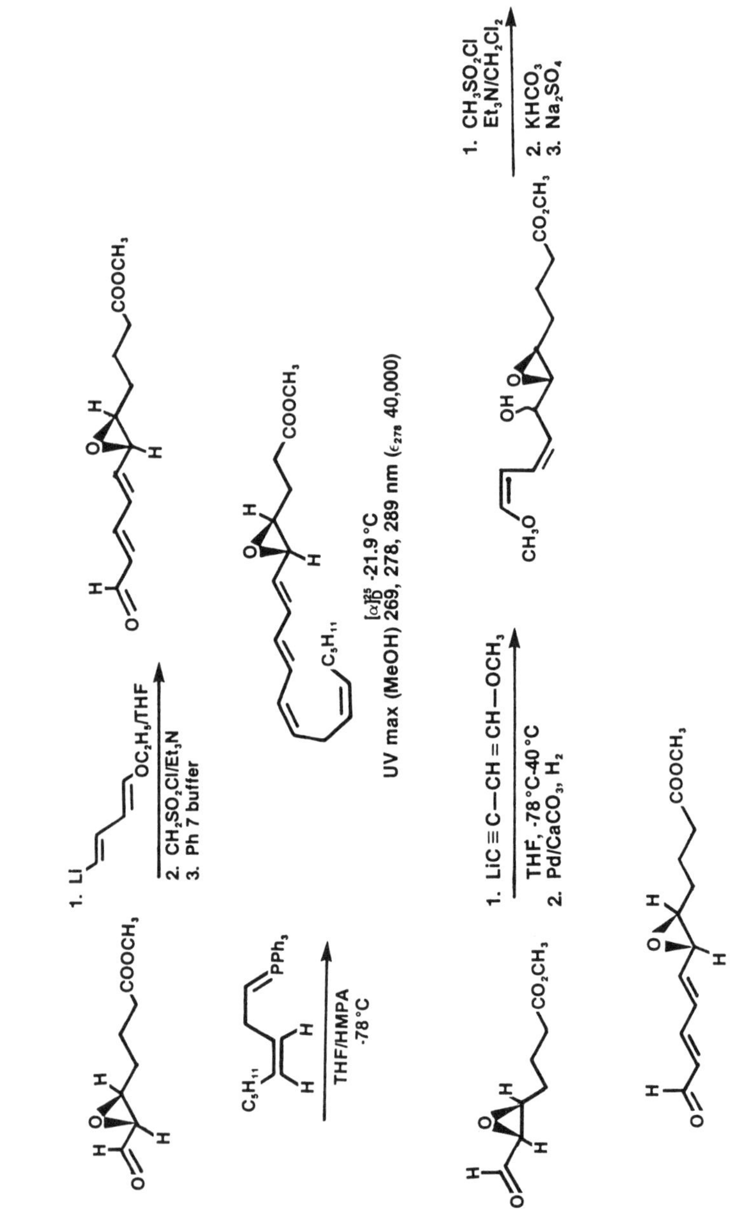

J. Org. Chem., **48**, 2114 (1983)

Figure 12. Stereospecific synthesis of leukotrienes A and C-1

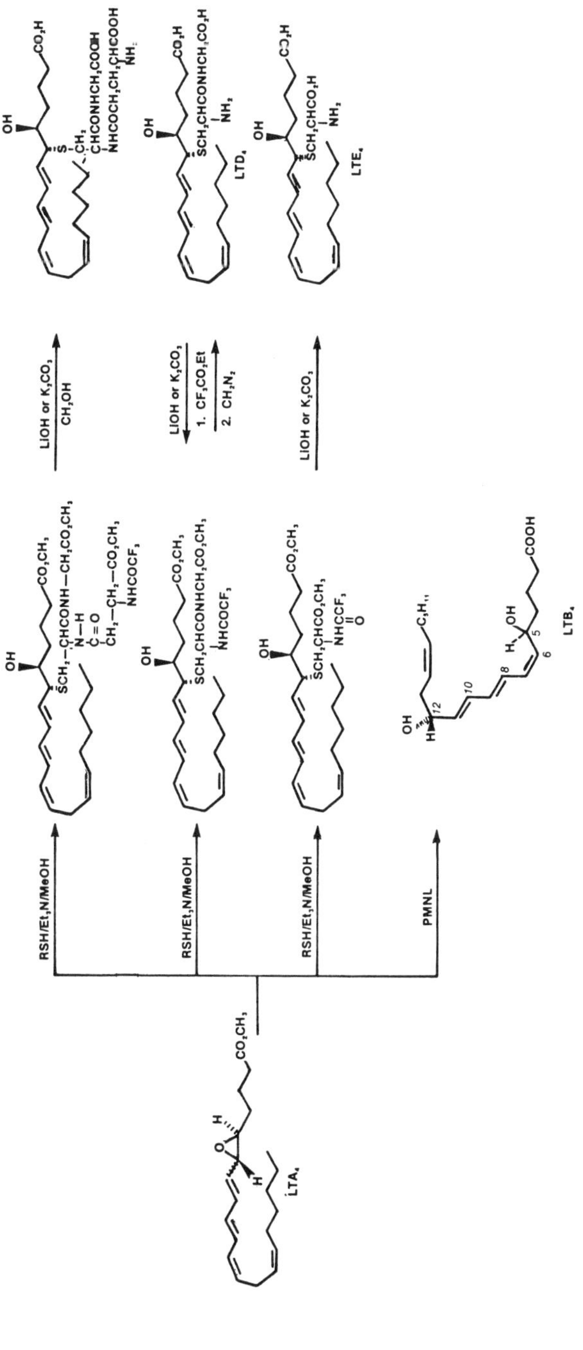

Figure 13. Stereospecific synthesis of leukotrienes

leukotriene B have been developed[23] which not only have made this substance available in quantity but which have also allowed clarification of the stereochemistry of the conjugated triene unit. An especially simple synthesis utilized as a key step the novel internal elimination depicted in Figure 14 which illustrates the terminal stage of the synthesis. The synthesis of the ester aldehyde corresponding to the C(1)-C(6) segment is shown in Figure 15 and that of the C(7)-C(20) segment is outlined in Figure 16. The trihydroxy ester monotosylate intermediate shown in Figure 15 can also serve as an intermediate in the synthesis of leukotriene A itself, being readily convertible by treatment with methoxide in methanol into the epoxy alcohol (Figure 11) precursor of LTA. Another route to the same epoxy alcohol which has been developed takes advantage of the Sharpless method for asymmetric epoxidation of allylic alcohols;[25] this process for the synthesis of the key LTA precursor is outlined in Figure 17.

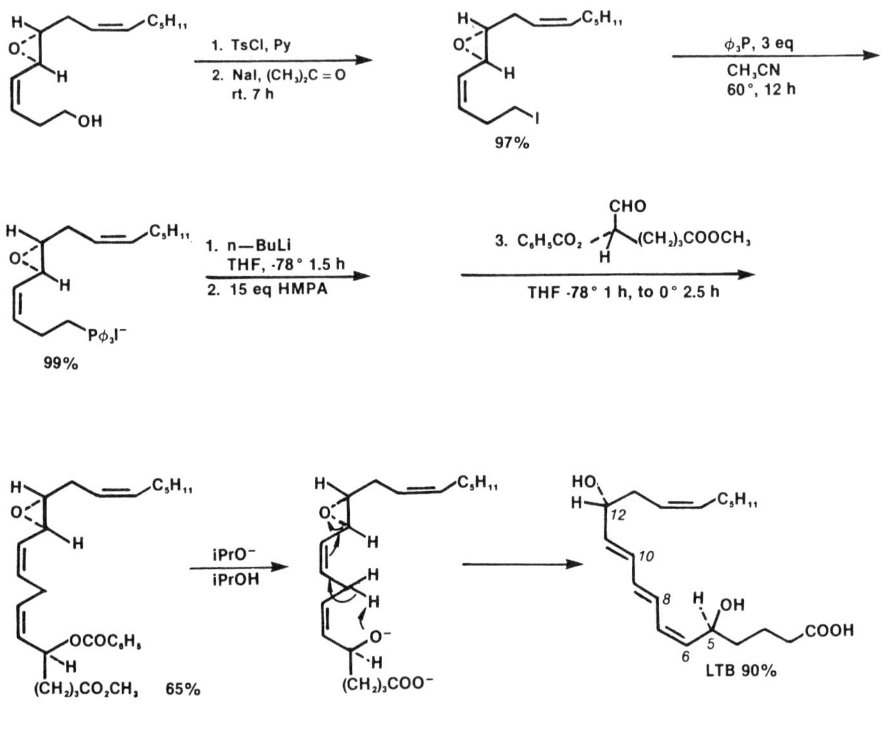

Figure 14. Synthesis of LTB

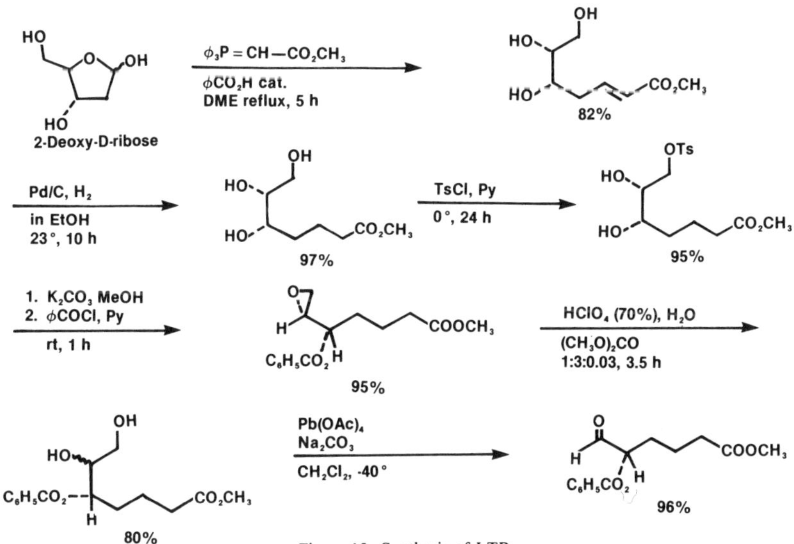

Figure 15. Synthesis of LTB

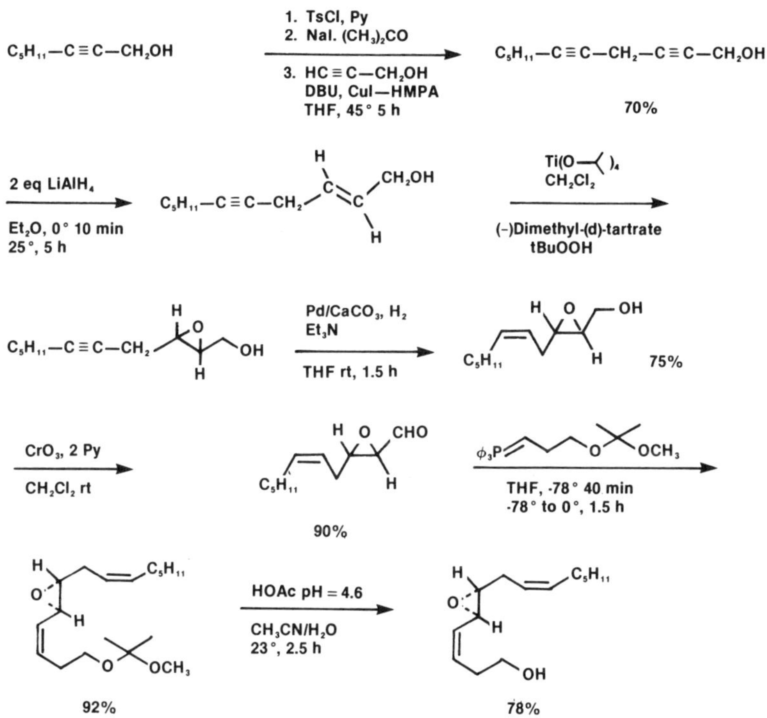

Figure 16. Synthesis of LTB

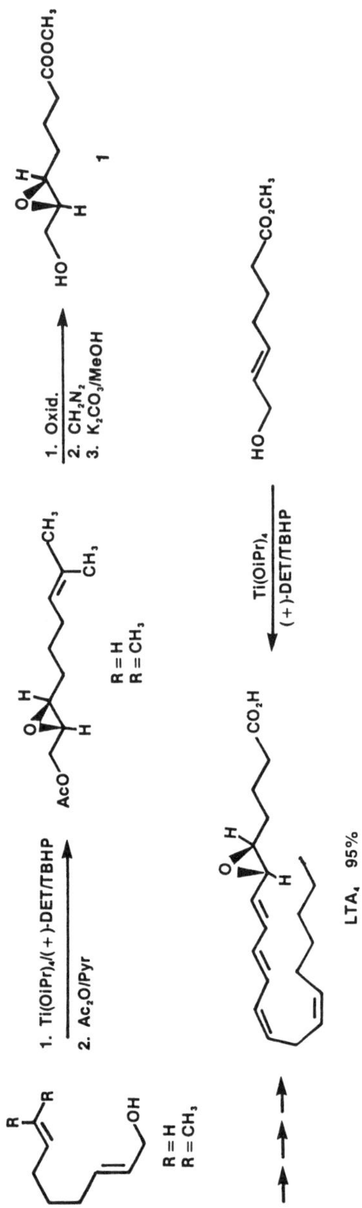

Figure 17. Chirally directed, enantioselective synthesis of intermediate (I) for LTA synthesis

All the leukotrienes (A through E) are now readily available in pure form by synthesis, and in consequence, biological and medical research on the role of these agents in hypersensitivity, allergy, and inflammation is proceeding at a rapidly accelerating pace.

The establishment of identity of synthetic leukotrienes and native SRSs was far from trivial, mainly because only minute amounts of the naturally occurring SRSs could be obtained (in fact, all of our work at Harvard was done without the benefit of any native SRS and neither pmr nor infrared spectra were available). In order to rule out certain stereochemical possibilities for the leukotrienes we therefore synthesized various isomers to compare with naturally derived material. Thus we identified LTC-2 as 11-*trans*-LTC,[20,26] and we also showed that the synthetic 6-*epi*-LTC and LTD[27] showed less than 1% of the activity of LTC and LTD and did not correspond to any detectable native SRS material (Figure 18). The formation of 11-*trans*-LTC from LTC appears to occur with great facility in the presence of certain unidentified impurities which can promote the isomerization catalytically. It seems likely that this 11-*cis* → 11-*trans* conversion is a free radical process involving addition elimination of a thiyl (RS) radical at C(12).[20] Carefully purified LTC (RP-HPLC) does not readily undergo isomerization.

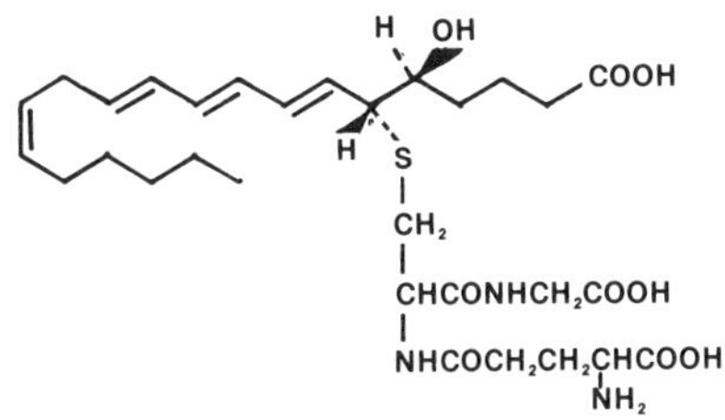

LTC-2 λmax 278

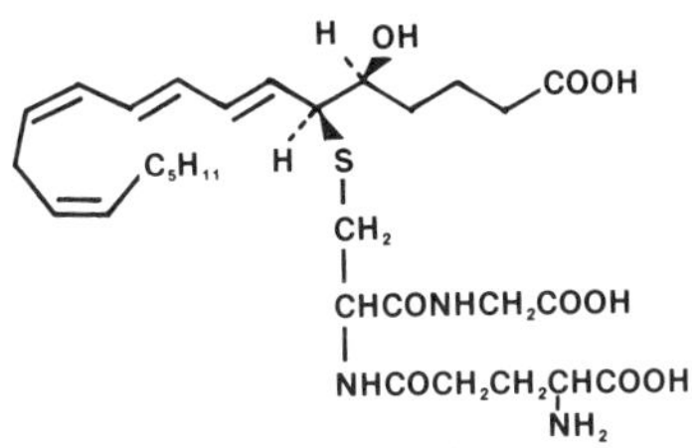

6 epi LTC-1 λmax 280

Figure 18. Other totally synthetic LTC's

We now turn to another aspect of these studies, namely the synthesis of the predecessor of the leukotrienes, 5-(*S*)-hydroperoxyeicosa-*trans*-6-*cis*-8,11,14-tetraenoic acid (5-HPETE), and its conversion by a biomimetic process into leukotriene A. Outlined in Figure 19a is the scheme successfully employed for the synthesis of (±)-5-HPETE and (±)-5-HETE. This approach has been adapted to the large scale synthesis of either 5-(*S*)-or 5-(*R*)-HETE[28,29] (Figure 19b outlines the method of resolution).

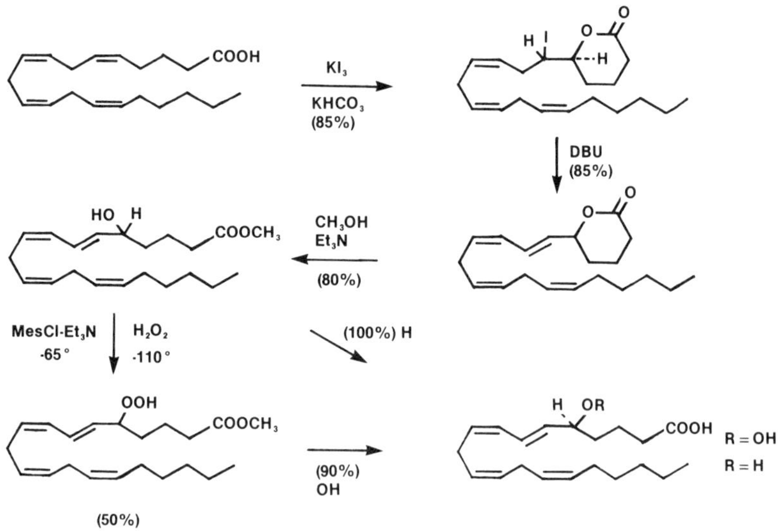

Figure 19a. Chemical synthesis of 5-HETE and 5-HPETE

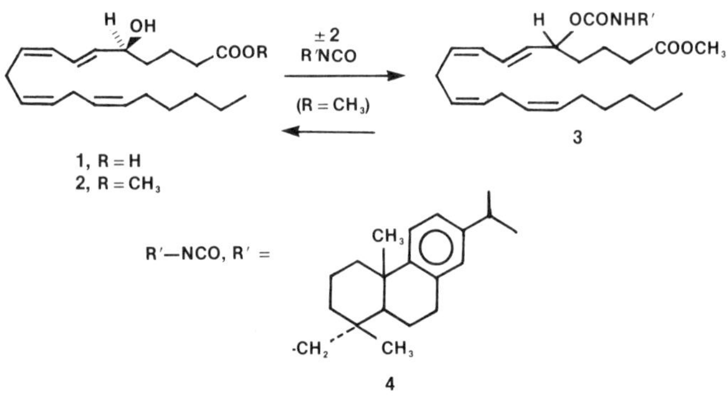

Figure 19b. Resolution of (±)-5-HETE

An enzymic synthesis of 5-(*S*)-HPETE was also developed. Advantage was taken of the availability of synthetic 5-HETE and 5-HPETE and a knowledge of their chromatographic behavior to test the ability of various plant lipoxygenases to convert arachidonic acid to 5-HPETE. The first system selected for study was the lipoxygenase of potato since this had been reported to afford an unusually high ratio of 9- to 13-hydroperoxides from linoleic acid. It was a fortunate choice because the potato lipoxygenase did indeed convert arachidonic acid to 5-(*S*)-HPETE![28]

As shown in Figure 20 it was also possible to transform 5-(*S*)-HPETE into LTA by a rational chemical process which followed an original mechanistic hypothesis. Reaction of the LTA so produced with glutathione gave LTC identical in all respects with an authentic sample. This final synthetic conversion brings us full circle to the original concept for leukotriene A structure and biosynthesis from which all of the subsequent chemical work issued. Figure 21 summarizes current knowledge of leukotriene biosynthesis from arachidonic acid.

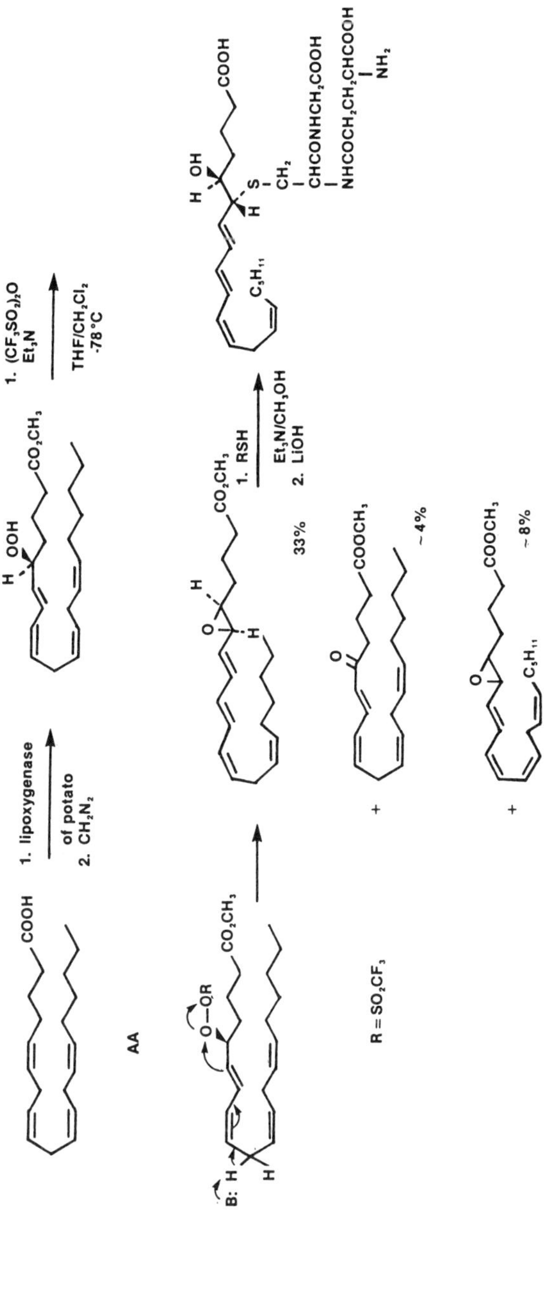

Figure 20. Conversion of arachidonic acid to 5-HPETE, LTA and LTC-1

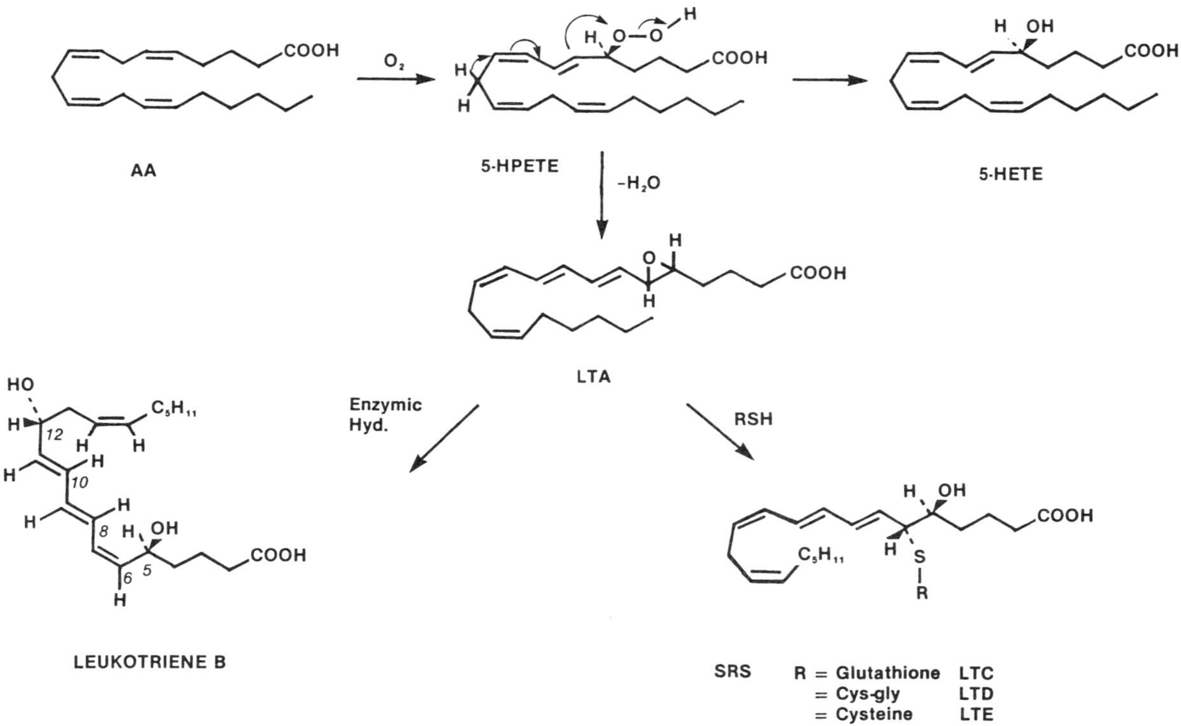

Figure 21. Biosynthesis of Leukotrienes

The likely role of leukotrienes in disease states such as asthma, allergy, and inflammation dictates the desirability of finding agents which inhibit this biosynthetic pathway, particularly at the first step, the 5-lipoxygenase (5-LO) catalyzed conversion of arachidonic acid to 5-HPETE. Studies directed at this objective have already yielded promising results using a rational approach. It was conjectured that if an acetylenic dehydroarachidonic acid (DHA; Figure 22 shows the four possible acetylenic DHAs) could serve as a substrate for a lipoxygenase, delivery of oxygen to the triple bond would result in an unstable vinylic hydroperoxide (estimated O-O bond dissociation energy *ca.* 13 kcal/mole) which could decompose to radicals before dissociation from the enzyme. Such homolysis could result in attack on the enzyme at the catalytic site and could cause inactivation. Such a process is outlined for soybean lipoxygenase, a 15-LO for arachidonic acid, with 14,15-DHA as the inactivator in Figure 23. In Figure 24 is shown a possible mechanism for the lipoxygenase reaction and the specific oxidation which has been observed between soybean lipoxygenase, oxygen, and arachidonic acid. The soybean lipoxygenase, i.e., 15-lipoxygenase (15-LO) system was chosen for initial studies because of the ready availability, purity, and stability of this enzyme. The following results were obtained:[31]

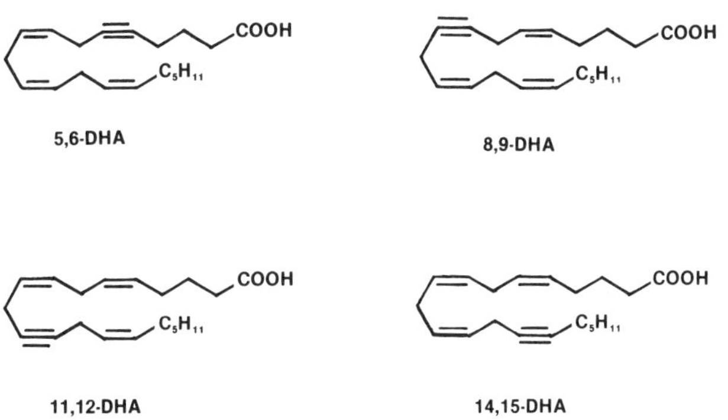

Figure 22. Four monoacetylenic dehydroarachidonic acids (DHA's)

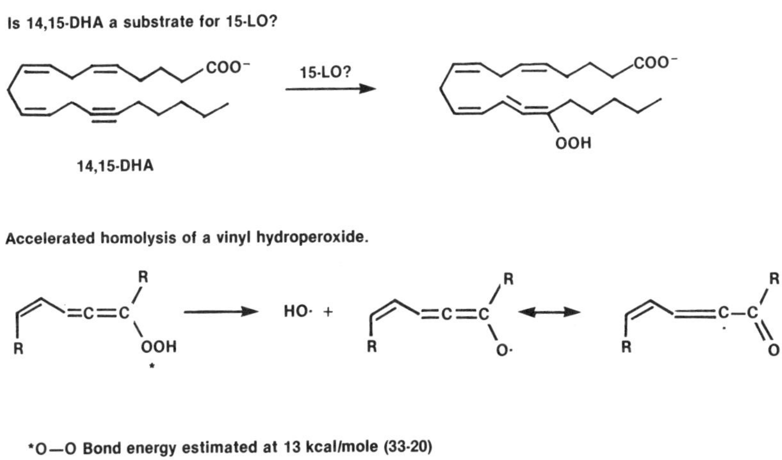

Figure 23. Development of an irreversible inhibitor of soybean lipoxygenase (15-LO)

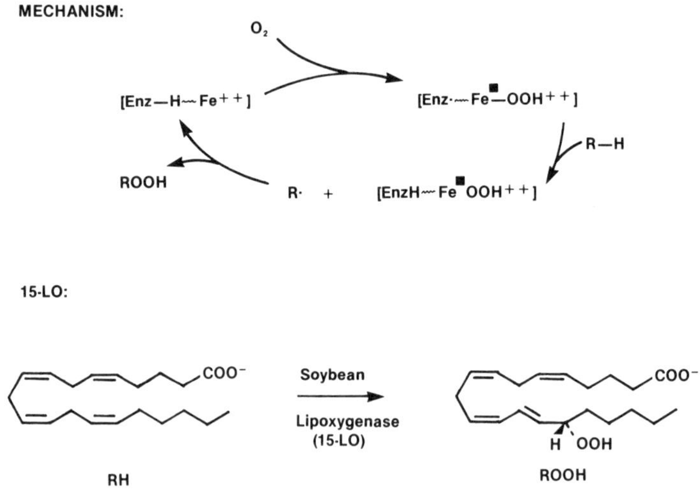

Figure 24. Inhibition of lipoxygenase (LO) reactions

1. 14,15-DHA irreversibly deactivates soybean 15-LO in a time-dependent manner in the presence of O_2 (but not in absence of O_2).
2. Complete deactivation of the 15-LO by 14,15-DHA occurs with an efficiency of 0.4% (260 14,15-DHA per enzyme molecule).
3. Upon deactivation of the 15-LO by tritiated 14,15-DHA 0.9 molecule of inhibitor becomes bound to the enzyme.
4. 5,6-DHA, 8,9-DHA, and 11,12-DHA + O_2 do not cause irreversible inhibition of the 15-LO even at 100 μM (14,15-DHA inhibition observed below 1 μM) and are not bound covalently to the enzyme.

Similar studies were also conducted with the prostaglandin synthetase (cyclooxygenase) from ram seminal vesicles. A summary of these results is as follows:[32]

1. 11,12-DHA irreversibly deactivates cyclooxygenase in a time-dependent manner in the presence of O_2 but not without O_2.
2. 14,15-DHA also irreversibly deactivates cyclooxygenase, but is much less effective than 11,12-DHA.
3. 5,6-DHA and 8,9-DHA behave like weak competitive inhibitors (substrate analogs); 5,6-DHA serves as a substrate undergoing conversion to 5,6-dehydro PGE_2.
4. In the absence of hematin, a necessary cofactor for PG biosynthesis, the PG cyclooxygenase is *not* inactivated by 11,12-DHA *and* O_2.

Finally, preliminary studies have been made with an unpurified preparation of 5-lipoxygenase from rat basophilic leukemia cells which have produced the following observations:[32]

1. Formation of 5-HETE, LTA, and LTB inhibited by preincubation of the 5-LO from rat basophilic leukemia cells with 5,6-DHA and O_2 in a time-dependent irreversible manner.
2. Inhibition of 5-LO by 5,6-DHA requires not only O_2 but also Ca^{++}. Note conversion of 5-HPETE from arachidonate requires O_2 and Ca^{++}.

It is concluded from these studies[31,32] that replacement of a *cis* HC=CH unit of arachidonic acid by a C≡C unit at the site of lipoxygenation leads to a dehydroarachidonic acid which effectively deactivates the enzyme in the presence of oxygen in a time-dependent irreversible way. Clearly 5,6-DHA effectively blocks the first step of the leukotriene pathway.

The four monoacetylenic DHAs shown in Figure 24 are accessible by efficient synthetic routes.[31–33] Studies on the use of 5,6-DHA in biological systems are in progress. Eicosa-5,8,11,14-tetraynoic acid, a long known competitive inhibitor of the prostaglandin synthetase, does not appear to block the leukotriene pathway.

It has recently been found that a biologically active isomer of LTB is formed from arachidonic acid upon incubation with mixed peripheral blood leukocytes.[34] The proposal that this substance is the product of successive 5- and 12-lipoxygenation reactions, i.e., is a 5-(*S*), 12-(*S*)-diHETE, has been confirmed by the unambiguous total synthesis of the same compound (Figure 25, compound 1) by way of intermediates 2 to 7 in Figure 25.[35] The biological role of 5-(*S*), 12-(*S*)-diHETE is not known, but it is of great interest that this substance can antagonize the action of LTB on neutrophils.

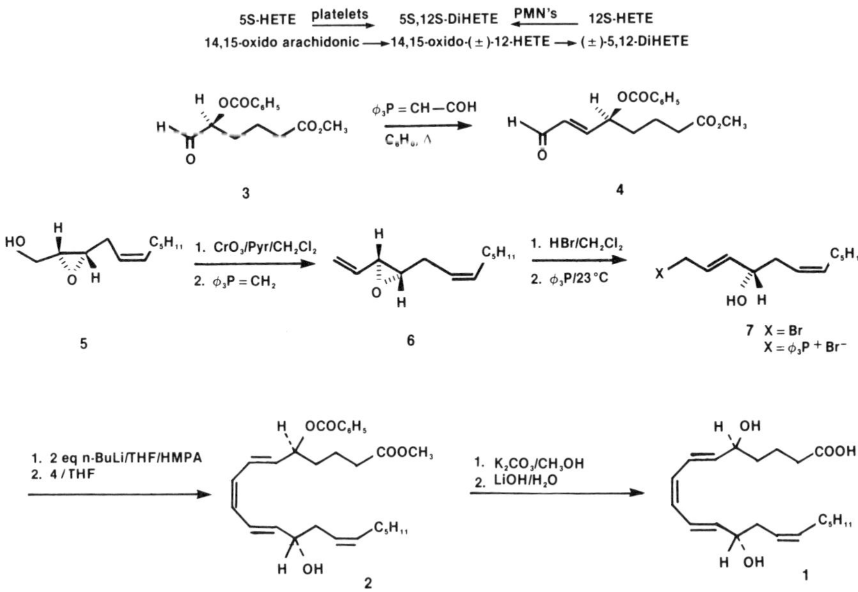

Figure 25. Total synthesis of 5S,12S DiHETE (12-*epi*-6-*trans*-8-*cis* LTB_4)

In 1980 it was pointed out that in principle compounds analogous to leukotriene A could be formed from 8-, 12-, or 15-HPETE by a mechanism paralleling the biosynthesis of LTA from 5-HPETE.[36] The three possible isomers of LTA, designated EPETEs, were synthesized[30b,36] and converted to conjugates with glutathione which are isomeric with leukotriene C; Figure 26 displays the structures of 11,12-EPETE and 14,15-EPETE and the corresponding peptide conjugates (RS = glutathionyl). The glutathione conjugates were found to be biologically inactive (10^{-3}) relative to leukotriene C.[37] Subsequent to this work the formation of hydration products[38] and also a glutathione conjugate[39] of 14,15-EPETE have been observed upon incubation of 15-(*S*)-HPETE with human leukocytes. The biological significance of these findings is unclear, though it should be mentioned in this context that 15-HETE and 5,15-di-HETE both appear to inhibit enzymic synthesis of 5-HPETE from arachidonic acid. The 15-EPETE pathway appears to be dominant in the case of eosinophiles, at least for porcine eos.[40] It seems likely that future research in this area will yield exciting new information.

The slow reacting substances LTC, LTD, and LTE have been termed "mediators in the regulation of the microenvironment" of cells or "hormones of the microenvironment" and are thought to play a role in maintaining local homeostasis.[4C] They occur in exceedingly low concentrations under physiological conditions and, as a consequence, the study of their regulatory action requires highly sensitive methods of chemical analysis. Toward this end radioimmuno assays have been developed which allow analysis for LTC or for LTC, LTD, and LTE as a group,[42] or specifically for LTB. Suitable immunogens were obtained by the chemical coupling of synthetic leukotrienes via the C(1) carboxyl or C(5) hydroxyl to a fraction of the free amino functions of bovine serum albumin.[42] (Figures 27, 28) One important advance which has resulted from the availability of the leukotriene radioimmuno assay has been the discovery of a new class of bone marrow-derived mast cells of unique biochemistry. These cells generate large amounts of leukotrienes relative to prostaglandins.[43]

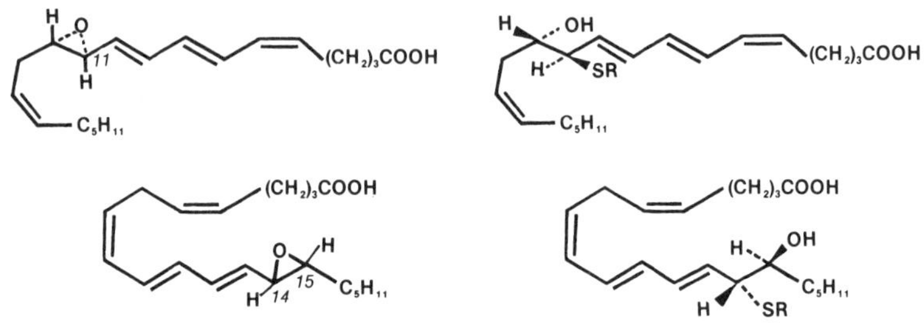

Figure 26. Biosynthetic analogs of LTA, LTC, LTD

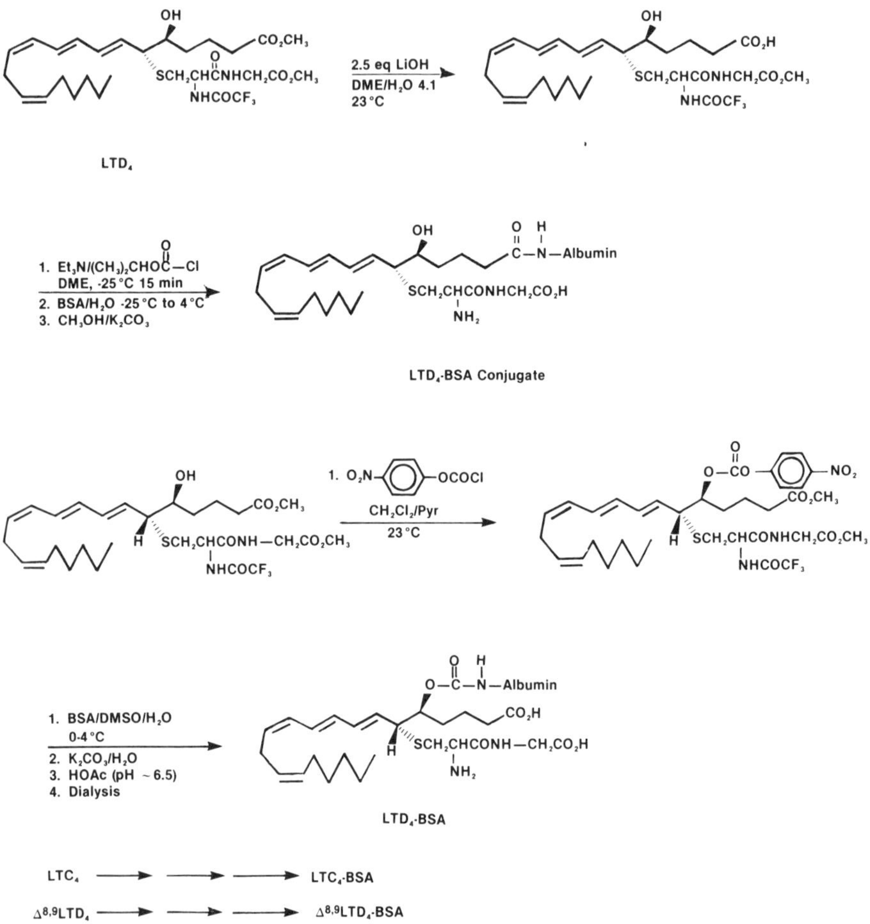

Figure 27. Conjugation of leukotrienes and bovine albumin

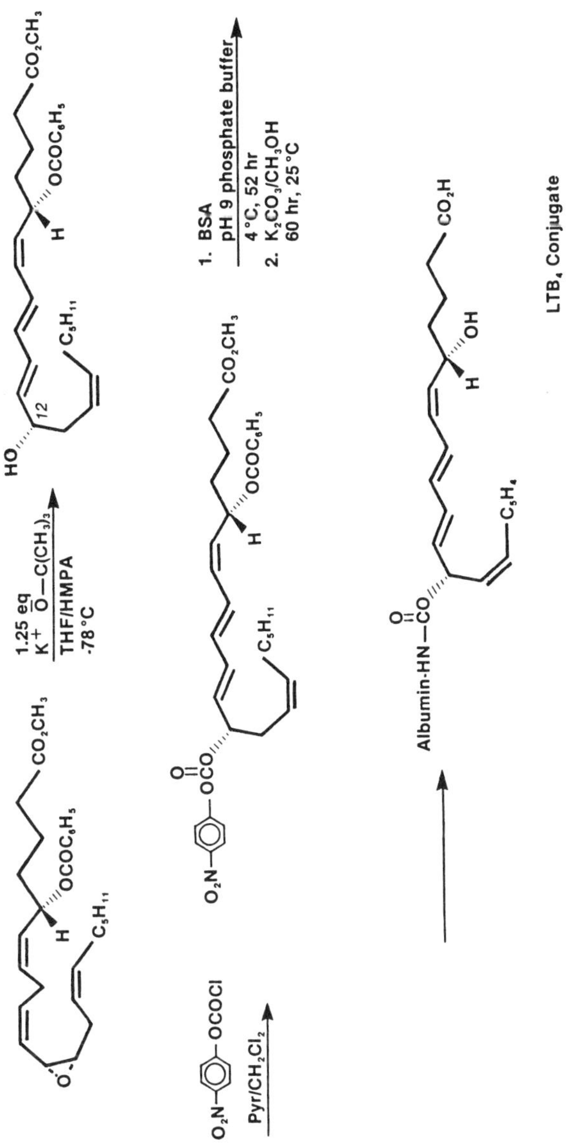

Figure 28. Conjugation of leukotriene B_4 with bovine albumin

The development of antagonists of the leukotrienes (anti-LTs) could be of considerable medical value. For this reason and also to explore the structural factors within the leukotrienes which are essential to the expression of biological activity, a substantial number of carefully chosen analogs of the leukotrienes has been synthesized. Some of the data which have been obtained are shown in Tables 4 & 5.[37,44] From these and more recent results a number of important conclusions have emerged with regard to the structural features which are crucial for biological activity. The crucial structural elements include the following: (1) lipophilic chain attached to C(6); (2) 5-hydroxyl group; (3) a single ionized carboxyl, *either* C(1) or gly. The sulfur atom need not be divalent. Considerable substitution can be tolerated in the peptide moiety of LTD; however, in the case of LTC the free α-amino carboxylic unit is necessary. Although some progress has been made in the development of an anti-LT, the goal of achieving high potency is still elusive.[45]

Table 4. Contractile activities of leukotriene analogs with various saturation of ethylenic bonds.

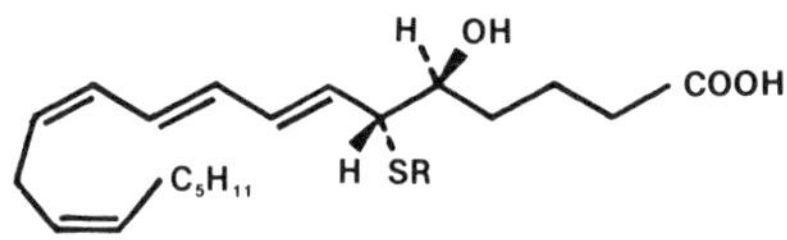

LTC: R = glutathione
LTD: R = Cys-Gly
LTE: R = Cys

	Pulmonary parenchymal strip	Ileum
Parent compound		
LTC	1	1
LTD	1	1
Analog		
14,15-dihydro-LTC	0.19	1.0
14,15-dihydro-LTD	0.66	0.66
7-*trans*-hexahydro-LTC	0.12	0.35
7-*trans*-hexahydro-LTD	0.019	0.005
7-*cis*-hexahydro-LTC	0.06	0.35
7-*cis*-hexahydro-LTD	0.03	0.007

Activities of the analogs are expressed relative to those of the parent compounds on guinea pig uterus.

Proc. Natl. Acad. Sci. USA Vol. 78, No. 5, pp. 3195-3198, May 1981
Medical Sciences

Table 5. Contractile activities of leukotriene analogs; activity ratios of LTD on guinea pig tissues

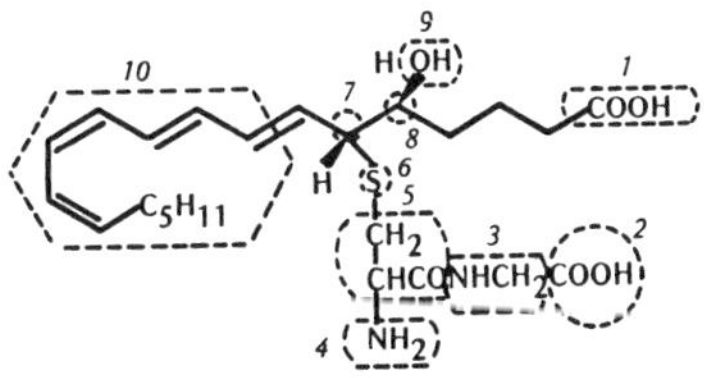

Entry	*Compound*	*Struotural variation*	*Pulmonary parenchymal strip, EC₅₀ ratio**	*Ileum dose ratio†*
1	LTD	None	1	1
2	LTD monoamide (C-1)	1	1	0.25
3	LTD bisamide (C-1, Gly)	1,2	<0.001	<0.002
4	Deamino LTD bisamide (C-1, Gly)	1,2,4	<0.001	<0.002
5	LTD monoamide (Gly)	2	0.10	ND
6	LTD monodimethylamide (Gly)	2	0.006	<0.002
7	D-Ala‡ LTD	3	0.11	0.33
8	L-Ala‡ LTD	3	0.09	0.05
9	Pro‡ LTD	3	0.03	0.03
10	Glu‡ LTD	3	0.02	0.02
11	Val‡ LTD	3	0.008	0.01
12	Homocys§ LTD	5	0.26	0.20
13	D-Pen§ LTD	5,6	<0.001	<0.002
14	D-Cys§ LTD	6	0.11	0.02
15	D-Pen LTE	5,6	<0.001	0.005
16	Deamino LTD	4	0.05	0.20
17	N-Acetyl-LTD	4	0.03	ND
18	N-Acetyl-LTE	4	0.04	0.04
19	LTD-sulfoxide, isomer 1	6	0.10	0.10
20	LTD-sulfoxide, isomer 2	6	<0.001	<0.002
21	LTC sulfone	6		
22	LTD sulfone			
23	6-Epi-LTD	7	0.004	0.005
24	6-Epi-LTC	7	0.005	0.002
25	5-Epi-LTD	8	0.004	0.005
26	5-Dehydroxy-9-12,14,15-hexahydro LTD	9,10	<0.001	<0.002

ND, not determined
*Analog/LTD EC_{50} ratio; EC_{50} of LTD = 0.6 nM
†Analog/LTD ratio; LTD reference dose = 0.4 nM
‡This amino acid replaced Gly of LTD.
§This amino acid replaced L-Cys of LTD.

PNAS (USA) **28,** 4579 (1981)

Acknowledgment

It is a pleasure to express admiration, appreciation, and gratitude to each of the collaborators whose names appear in the references to this article. The research has yielded great personal satisfaction as well as scientific excitement.

References

1. W. Feldberg and C.H. Kellaway, *J. Physiol.*, **94**, 187 (1938).
2. C.H. Kellaway and E.R. Trethewie, *Q.J. Physiol.*, **30**, 121 (1940).
3. (a) W.E. Brocklehurst, *J. Physiol.*, **120**, 16P (1953); (b) W.E. Brocklehurst, *Prog. Allergy,* **6**, 539 (1962).
4. (a) R.P. Orange and K.F. Austen, *Adv. Immunol.*, **10**, 105 (1969); (b) R.P. Orange, D.J. Stechshulte and K.F. Austen, *Fed. Proc. Am. Soc. Exp. Biol.*, **28**, 1710 (1969); (c) R.A. Lewis and K.F. Austen, *Nature,* **293**, 103 (1981).
5. (a) K. Strandberg and B. Uvnar, *Acta Physiol. Scand.*, **82**, 358 (1971); (b) W. Dawson and R. Tomlinson, *Brit. J. Pharmacol.*, **52**, 107P (1974); (c) M.K. Bach, J.R. Brashler and R.R. Gorman, *Prostaglandins,* **14**, 21 (1977); (d) B.A. Jakschik, S. Falkenhein, and C.W. Parker, *Proc. Natl. Acad. Sci. USA,* **74**, 4577 (1977).
6. M.K. Bach and J.R. Brashler, *J. Immunol.*, **120**, 998 (1978).
7. R.P. Orange and E.G. Moore, *ibid.*, **116**, 392 (1976).
8. H.R. Morris, G.W. Taylor, P.J. Piper, P. Sirois, and J.R. Tippins, *FEBS Lett.*, **87**, 203 (1978).
9. P. Sirois, *Prostaglandins,* **17**, 395 (1979).
10. E.J. Corey and A.G. Hortmann, *J. Am. Chem. Soc.*, **87**, 5736 (1965).
11. I. Agata, E.J. Corey, A.G. Hortmann, J. Klein, S. Proskow and J.J. Ursprung, *J. Org. Chem.*, **30**, 1698 (1965).
12. E.J. Corey, Y. Arai, and C. Mioskowski, *J. Am. Chem. Soc.*, **101**, 6748 (1979).
13. P. Borgeat and B. Samuelsson, *Proc. Natl. Acad. Sci. USA,* **76**, 3213 (1979).
14. R.C. Murphy, S. Hammarstrom, and B. Samuelsson, *Proc. Natl. Acad. Sci. USA,* **76**, 4275 (1979).
15. S. Hammarstrom, R.C. Murphy, B. Samuelsson, D.A. Clark, C. Mioskowski, and E.J. Corey, *Biochem. Biophys. Res. Commun.*, **91**, 1266 (1979).
16. S. Hammarstrom, B. Samuelsson, D.A. Clark, G. Goto, A. Marfat, C. Mioskowski and E.J. Corey, *ibid.*, **92**, 946 (1980).
17. E.J. Corey, D.A. Clark, G. Goto, A. Marfat, C. Mioskowski, B. Samuelsson, and S. Hammarstrom, *J. Am. Chem. Soc.*, **102**, 1436, 3663 (1980).
18. For a more recent synthesis of one of the early intermediates in this route see, E.J. Corey, S. Hashimoto, and A.E. Barton, *J. Am. Chem. Soc.*, **103**, 721 (1981).
19. See, for example, (a) R.A. Lewis, K.F. Austen, J.M. Drazen, D.A. Clark, A. Marfat and E.J. Corey, *Proc. Natl. Acad. Sci. USA,* **77**, 3710 (1980) (LTD); and (b) R.A. Lewis, J.M. Drazen, K.F. Austen, D.A. Clark, and E.J. Corey, *Biochem. Biophys. Res. Commun.*, **96**, 271 (1980).
20. For syntheses of LTD and LTC-2 see, E.J. Corey, D.A. Clark, A. Marfat and G. Goto, *Tetrahedron Letters,* **21**, 3143 (1980).
21. O. Radmark, C. Malmsten, B. Samuelsson, D.A. Clark, G. Goto, A. Marfat and E.J. Corey, *Biochem. Biophys. Res. Commun.*, **92**, 954 (1980).
22. For nomenclature see, B.Samuelsson and S. Hammarstrom, *Prostaglandins,* **19**. 645 (1980).
23. (a) E.J. Corey, A. Marfat, G. Goto and F. Brion, *J. Am. Chem. Soc.*, **102**, 7984 (1980); and (b) E.J. Corey, A. Marfat, J. Munroe, K.S. Kim, P.B. Hopkins and F. Brion, *Tetrahedron Letters,* **22**, 1077 (1981).
24. O. Radmark, C. Malmsten, B. Samuelsson, G. Goto, A. Marfat and E.J. Corey, *J. Biol. Chem.*, **255**, 11828 (1980).
25. (a) E.J. Corey, S. Hashimoto, and A.E. Barton, *J. Am. Chem. Soc.*, **103**, 721 (1981); (b) B.E. Rossiter, T. Katsuki, and K.B. Sharpless, *J. Am. Chem. Soc.*, **103**, 464 (1981).
26. D.A. Clark, G. Goto, A. Marfat, E.J. Corey, S. Hammarstrom, and B. Samuelsson, *Biochem. Biophys. Res. Commun.*, **94**, 1133 (1980).
27. E.J. Corey and G. Goto, *Tetrahedron Letters,* **21**, 3463 (1980).
28. E.J. Corey, J.O. Albright, A.E. Barton and S. Hashimoto, *J. Am. Chem. Soc.*, **102**, 1435 (1980).
29. E.J. Corey and S. Hashimoto, *Tetrahedron Letters,* **22**, 299 (1981).
30. (a) E.J. Corey, A.E. Barton, and D.A. Clark, *J. Am. Chem. Soc.*, **102**, 4278 (1980); (b) E.J. Corey and A.E. Barton, *Tetrahedron Letters,* **23**, 2351 (1982).
31. E.J. Corey and H. Park, *J. Am. Chem. Soc.*, **104**, 1750 (1982).
32. E.J. Corey and J.E. Munroe, *J. Am. Chem. Soc.*, **104**, 1752 (1982).
33. E.J. Corey and J. Kang, *Tetrahedron Letters,* **23**, 1651 (1982).
34. P. Borgeat, S. Picard, P. Vallerand, and P. Sirois, *Prostaglandins Med.*, **6**, 557 (1981).
35. E.J. Corey, A. Marfat, and B.C. Laguzza, *Tetrahedron Letters,* **22**, 3339 (1981).

36. (a) E.J. Corey, A. Marfat, and G. Goto, *J. Am. Chem. Soc.*, **102**, 6607 (1980); (b) E.J. Corey and Y. Nii, unpublished synthesis of 8,9-EPETE.
37. J.M. Drazen, R.A. Lewis, K.F. Austen, M. Toda, F. Brion, A. Marfat, and E.J. Corey, *Proc. Natl. Acad. Sci. USA,* **78**, 3195 (1981).
38. U. Lundberg, O. Radmark, C. Malmstem, and B. Samuelsson, *FEBS Letters,* **126**, 127 (1981).
39. D. Sok, C. Han, W. Shieh, B. Zhou, and C.J. Sih, *Biochem. Biophys. Res. Commun.*, **104**, 1363 (1982).
40. J.M. Bailey, R.W. Bryant, C.E. Low, M.B. Pupillo, and J.Y. Vanderhoek, *Adv. Prostaglandin Thromboxane Res.*, **9**, 341 (1982).
41. J.A. Oates, paper presented at the 5th International Conference on Prostaglandins, Florence, May 1982.
42. L. Levine, R. A. Morgan, R.A. Lewis, K.F. Austen, D.A. Clark, A. Marfat, and E.J. Corey, *Proc. Natl. Acad. Sci. USA,* **78**, 7692 (1981).
43. E. Razin, J.M. Mencia-Huerta, R.A. Lewis, E.J. Corey, and K.F. Austen, *Proc. Natl. Acad. Sci. USA,* **79**, August (1982).
44. R.A. Lewis, J.M. Drazen, K.F. Austen, M. Toda, F. Brion, A. Marfat, and E.J. Corey, *Proc. Natl. Acad. Sci. USA,* **78**, 4579 (1981).
45. We are grateful to the U.S. National Science Foundation and the National Institutes of Health for financial support of the research program on leukotrienes at Harvard.

The Role of Arachidonic Acid Products in the Respiratory System

By W. Dawson

LILLY RESEARCH CENTRE LIMITED, ERL WOOD MANOR, WINDLESHAM, SURREY GU20 6PH, U.K.

Introduction

In any review of the respiratory system it must be recognised that the lung is not an homogeneous organ, nor does it function in isolation. Large or small airways, vasculature and parenchymal tissue are all capable of synthesising and subsequently releasing different patterns of pharmacologically active substances and, indeed, of responding differently to the same stimulus. In addition, the lung is a secretory organ and these secretions are susceptible to variation as a result of disease or the action of pharmacological agents.

The impact of autonomic nervous system regulation of various types of smooth muscle in the lung undoubtedly influences the basic state of muscle tone and this can modify the subsequent response to agonists. This certainly applies to the action of products of arachidonic acid, notably those synthesised *via* the cyclooxygenase pathway.

Whilst it is clear that many products of arachidonic acid metabolism are potent pharmacological agents, the link between these activities and various pathophysiological states is important and must be clearly demonstrated before hypotheses on biological roles can be suggested.

An earlier review[1] raises many of the topics discussed here.

The lung

Although the major function of the lung is to exchange gases, it is also a primary organ in terms of contact with the environment. Many of its components relate to host defence, therefore, and it is important to consider this aspect of function in an assessment of the activity of pharmacological agents. The interrelationship of these systems is detailed in Figure 1.

Figure 1

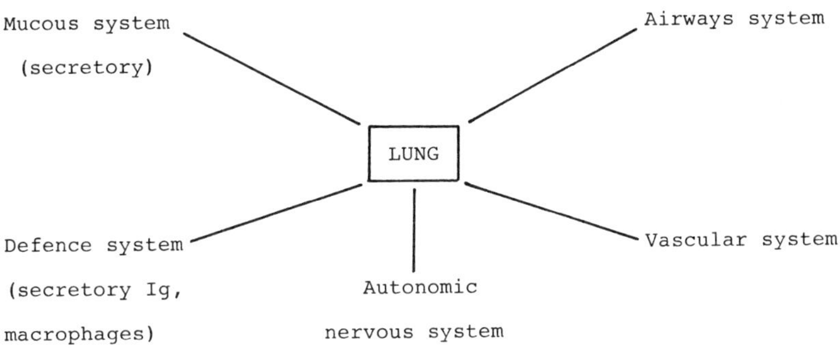

The airways system of the lung is conveniently divided into the central or large airways, consisting of trachea, bronchi and the larger bronchioles and the peripheral airways[2]. Asthma is defined as reversible obstructive airways disease and is considered to affect the peripheral airways to the greatest extent[3]. To the pharmacologist, the airways are examples of smooth muscle and much of the earlier research was conducted on tracheal chains, spirally cut strips of bronchial muscle and the like[4]. More recently, preparations using peripheral airway smooth muscle have been developed[5] and comparison of these two assay systems have revealed significant differences in response to spasmogens, particularly with respect to arachidonate metabolites[6].

The vascular system within the lung is complex and the organ's ability to change regional blood flow within itself is not clearly understood. It seems probable that such local change in blood flow is in response to chemoreceptor stimulation by minute changes in blood pH or in partial pressures of blood gases[7]. As these same parameters are capable of affecting both mediator release and the response of smooth muscle to mediators[8], it follows that careful consideration be given to the transposition of data from <u>in vitro</u> assay systems to the <u>in vivo</u> situation.

The autonomic nervous system has the ability to induce gross changes in both airways and vascular systems in the lung[9]. Indeed, many workers believe that asthma is primarily a central nervous system disorder with excessive parasympathetic stimulation[10] causing the observed bronchospasm, although others take an alternate view[11]. In the normal physiological state, it seems probable that major adjustments in lung function will be provided by the autonomic nervous sytem but fine control will be exerted by local mechanisms where products of the arachidonate metabolism would seem to be prime candidates.

In response to particulate challenge from the environment the lung is capable of providing phagocytic cells in the form of alveolar macrophages to remove such stimuli and, in addition, the presence of secretory immunoglobulins of the IgA class will facilitate the removal of common antigenic materials. The mechanism by which antigen is recognised and secretory IgA provided to it has not been extensively studied, but the alveolar macrophage probably provides an accessory cell function for the immune system in the lung.

The mucous system is also part of the lung's defence mechanism and includes ciliated epithelium throughout much of the large airway system to facilitate removal of the secreted material. The formation of mucus plugs in smaller airways is a recognised phenomenon in asthma in man and results from both increased secretion and viscosity of the mucus and also the constriction of the airways, particularly peripherally[12]. Recent work has shown that both prostaglandin and lipoxygenase products are capable of influencing these parameters.

Arachidonic acid metabolites

Arachidonic acid derives from phospholipids in cell membranes and is acted on by two complex enzyme systems, fatty acid cyclooxygenase (prostaglandin synthetase) and fatty acid lipoxygenase.

The cyclooxygenase enzyme inserts two molecules of oxygen, one by an endoperoxide bridge across C_9 and C_{11} of the arachidonic acid backbone and the other by hydroperoxidation at C_{15}[13] to form PGG_2, which is converted to PGH_2 by reduction at C_{15} to 15-hydroxyl[13]. PGH_2 is the substrate for a number of synthetase and isomerase enzymes which result in PGE_2, $PGF_{2\alpha}$, PGD_2, PGI_2 and thromboxane A_2 (TxA_2)(Figure 2). The inactivation of PGs and TxA_2 follows a common pathway with reduction of the 13:14 double bond, oxidation at C_{15} to ketone, various steps of ω-oxidation and β-oxidation (two steps) of the α-side chain[14].

Figure 2

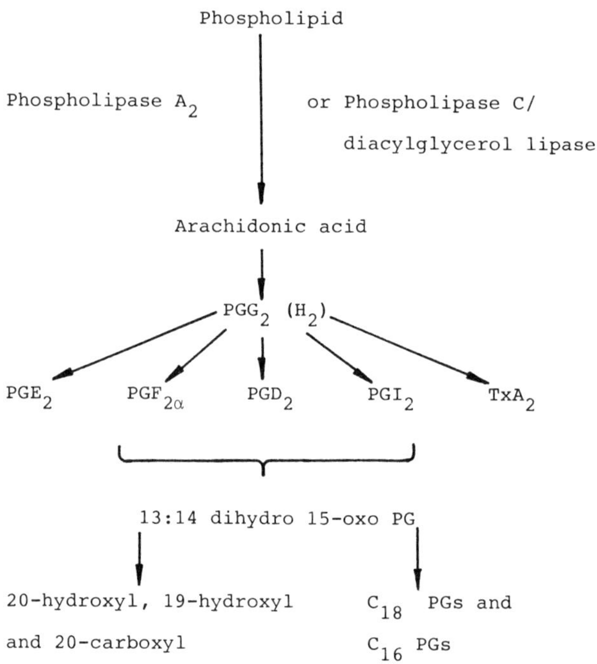

In addition to the enzymic reactions mentioned, some chemical conversions do occur and this is still an active area of research.

Various positional lipoxygenases are capable of using arachidonic acid as substrate (Figure 3)[15]. The most important seem to be the 5-, 12- and 15-lipoxygenases, with the number referring to the C atom of arachidonic acid. In addition there are certainly 8- and an 11-lipoxygenase. Mono and disubstitution can take place forming various HETEs (hydroxyeicosa tetranenoic acids). Epoxidation at 5:6 and 14:15 positions can occur, when hydrolysis leads to diHETEs, and thiolysis to a range of potent pharmacological agents, the leukotrienes, when the thiol is glutathione.

Figure 3

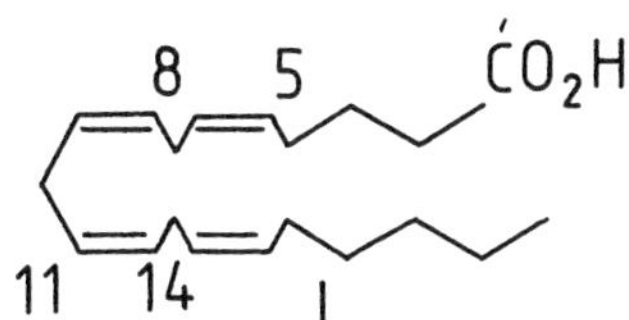

Arachidonic acid

↓ HYDROPEROXIDATION, FOLLOWED BY REDUCTION

5-, 8-, 11-, 12-, 15- H(P)ETE

↓

LTA_4 (5:6 EPOXIDE)

Enzymic hydrolysis

5S 12R diHETE (leukotriene B_4) → 20-hydroxyl, 20-carboxyl

LTC_4 (Glutathione) → LTD_4 (cysteinyl glycine) → LTE_4 (cysteine) → ?

Di-HETEs can also be synthesised in vitro and probably in vivo by chemical or biochemical routes. Certainly, 5HETE is a substrate for the 12-lipoxygenase and vice versa.

The tissue distribution of the positional lipoxygenases varies; for example, platelets contain 12-lipoxygenase but not the 5-enzyme. Polymorphonuclear leucocytes (PMN) appear to have mostly 5-lipoxygenase but may also contain the 15-variant. Macrophages contain both 5- and 12- enzymes. The distribution of glutathione-S-transferase is a subject of active research at the present time. As arachidonate is the substrate for both cyclooxygenase and lipoxygenase enzymes, the relative distribution of both is of considerable importance and may well relate to the functional aspects of arachidonate metabolites in a given tissue or organ.

Cyclooxygenase products

Effects on airways. PGF_2 and PGD_2 are potent bronchoconstrictor agents[16-20] and there is evidence that asthmatic patients are more sensitive to PGF_2 challenge than normal individuals[20,21]. PGE_2 and PGI_2, on the other hand are active bronchodilator compounds[22-25]. The metabolites of these PGs have less or no bronchoreactivity although one, 15-oxo PGF_2 seemed to sensitise tissues to other spasmogens[26]. TxA_2 seems to have spasmogenic activity but has an extremely short half-life and its stable metabolite TxB_2 is without activity.

The concept of maintenance of airway homeostasis by PGs is an attractive one (Figure 4).

Figure 4

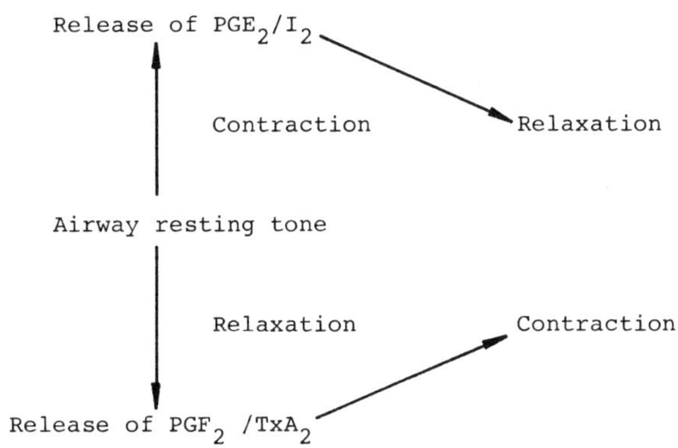

Early work[27] has been confirmed[28] showing that in airway preparations with high tone PGE_2 is released and causes relaxation whilst the converse is true and PGF_2 is released to increase tone. This interaction is similar to that proposed in the cardiovascular area where TxA_2 causes adherence of platelets to endothelium and this effect is counteracted by the synthesis of PGI_2 by endothelial cells.

Effect on mucus secretion. PGD_2 and PGF_2 increase mucus production at high concentrations[29], whilst PGE_2 reduces its production at relatively low concentrations. In the same system PGI_2 and TxB_2 were without effect. In contrast other workers showed that PGE_1 stimulated mucus secretion and mucociliary transport[30,31]. The development of more definitive testing systems in both animals and man should provide a clearer understanding of this complex but important area.

Effect on mediator release. PGs which elevate intracellular levels of cAMP (PGE_2, PGI_2) also reduce the release of histamine, PG's and LTs in allergic challenge situations. $PGF_{2\alpha}$, which increases cGMP, has the reverse effect[32].

Lipoxygenase products

Effect on airways. The potent bronchoconstriction evoked in sensitised lungs in response to challenge was ascribed to slow reacting substance of anaphylaxis (SRS-A)[33]. The identification of 5-lipoxygenase products in SRS-A[34,35] and their naming as leukotrienes[36] has stimulated much research in this area. The leukotrienes (C_4, D_4 and E_4) are potent bronchoconstrictor agents being perhaps 1000 times more potent than histamine[37-39]. Clearly LTE_4 is the least active of these three compounds. It is of considerable interest that they are more effective in reducing compliance, thought to be a peripheral airways effect, than in increasing resistance, a response associated with the large airways. This correlation with airway events in asthma has stimulated the pharmaceutical industry to explore the various avenues possible to limit either the formation of, or the actions of these agents, and should ultimately confirm or deny the clinical importance of 5-lipoxygenase products. The biological activities of LTs have been found to reside in structures with considerable stereochemical and geometric specificity[40], which encourages the belief that selective antagonists may be found.

LTB_4, a potent chemotaxin[41-43] has been implicated in inflammatory responses but it is also spasmogenic and may also increase vascular permeability. These properties suggest that it may be involved in the cellular events associated with chronic asthma but further research is clearly required.

Effects on mucus secretion. LTC_4 and LTD_4 have been shown to increase mucus production and viscosity at low concentrations in vitro[44], and in vivo may reduce ciliary transport of mucus in man[45]. The pharmacology of the mucociliary system has been reviewed[46] and although the LTs were not used directly, the effects of challenge and treatment with anti-allergic agents provide thoughtful reading.

Effects on mediator release. LTC_4 and LTD_4 enhance mediator release[47,48] from sensitised, challenged lung tissue and it is clear, particularly in the guinea pig, that the biological effects of LTs are modulated via the release of cyclooxygenase products, especially TxA_2 (B_2). This appears not to be so in human tissue and careful interpretation of the data from various species is essential.

In this section on lipoxygenase products, no mention has been made of the biological activity of the various HETEs, mainly because, to date, they appear to have considerably less activity than the leukotrienes. However, it is important to remember that, quantitatively, they may occur at concentrations 1000 times in excess of the leukotrienes and a modest intrinsic activity may be of considerable relevance in a pathophysiological state. An example is that of psoriatic skin, characterised by PMN infiltration, high levels of 12-HETE and lower levels of 5-lipoxygenase products[49]. 12-HETE has slight chemoattractant properties and may well be important in this condition.

Effect of drugs

A detailed literature exists describing the induction of severe asthmatic attacks by inhibitors of the cyclooxygenase enzyme. Aspirin and aspirin-like drugs have been reported to induce attacks in 10% of adult asthmatics[50]. Mechanistically, it is attractive to hypothesise that arachidonate, released on challenge, is diverted to lipoxygenase enzymes rather than the more prevalent cyclooxygenase system which has been inhibited, a view supported by work over many years[51-53]. Alternatively, the reduced biosynthesis of bronchodilator PGs could provide the explanation.

No compounds have been claimed to be selective PG or LT antagonists, although FPL 55712 is a selective inhibitor of SRS-A responses[54] and also antagonises contractions induced by LTD_4 *in vitro*. It appears to have less activity against LTC_4. Neither are there compounds which selectively inhibit the 5-lipoxygenase enzyme, although one compound[55] is claimed to inhibit glutamyl transpeptidase and thus prevents the conversion of LTC_4 to LTD_4. Clearly, we must await the efforts of the synthetic organic chemists to provide molecules with the required selectivity for extensive biological evaluation.

The available "antiallergic drugs" include B_2 agonists, phosphodiesterase inhibitors, disodium cromoglycate and glucocorticoid steroids. Apart from the latter group of compounds, none appear to modify the synthesis, release or action of the arachidonate metabolites in a selective way. The steroids reduce the availability of arachidonic acid by causing the synthesis of a peptide, lipomodulin (macrocortin) which inhibits phospholipase A_2 but this is only one possible explanation of steroid action.

It would seem, therefore, that compounds modifying the lipoxygenase pathway(s) would certainly provide a novel approach to therapy of lung disease.

Concluding remarks

This brief review has attempted to describe the wide ranging pharmacological activities of arachidonate metabolites in the lung and to make some assessment of the present relevance of these activities to the pathogenesis of asthma in particular. The lung is heir to a variety of other disease states and in many of them changed levels of prostaglandins, in particular, have been observed. The clinical evidence, albeit indirect from cyclooxygenase inhibition, suggests that prostaglandins may be more beneficial than harmful. Data is not available to comment on the clinical aspects of lipoxygenase products with any certainty, although the preclinical evidence is strong that the inhibition of release or antagonism of action of the leukotrienes should be therapeutically beneficial.

One aspect has not been discussed. Some reports[56,57] have suggested that the process of sensitisation can modify that ability of lung tissue to synthesise at least cyclooxygenase products. If this is so in man, the potential exists to restore this deregulated system to normal, perhaps by influencing biochemical or immunological messengers and this could be an attractive direction for novel therapeutic agents.

No attempt has been made to describe the complex interactions between combination of antigen with antibody on the mast cell surface, the subsequent activation processes releasing histamine and causing the biosynthesis of arachidonate metabolites. It seems probable that there are many feedback systems with these potent pharmacological agents facilitating or inhibiting their own synthesis, release and action. Specific tools are required by the biologist to untangle these events and the evidence is strong that research in this area will give us the means to undertake the task, from which, hopefully, will derive useful therapeutic agents.

References

[1]E.W. Spannhake, A.L. Hyman and P.J. Kadowitz, Prostaglandins, 1981, 22, 1013.

[2]A.W. Ham and T.S. Leeson, "Histology", Pitman Medical, London, 1961, p.573.

[3]R. Porter and J. Birch. "Identification of asthma: Ciba Foundation Study Group No38", Churchill Livingstone, Edinburgh and London, 1971, p.13

[4]J.W. Constantine, J. Pharm. Pharmac., 1965, 17, 387.

[5]H.W. Mitchell and M.A. Denborough, Eur. J. Pharmac., 1979, 54, 69.

[6]P.J. Piper, M.N. Samhoun, J.R. Tippins, T.J. Williams, M.A. Palmer and M.J. Peck, "SRS-A and Leukotrienes", Research Studies Press, Chichester, 1980, p.81.

[7]J.H. Comroe, "Physiology of respiration, an introductory text", Year Book Medical Publishers, Chicago, 1974, p.1.

[8]J. Widdicombe. "International Review of Physiology", University Park Press, London, 1977, Vol.14, p.291.

[9]J.B. Richardson (1977) "Asthma", Academic Press, New York, 1977, p.237.

[10]J.G. Widdicombe, "Asthma", Academic Press, New York, 1977, p.225.

[11]R. Bookman, Ann. Allergy, 1982, 48, 65

[12]M.S. Dunnill, J. Clin. Path., 1960, 13, 27.

[13]M. Hamberg and B. Samuelsson, J. Amer. Chem. Soc., 1966, 88, 2349.

[14]E. Anggard and C. Larson, Eur. J. Pharmac., 1971, 14, 66.

[15]B. Samuelsson, "SRS-A and Leukotrienes", Research Studies Press, Chichester, 1981, p.45.

[16]A.A. Mathe and P. Hedquist, Am. Rev. Resp. Dis., 1975, 3, 313.

[17]R. Brown, R.H. Ingram, Jr. and E.R. McFadden, Jr., J. Appl. Physiol., 1978, 44, 150.

[18]K.R. Patel, Postgrad. Med. J., 1976, 52, 275.

[19]M.A. Wasserman, D.W. DuCharme, R.L. Griffin, G.L. DeGraaf and F.G. Robinson, Prostaglandins, 1977, 13, 255.

[20]H. Ankut, B.A. Peskar and W. Bemaner, Arch. Pharmac., 1978, 305, 247.

[21]A.P. Smith, M.F. Cuthbert and L.S. Dunlop, Clin. Sci. Mol. Med., 1975, 48, 421.

[22]M.E. Rosenthale, A. Dervinis and J. Kassarich, J. Pharmac. Exp. Ther., 1978, 207, 83.

[23]M.A. Wasserman, R.L. Griffin and F.B. Marsalisi, J. Pharmac. Exp. Ther., 1980, 214, 68.

[24]M.A. Wasserman, D.W. Ducharme, R.L. Griffin, M.G. Wendling and G.L. DeGraaf, Pharmacologist, 1977, 19, 145.

[25]E.W. Spannhake, J.L. Levin, B.T. Mellion, C.A. Gruetter, A.L. Hyman and P.J. Kadowitz, J. Appl. Physiol., 1980, 49, 521.

[26]W. Dawson, R.L. Lewis, R.E. McMahon and W.J.F. Sweatman, Nature, 1974, 250, 331.

[27]J. Orehek, J.S. Douglas, A.J. Lewis and A. Bouhuys, Nature, 1973, 245, 84.

[28]W. Dawson and J.R. Boot, Fed. Proc., 1977, 36, Abstract No. 3550.

[29]Z. Maron, J.H. Shelhamer, M. Kaliner, J. Clin. Invest., 1981, 67, 1695

[30]J. Iravani and C.N. Melville, Respiration, 1975, 32, 305.

[31]D.R. Maurer, J. Schor, M. Sielzak, A. Wanner and W.M. Abraham, Cell Motility, 1, 67.

[32]L.M. Lichtenstein, "Asthma", Academic Press, New York, 1973, p91

[33]C.H. Kellaway and E.R. Trethewie, Q.J. Exp. Physiol., 1940, 30, 121.

[34]B. Sammuelsson, S. Hammarstrom and P. Borgeat, "Advances in Inflammation Research", Raven Press, New York, 1979 Vol.1, p405.

[35]H.R. Morris, P.J. Piper, G.W. Taylor and J.R. Tippins, Prostaglandins, 1980, 19, 371.

[36]B. Samuelsson and S. Hammarstrom, Prostaglandins, 1980, 19, 645.

[37]S.-E. Dahlen, P. Hedquist, S. Hammarstrom and B. Samuelsson, Nature, 1980, 288, 484.

[38]J.M. Drazen, K.F. Austen, R.A. Lewis, D.A. Clark, G. Goto, A. Marfat and E.J. Corey, Proc. Nat. Acad. Sci. (U.S.A.), 1980, 77, 4354.

[39]J.W. Weiss, J.M. Drazen, E.R. McFadden and P. Weller, E.J. Corey, R.A. Lewis and K.F.Austen, J.A.M.A., 1983, 249, p.2814

[40]S.R. Baker, J.R. Boot, W. Dawson, W.B. Jamieson, D.J. Osborne and W.J.F. Sweatman, "Advances in Prostaglandin, Thromboxane and Leukotriene Research", 1982, Raven Press, New York, p.223.

[41]A.W. Ford-Hutchinson, M.A. Bray, M.V. Doig, M.E. Shipley and M.J.H. Smith, Nature, 1980, 286, 264.

[42]E.J. Goetzl and W.C. Pickett, J. Exp. Med., 1981, 153, 482.

[43]C.L. Malmsten, J. Palmblad, A-M. Uden, O. Radmark, L. Engstedt and B. Samuelsson, Acta Physiol. Scand., 1980, 110, 449.

[44]Z. Marom, J.H. Shelhamer, M.K. Bach, D.R. Morton and M. Kaliner, Am. Rev. Resp. Dis., 1982, 126, 449.

[45]T. Ahmed, D.W. Greenblatt, S. Birch, B. Marchette and A. Warner, Am. Rev. Resp. Dis., 1981, 124, 110.

[46]A. Wanner and T. Ahmed, "Immunopharmacology of the lung", Dekker, New York, 1983, p.369.

[47]D.M. Engineer, P.J. Piper and P. Sirois, Br. J. Pharmac., 1977, 59, 444P.

[48]A.A. Mathe, K. Strandberg and S.-S. Yen, Prostaglandins, 1977, 14, 1105.

[49]S. Hammarstrom, M. Hamberg, B. Samuelsson, E. Duell, M. Stawiski and J. Voorhees, Proc. Natl. Acac. Sci. (U.S.A.), 1975, 72, 5730.

[50]M. Samter and R.F. Beers, Jr., Ann. Intern. Med., 1968, 68, 975.

[51]J.L. Walker, "Advances in the Biosciences", Pergamon Press, New York, 1977, Vol.9, p.235.

[52]J.F. Burka and R.J. Flower, Br. J. Pharmac., 1979, 65, 35.

[53]N.A.M. Paterson, J.R. Burka and I.D. Craig, J. Allergy Clin. Immunol., 1981, 61, 427.

[54]P. Sheard, T.B. Lee and M.L. Tattersall, Mon. Allergy, 1977, 12, 245.

[55]L. Orning and S. Hammarstrom, J. Biol. Chem., 1980, 255, 8023.

[56]R. Benzie, J.R. Boot and W. Dawson, J. Physiol., 1975, 246, 280P.

[57]A.A. Mathe, S.K. Puri, and L. Volicer, Res. Commun. Chem. Path. Pharmac., 1974, 8, 635.

Peptidoleukotriene Agonists and Antagonists

By J. G. Gleason*, T. W. Ku, C. M. Kinzig, M. E. McCarthy, C. D. Perchonock, I. Kuzinskas, B. Berkowitz, and B. M. Weichman

DEPARTMENTS OF MEDICINAL CHEMISTRY AND PHARMACOLOGY, SMITH KLINE & FRENCH LABORATORIES, PHILADELPHIA, PA. 19101, U.S.A.

Slow reacting substance of anaphylaxis (SRS-A) is comprised of three structurally related, biologically active arachidonic acid metabolites: leukotriene C_4 (LTC_4), D_4 (LTD_4), and E_4 (LTE_4) (1,2,3). Release of these peptidoleukotrienes has been demonstrated following antigen provocation of sensitized human and animal lung tissue (4). In man, they are potent bronchoconstrictive agents both in vitro (5) and in vivo (6,7); in animal models, they exhibit potent contractile effects on airway and vascular smooth muscle (8,9,10), induce tracheal edema (11) and stimulate mucous hypersecretion (12,13).

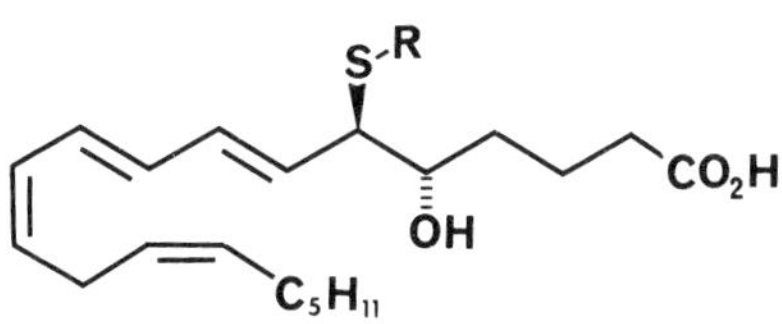

LTC_4, R = Cys(γ-Glu)(Gly)

LTD_4, R = Cys-Gly

LTE_4, R = Cys

Thus, the peptidoleukotrienes may function as mediators of anaphylactic reactions and could contribute to the pathophysiology of allergic asthma. It follows, therefore, that a potent and selective leukotriene receptor antagonist may have significant therapeutic value in the treatment of asthma and other immediate hypersensitivity diseases.

We and others have explored structural requirements for leukotriene agonist activity on guinea pig airway smooth muscle with a view to the design of a selective, potent receptor antagonist. A summary of some of our results is presented in Table 1. For this very limited series of compounds, it appears that the chirality at C-5 and C-6 and the length and degree of unsaturation of the lipid chain have considerable influence on affinity, but not on intrinsic agonist activity. The high preference for the 5S,6R absolute stereochemistry for agonist activity for both LTD_4 and LTE_4 suggests structural specificity expected of a receptor-mediated effect. The double bonds at C-9, C-11 and C-14 are not essential for agonist activity; 7Z-LTD_1 exhibits potent agonist activity and stereochemical selectivity on the guinea pig lung parenchymal strip, while its chemical stability is enhanced compared with LTD_4. Consequently, subsequent studies were carried out in the hexahydroleukotriene series. These early structure-activity studies also suggested that the polar C-1 to C-6 region of the eicosanoid molecule was critical for agonist activity (14,15,16). We have discovered (17,18) novel leukotriene analogs which, as a consequence of changes in this carboxyl region of LTD_1, possess significant leukotriene antagonist activity in several pharmacologically relevant models.

A series of hexahydroleukotriene analogs was prepared to assess the effect of altering the distance between the apparently critical C-1 carboxyl and C-5 hydroxyl groups which might be critical for the agonist activity of LTD_4. The synthesis of these compounds is exemplified in Figure I by the synthesis of 4R,5S,6Z-2-nor-LTD_1, 4b (SK&F 101132). The geometry of the cis-olefin and the trans-epoxide of 2-nor-LTA_1 methyl ester were established by analysis of nmr coupling constants and nuclear Overhauser experiments. To

TABLE I Relative Agonist Potential of C_{20} Leukotriene Analogs on Guinea Pig Lung Parenchymal Strips

Compounds	Agonist Potency[a] (LTD_4 = 1.0)[b]
LTD_4	1.0
5R,6S-LTD_4	0.01
7Z-LTD_1	0.1
7Z,5R,6S-LTD_1	0.02
LTE_4	0.15
5R,6S-LTE_4	0.01
7Z-LTE_1	0.03
7Z-LTC_1 [c]	0.2
7Z-desamino-LTE_1 [d]	1.4
LTE_1[d]	0.002
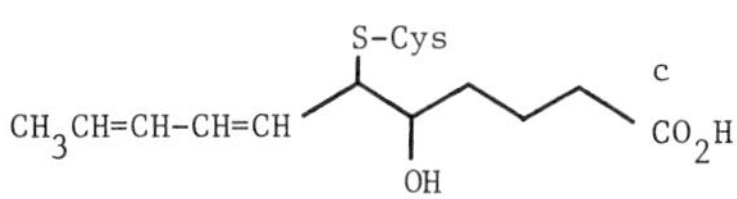	$<$ 0.001

a) maximal contractions for all compounds were at least 80% that observed for LTD_4 b) LTD_4: EC_{50}=1.4 nM on guinea lung parenchymal strips

c) diastereomeric d) racemic

facilitate separation of diastereomers, the intermediate cysteinyl-glycyl adduct was lactonized and separated by prep. HPLC. Hydrolytic removal of the amino and carboxyl protecting groups afforded the target compounds. The absolute configuration of the 2-nor-LTD_1 isomers was established by Raney nickel desulfurization of the chirally pure lactones and comparison with the known γ-lactones of 4-hydroxynonadecanoic acid (19). Analogous methodology was employed for the synthesis of LTD_1 and 2-homo-LTD_1 analogs with the exception that diastereomer separation was achieved without lactonization.

FIGURE I. SYNTHESIS OF 2-NOR-LEUKOTRIENES

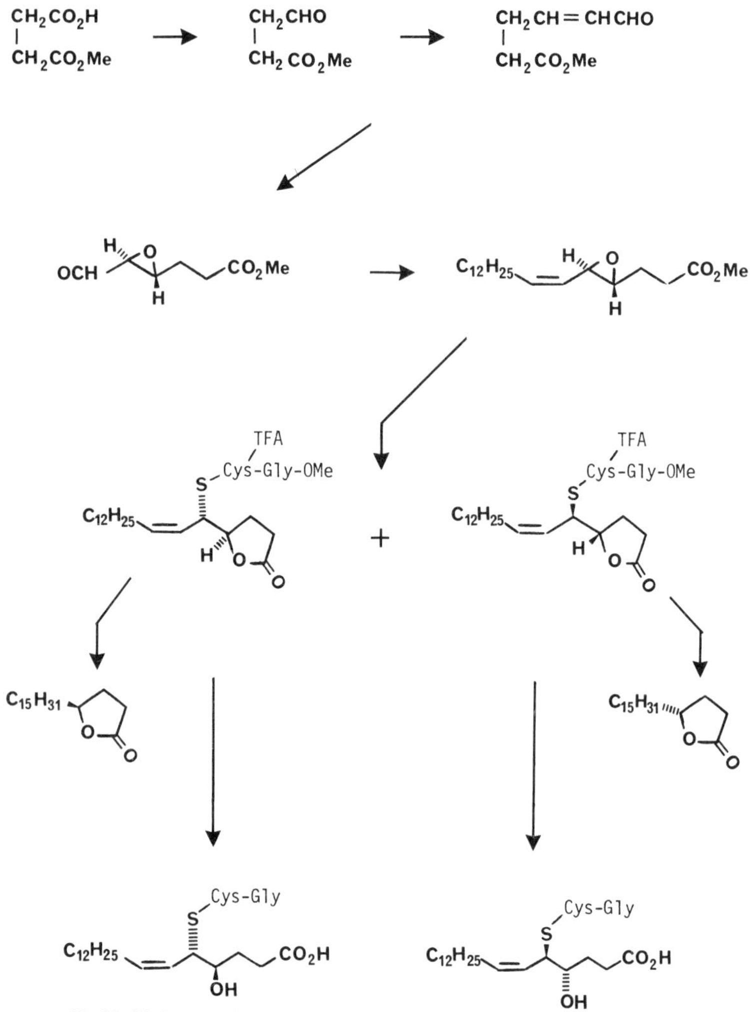

4R,5S,6Z-2-nor-LTD_1; 4b

$[\alpha]_D^{22}$ -34.3 (1,MeOH)

SK&F 101132

4S,5R,6Z-2-nor-LTD_1; 4a

$[\alpha]_D^{22}$ +59.4 (1,MeOH)

Altering the separation of the C-1 carboxyl and the C-5 hydroxyl has a dramatic effect on intrinsic agonist activity, as can be seen from Table 2. Lengthening the propylene bridge by one methylene residue (2a, 2b) has little or no effect on the agonist activity as measured on guinea pig lung parenchymal strips. In contrast, deletion of a methylene group between C-1 and C-4, particularly in the unnatural 4R,5S series, affords 4b, SK&F 101132, which only weakly contracts lung parenchymal tissue, but which, on pretreatment at 100μM, significantly antagonizes the subsequent contraction elicited by LTD_4 (17,18). Inhibition of LTD_4-induced contractions of guinea pig trachea and pulmonary arterial rings by SK&F 101132, was also demonstrated (Figure 2). This compound is a selective leukotriene antagonist; no antagonism of histamine-, carbachol-, or KCl-induced contractions of guinea

TABLE 2: BIOLOGICAL COMPARISON OF LTD_4 ANALOGS

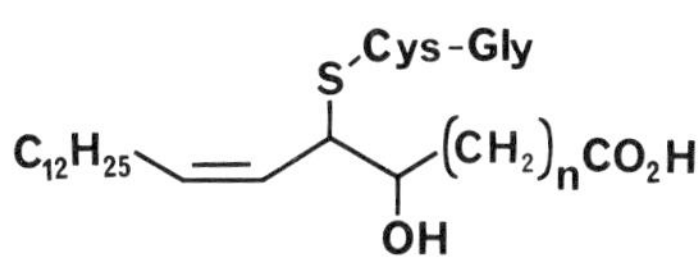

Compound	n	Stereochemistry	G.P. Lung Parenchyma Agonist Activity EC_{50}(nM)	Relative Contractile Activity*
1a	LTD_4	5S, 6R	1.4	1
2a	4	6S, 7R	6.5	1
2b	4	6R, 7S	4.5	1
3a	3	5S, 6R	15	1
3b	3	5R, 6S	610	1
4a	2	4S, 5R	630	0.5
4b	2	4R, 5S	500	0.1

*Contractile activity is expressed as the maximal contraction achieved relative to natural LTD_4.

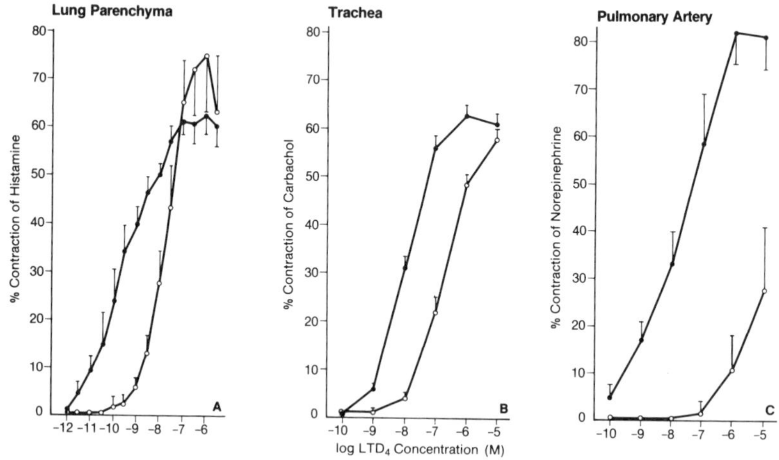

Figure 2. Antagonism of the LTD_4-induced contraction of guinea pig a) lung parenchyma, b) trachea, and c) pulmonary arterial tissues by 4b (10^4M). Tissues were incubated with 4b (●) or vehicle (0) or vehicle (0) for 30 minutes prior to construction of cumulative LTD_4 concentration response curves. Each point represents the mean percentage of contraction ± S.E.M.

pig trachea was observed (18). The ability of SK&F 101132 to block an LTD_4-induced bronchospasm *in vivo* was also examined. When administered i.v. (5 mg/kg) one minute prior to i.v. challenge of LTD_4 (1 nM/kg), greater than 80% reduction of the resulting bronchospasm, as measured by changes in both pulmonary resistance and dynamic lung compliance, was observed (18). Thus, SK&F 101132 appears to be a selective peptidoleukotriene antagonist both *in vitro* and *in vivo*.

The antagonist, SK&F 101132, exhibits a number of significant structural differences from the natural agonists. One difference, which might have significant impact on receptor affinity, is the C-6 double bond. In 4b, this group has Z geometry, which is in contrast with the E stereochemistry observed for the corresponding C-7 bond in LTD_4. Surprisingly, the corresponding E isomers, 5a and 5b, exhibited weak intrinsic agonist activity, and at 100μM were less potent as antagonists of the LTD_4-induced contraction than the Z isomer 4b (Figure 3).

Figure 3

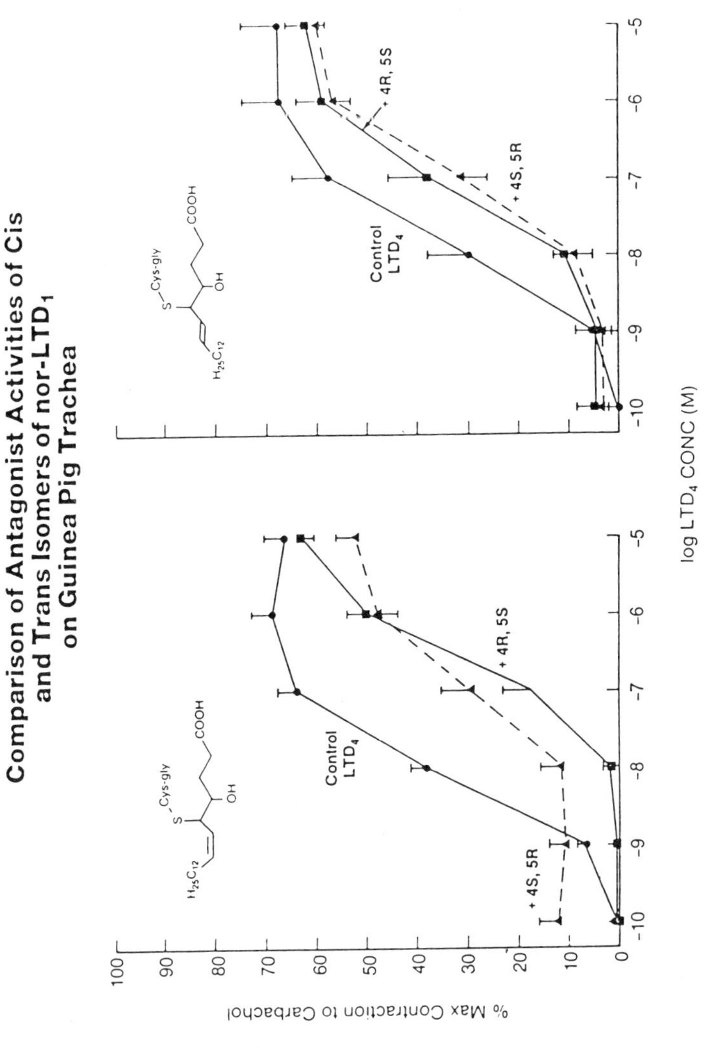
Comparison of Antagonist Activities of Cis
and Trans Isomers of nor-LTD$_1$
on Guinea Pig Trachea
% Max Contraction to Carbachol
100
90
80
70
60
50
40
30
20
10
0
-10
-9
-8
-7
-6
-5
log LTD$_4$ CONC (M)
Control
LTD$_4$
+ 4S, 5R
+ 4R, 5S
S
Cys-gly
COOH
OH
$H_{25}C_{12}$

TABLE 3: COMPARATIVE ANTAGONIST ACTIVITY OF NOR-LEUKOTRIENE ANALOGS ON GUINEA PIG TRACHEAL SPIRAL STRIPS

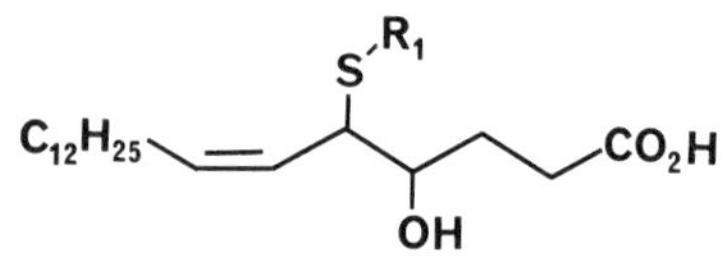

	R_1	R_2	Stereochemistry	Antagonist Activity* K_B** (μM)
4a	Cys-Gly	OH	4S, 5R, 6Z	10.0 †
4b	Cys-Gly	OH	4R, 5S, 6Z	7.0
5a	Cys-Gly	OH	4S, 5R, 6E	20.4 †
5b	Cys-Gly	OH	4R, 5S, 6E	25.6 †
6	Cys-Gly	NH_2	4R, 5S, 6Z	3.3
7a	Cys	OH	4S, 5R, 6Z	3.1
7b	Cys	OH	4R, 5S, 6Z	14
8	Cys(γ-Glu)Gly	OH	4RS, 5SR, 6Z	Agonist
9	CH_2CH_2CO-Gly	OH	4RS, 5SR, 6Z	0.50

† partial agonist

* in comparison, the K_B for FPL 55712 is 0.1 μM.

**K_B, an estimate of antagonist potency, was calculated from the equation K_B=Antagonist concentration/(X-1) where X is the dose ratio of agonist required to elicit an equal contraction in the presence and absence of antagonist.

Several analogs of SK&F 101132 were prepared to explore the SAR of this unique series, and the results are summarized in Table 3. Thus, the Cys-Gly (4b), Cys (7b) and desamino-Cys-Gly (9) 2-nor-leukotriene analogs all exhibit significant antagonist activity, with the latter desamino analog being the

Figure 4

Removal of the Peptide Amino Group Enhances Antagonism of LTD_4 on Guinea Pig Trachea

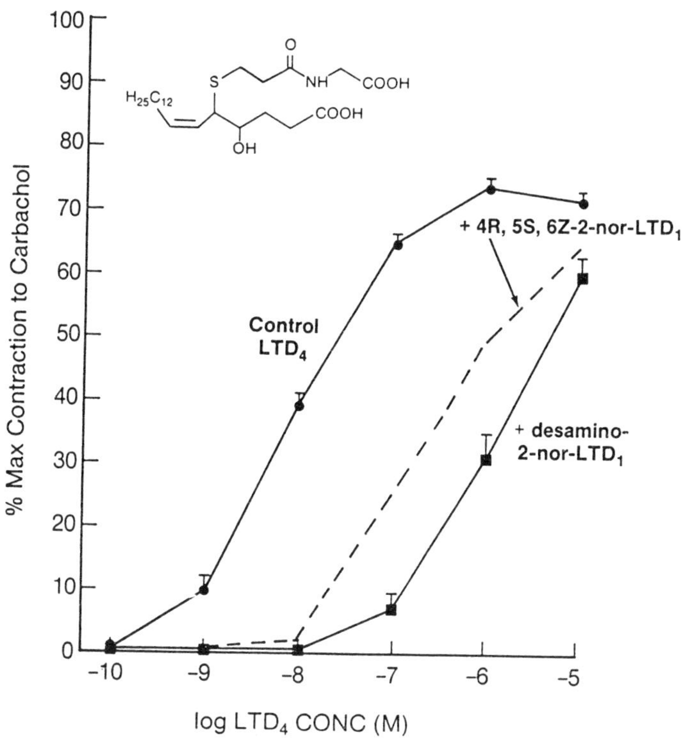

most potent of the series (Figure 4). In contrast, the 2-nor-LTC_1 analog 8 is a full agonist. The observation of agonist, rather than antagonist, activity for the nor-LTC_1 analog is somewhat surprising. Distinct receptor subtypes for LTC_4 and LTD_4 have been suggested by ligand binding studies (20,21). It is possible that the LTC_4 receptor possesses distinctly different structure-activity requirements and that structural modifications which result in antagonism at the LTD_4 receptor do not necessarily result in similar antagonism at the LTC_4 receptor.

Our results demonstrate that SK&F 101132 and related 2-nor leukotriene analogs possess significant leukotriene antagonist activity, both in vitro and in vivo. While the precise role of the peptidoleukotrienes in the pathophysiology of asthma remains to be determined, a leukotriene antagonist may offer new and useful therapeutic opportunities for immediate hypersensitivity diseases as well as provide tools for further exploration of the role of leukotrienes in mammalian physiology.

References

1. Hammarstrom, S., R.C. Murphy, B. Samuelsson, D.A. Clark, C. Mioskowski, E.J. & Corey, (1979), Biochem. Biophys. Res. Commun., 91, 1266.

2. Lewis, R.A., K.F. Austen, J.M. Drazen, D.A. Clark, A. Marfat, & E.J. Corey, (1980), Proc. Natl. Acad. Sci., USA., 77, 3710.

3. Morris, H.R., G.W. Taylor, P.J. Piper, M.N. Samhoun, & J.R. Tippins, (1980), Prostaglandins, 19, 185-201.

4. Lewis, R.A., K.F. Austen, J.M. Drazen, D.A. Clark, A. Marfat, & E.J. Corey, (1980) Proc. Natl. Acad. Sci., USA., 77, 3710.

5. Hanna, C.J., M.K. Bach, P.D. Pare, & R.R. Schellenberg, (1981), Nature, 290, 343.

6. Holroyde, M.C., R.E.C. Altounyan, M. Cole, M. Dixon, & E.V. Elliott, (1981), Lancet, 1981, 17.

7. Weiss, J.W., J.M. Drazen, N. Coles, E.R. McFadden, P.F. Weller, E.J. Corey, R.A. Lewis, & K.F. Austen, (1982), Science, 216, 196.

8. Gleason, J.G., R.D. Krell, B.M. Weichman, F. Ali, & B. Berkowitz, (1982) Adv. Prost. Thromb. and Leukotriene Res., 9, 243.

9. Dahlen, S.E., P. Hedquist, S. Hammarstrom, & B. Samuelsson, (1980), Nature, 288, 484.

10. Weichman, B.M., R.M. Muccitelli, R.R. Osborn, D.A. Holden, J.G. Gleason, & M.A. Wasserman, (1982), J. Pharmacol. Exp. Ther., 222, 202.

11. Woodward, D.F., B.M. Weichman, C.G. Gill, M.A. Wasserman, (1983), Prostaglandins, 25, 131.

12. Marom, Z., J.H. Shelhamer, M.K. Bach, D.R. Morton, & M. Kaliner, (1982), Amer. Rev. Respir. Dis., 126, 449.

13. Peatfield, A.C., P.J. Piper, & P.S. Richardson, (1982), Br. J. Pharmac., 77, 391.

14. Lewis, R.A., J.M. Drazen, K.F. Austen, M Toda, F. Brion, A. Marfat, & E.J. Corey, (1981), Proc. Natl. Acad. Sci. USA., 78, 4579.

15. Drazen, J.M., R.A. Lewis, K.F. Austen, M. Toda, F. Brion, A. Marfat, & E.J. Corey, (1981), Proc. Natl. Acad. Sci. USA.,78, 3195.

16. Okuyama, S., S. Miyamoto, K. Shimoji, Y. Konishi, D. Fukushima, H. Niwa, Y. Arai, M. Toda, M. Hayaski, (1982), Chem. Pharm. Bull., 30, 2453.

17. Gleason, J.G., Ku, T.W., Mccarthy, M.E., Weichman, B.M., Holden, D., Osborn, R.R., Zabko-Potapovich, B., Berkowitz, B., & Wasserman, B.A., (1983), Biochem. Biophys. Res. Commun., 117, 732.

18. B.M. Weichman, M.A. Wasserman, D.A. Holden, R.R. Osborn, D.F. Woodward, T.W. Ku & J.G. Gleason, (1983), J. Pharm. Exp. Therap., 227, 700.

19. Nishizawa, M., M. Yamada, & R. Noyori, (1981), Tet. Lett., 22, 4933.

20. Hogaboom, G.K., Mong, S., Wu, H., and Crooke, S.T., (1983), Biochem. Biophys. Res. Commun., 116, 1136.

21. Krilis, S., Lewis, R.A., Corey, E.J., and Austen, F.K., (1983), J. Clin. Inv., 72, 1516.

The Design of Selective Thromboxane Synthetase Inhibitors

By P. E. Cross and R. P. Dickinson*

PFIZER CENTRAL RESEARCH, SANDWICH, KENT CT13 9NJ, U.K.

The last few years have seen major advances in our knowledge of arachidonic acid (AA) metabolism. Until 1974, only the pathways leading to formation of the prostaglandins PGD_2, PGE_2 and $PGF_{2\alpha}$ had been elucidated, but then an important new pathway was reported by Samuelsson and co-workers.[1,2] It was found that an enzyme present in blood platelets, thromboxane synthetase, converts the prostaglandin endoperoxide intermediate PGH_2 into the unstable thromboxane A_2 (TxA_2) which rapidly hydrolyses under physiological conditions to TxB_2. Thromboxane A_2 was found to be a potent vasoconstrictor and platelet aggregating agent. It was suggested[3] that it is probably involved in normal haemostasis but that, in certain circumstances, it may be involved in pathological conditions in which there is an increased tendency to vasospasm and/or thrombosis. These findings suggested that a compound which would inhibit TxA_2 synthetase selectively without affecting the formation of antiaggregatory or vasodilator prostaglandins, might be useful in the treatment or prophylaxis of several cardiovascular diseases. A cyclo-oxygenase inhibitor would also block formation of TxA_2 but the formation of all other prostaglandins would be prevented.

Another major advance in prostaglandin research was the discovery by Vane and his group at Wellcome that an enzyme present in vascular endothelium, PGI_2 synthetase, converts PGH_2 to the potent anti-aggregatory vasodilator PGI_2.[4-9] It was suggested that, under normal circumstances, there is a balance between TxA_2 and PGI_2 but that, in pathological situations, this balance may be tilted in favour of TxA_2. The discovery of PGI_2 synthetase highlighted a possible additional advantage of a selective TxA_2 synthetase inhibitor since, by blocking only the formation of TxA_2, there is the possibility of diversion of the PGH_2 substrate down the PGI_2 pathway.[10,11]

Subsequent work from many groups has suggested that an imbalance in TxA_2 and PGI_2 production may be a causative factor in a variety of conditions such as heart attack, thrombotic stroke, angina, peripheral vascular disease, the cardiovascular consequences of endotoxin shock and in tumour cell metastasis.[12]

The starting point for our work was the observation by the groups of Vane[13] and Needleman[14] that imidazole is a weak, though selective, inhibitor of TxA_2 synthetase. We investigated a range of substituted imidazoles with the aim of identifying more potent compounds. In all our work, the inhibitory activity of compounds against TxA_2 synthetase was determined using a human platelet microsomal preparation and enzymically generated PGH_2 as the substrate. The extent of conversion to TxA_2 was determined initially using a bioassay technique,[15] but in later work a specific radioimmunoassay for TxB_2 was used.[16] It was found that a lipophilic substituent at the 1-position increases activity, whereas substitution at the 2-position is detrimental (Table 1). Identical findings have been reported by other groups.[17,18]

INHIBITION OF TxA_2 SYNTHETASE BY IMIDAZOLES

COMPOUND	IC_{50}(M)	COMPOUND	IC_{50}(M)
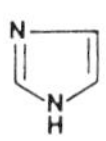	1.1×10^{-4} a	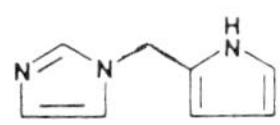	2.4×10^{-7} a
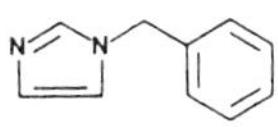	7.2×10^{-7} b	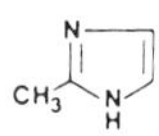	$> 10^{-4}$ a
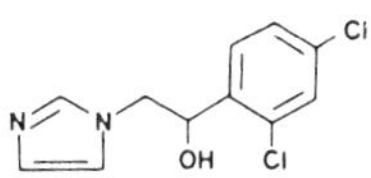	2.0×10^{-6} a	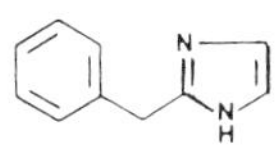	$> 10^{-4}$ a
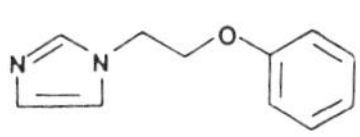	6.6×10^{-7} b	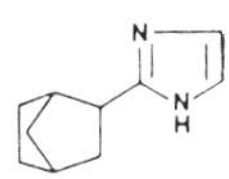	$> 10^{-4}$ a

a : BIOASSAY
b : RADIOIMMUNOASSAY

Table 1

A similar structure-activity relationship has been observed for imidazoles which inhibit oxidation reactions involving the haem enzyme cytochrome P-450, e.g. liver microsomal drug oxidations and adrenal steroid hydroxylation reactions.[19-24] Imidazoles are thought to inhibit cytochrome P-450-dependent reactions via co-ordination of the basic nitrogen to the fifth or sixth ligand site of the haem moiety. Binding of imidazoles in this way induces the characteristic so-called type II spectral change. Substitution at the 2-

position of the imidazole would be expected to sterically hinder interaction with the haem iron with a consequent reduction in inhibitory activity.[24]

In view of the apparent similarity in the SAR trends, it occurred to us that TxA_2 synthetase may be a haem-dependent enzyme. Further support for a possible link between TxA_2 synthetase and cytochrome P-450 came from the work of Cinti and Feinstein[25] who showed the presence of cytochrome P-450 in the platelet microsomal fraction. The cytochrome P-450 inhibitor metyrapone (Table 2) was found to inhibit AA-induced platelet aggregation and the formation of platelet aggregating factors on incubation of AA with platelet microsomes. A typical type II spectral change was found on incubation of metyrapone with platelet microsomes. It was concluded that cytochrome P-450 may play a role in the complex enzyme systems which convert AA to platelet aggregating factors, but it was not possible to decide from these results whether metyrapone was acting at the cyclo-oxygenase or TxA_2 synthetase steps. We found that it is in fact a moderately potent TxA_2 synthetase inhibitor having no effect on cyclo-oxygenase.

The possible involvement of iron in the transformation of PGH_2 to TxA_2 was also suggested by Turner and Herz.[26] Based on known ferrous ion-promoted reactions of organic peroxides, they proposed a chemical model for TxA_2 formation involving a ferrous-ferric redox system (Fig. 1).

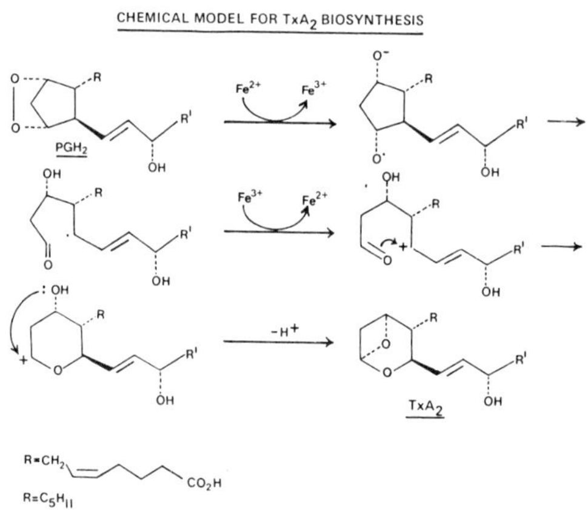

Fig. 1

Other workers have also suggested the possible dependence of TxA_2 synthetase on haem iron[17,27] and, very recently, Ullrich and Haurand[28] have shown that partially purified human platelet cytochrome P-450 transforms PGH_2 to TxA_2. A close similarity was found between the cytochrome P-450 spectral dissociation constants and IC_{50} values for inhibition of TxA_2 synthesis for a series of inhibitors. Purification of the cytochrome P-450 increased the specific activity for TxA_2 synthesis and it was concluded that platelet cytochrome P-450 and TxA_2 synthetase are identical.

In addition to suggesting a chemical model for TxA_2 formation based on a ferrous-ferric redox system, Turner and Herz[26] also proposed a related mechanism for the conversion of PGH_2 to PGI_2. More recently Ullrich et al[29,30] have obtained spectral evidence for the cytochrome P-450 nature of PGI_2 synthetase, and DeWitt and Smith[31] have obtained evidence to suggest that a haem moiety at the active site of the enzyme participates in the transformation of PGH_2 to PGI_2.

Thus, although not all this information was available at the start of our work, it appeared that if TxA_2 synthetase inhibitors were to be designed on the basis of a possible cytochrome P-450 mechanism, then selectivity problems could arise. However, we decided to concentrate initially on the design of potent inhibitors of TxA_2 synthetase and to tackle any selectivity problems subsequently.

In assessing points for possible synthetic entry we noted that the structure of metyrapone bears some resemblance to that of the anti-inflammatory agent nictindole which Gryglewski[32,33] had reported to be a much more potent TxA_2 synthetase inhibitor than imidazole (Table 2). Nictindole is less selective than metyrapone since it also inhibits cyclo-oxygenase but it was reasoned that

METYRAPONE AND NICTINDOLE:ENZYME INHIBITION IN VITRO

COMPOUND	IC_{50} (M)		
	TxA_2 SYNTHETASE [a]	PGI_2 SYNTHETASE	CYCLO-OXYGENASE
METYRAPONE 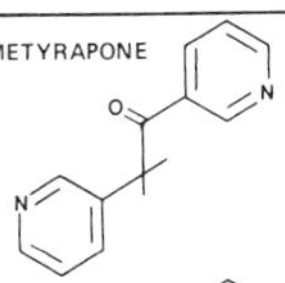	4.0×10^{-6}	1.0×10^{-4}	$> 10^{-3}$
NICTINDOLE 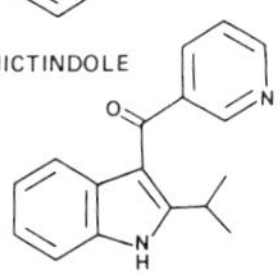	8.0×10^{-8}	4.5×10^{-3}	5.2×10^{-6}

a = Bioassay

Table 2

replacement of the nicotinoyl group by the more basic imidazolylmethyl group should maintain or even enhance activity against TxA_2 synthetase while hopefully retaining the selectivity of imidazole with respect to cyclo-oxygenase and PGI_2 synthetase. Therefore 3-(1-imidazolylmethyl)indole was prepared, together with some analogues in which modifications were made to the heterocyclic substituent (Table 3). The imidazolylmethyl analogue proved to be the most potent inhibitor. The marked detrimental effect on activity of substitution adjacent to the basic nitrogen atom is shown by the results for the 2-methylimidazole and benzimidazole analogues. Replacement of imidazole with pyrazole or 1,2,4-triazole results in loss of activity. Only the pyridylmethyl analogue retains activity.

3–SUBSTITUTED INDOLES : INHIBITION OF TxA_2 SYNTHETASE IN VITRO [a]

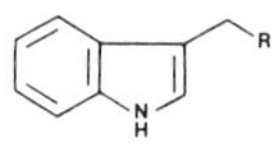

R	IC_{50}(M)	R	IC_{50}(M)
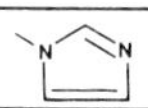	2.0 x 10^{-8}	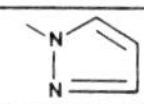	> 10^{-6}
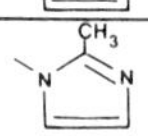	3.0 x 10^{-4}	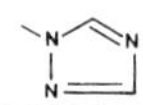	> 10^{-6}
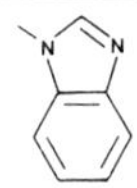	4.0 x 10^{-5}	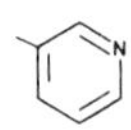	3.8 x 10^{-8}

a : Bioassay

Table 3

3 – IMIDAZOLYLMETHYLINDOLES
ENZYME INHIBITION IN VITRO

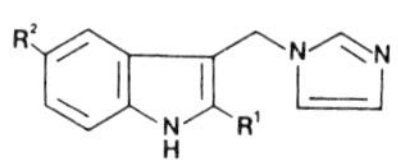

R^1	R^2	IC_{50}(M)			
		TxA_2 [a] SYNTHETASE	PGI_2 SYNTHETASE	CYCLO-OXYGENASE	11β–HYDROXYLASE
H	H	2.0 x 10^{-8}	6.5 x 10^{-4}	> 10^{-3}	1.0 x 10^{-5}
CH_3	H	1.2 x 10^{-8}	3.0 x 10^{-5}	> 10^{-4}	98% @ 10^{-4}
$CH(CH_3)_2$ [b]	H	1.7 x 10^{-8}	>10^{-4}	> 10^{-4}	2.0 x 10^{-6}
$C(CH_3)_3$	H	2.5 x 10^{-9}	9.5 x 10^{-7}	> 10^{-4}	N.D.
H	CH_3	5.8 x 10^{-8}	3.5 x 10^{-5}	> 10^{-4}	N.D.
H	OCH_3	2.0 x 10^{-7}	1.0 x 10^{-4}	> 10^{-4}	N.D.
H	OH	2.3 x 10^{-9}	8.0 x 10^{-6}	> 10^{-4}	N.D.
H	Br	1.0 x 10^{-7}	1.7 x 10^{-5}	> 10^{-4}	N.D.
H	$N(CH_3)_2$	6.4 x 10^{-8}	>10^{-4}	> 10^{-4}	N.D.

a : BIOASSAY b : UK-34,787
N.D. : NOT DETERMINED

Table 4

The effect of substitution in the indole ring of 3-(1-imidazolylmethyl)indole on both activity and selectivity was then examined (Table 4). 2-Alkyl substituted compounds retain activity against TxA_2 synthetase and, in the case of t-butyl, potency is increased. Substitution at the 5-position tends to reduce potency slightly with the exception of the hydroxy compound. In no case is there significant activity against cyclo-oxygenase, but activity against PGI_2 synthetase is variable. Thus, the unsubstituted compound and the isopropyl analogue, UK-34,787, show no significant activity but substantial activity is shown by the t-butyl analogue, although the selectivity is still in favour of TxA_2 synthetase. These results indicate that, as far as enzymes involved in prostaglandin formation are concerned, the design of selective TxA_2 synthetase inhibitors is possible, although some activity against PGI_2 synthetase may be encountered.

However, we were also concerned about possible side effects resulting from inhibition of other cytochrome-P450 enzymes. Thus, liver microsomal enzyme inhibition could affect the metabolism of co-administered drugs, and inhibition of adrenal enzymes could affect steroid biosynthesis. For example, metyrapone inhibits steroid 11β-hydroxylase which converts 11-deoxycortisol to cortisol. This causes an increase in ACTH secretion which can result ultimately in adrenal hyperplasia.[34] As a measure of activity against other cytochrome P-450 enzymes, we studied the effect of a number of the compounds against rat adrenal steroid 11β-hydroxylase _in vitro_ using 11-deoxycorticosterone as substrate. Although the compounds examined do inhibit the enzyme, higher concentrations than those needed to inhibit TxA_2 synthetase are required. However, further evaluation of UK-34,787 showed that it significantly prolongs pentobarbitone sleeping time in rats at a dose of 10mg/kg p.o. indicating liver microsomal enzyme inhibition. The dose is similar to that necessary for complete inhibition of thromboxane formation, indicating that it would be desirable to achieve a greater separation of activities against TxA_2 synthetase and other cytochrome P-450 enzymes.

As an approach to overcoming this selectivity problem we decided to examine whether introduction of hydrophilic substituents would decrease affinity for the liver and adrenal enzymes while maintaining activity against TxA_2 synthetase. Carboxylic acid groups were of particular interest since there is presumably a carboxyl binding site in the vicinity of the active site of the enzyme to accommodate the PGH_2 carboxyl group. Initial investigations were carried out on some arylalkyl and aryloxyalkyl substituted imidazoles (Table 5). One striking result is the marked increase in activity against TxA_2

1–SUBSTITUTED IMIDAZOLES: ENZYME INHIBITION IN VITRO

COMPOUND	IC_{50}(M)			
	TxA_2 SYNTHETASE [a]	PGI_2 SYNTHETASE	CYCLO–OXYGENASE	11β–HYDROXYLASE
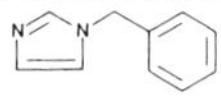	7.2×10^{-7}	$> 10^{-2}$	$> 10^{-2}$	72% @ 10^{-4}
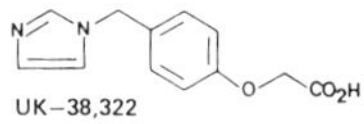 UK–38,322	3.1×10^{-8}	$> 10^{-4}$	$> 10^{-4}$	$> 10^{-4}$
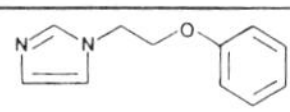	6.6×10^{-7}	$> 10^{-4}$	6.7×10^{-5}	72% @ 10^{-5}
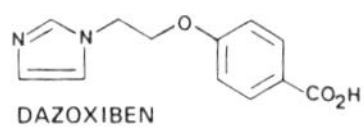 DAZOXIBEN	3.0×10^{-9}	$> 10^{-4}$	$> 10^{-4}$	$> 10^{-4}$

a: RADIOIMMUNOASSAY

Table 5

synthetase which results from introduction of the carboxyl substituents to give UK-38,322 and dazoxiben. Also, the reduction in activity against 11β-hydroxylase is notable. None of the compounds shows any activity against PGI_2 synthetase, although one compound, phenoxyethylimidazole, has weak activity against cyclo-oxygenase.

Further SAR studies were carried out in both series of compounds and results for some phenoxyethylimidazoles related to dazoxiben are summarised in Table 6. Conversion of dazoxiben to the corresponding amide leads to reduction in potency against TxA_2 synthetase, and alkylation of the amide group reduces the potency further. Moving the carboxyl group to other positions in the benzene ring has a detrimental effect, the reduction in potency being particularly marked for the 2-isomer. Introduction of linking groups (CH_2 and OCH_2) between

PHENOXYETHYLIMIDAZOLE ANALOGUES
ENZYME INHIBITION IN VITRO

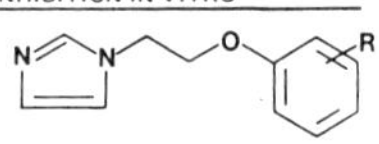

R	IC_{50}(M)	
	TxA_2 SYNTHETASE [a]	PGI_2 SYNTHETASE
H	6.6×10^{-7}	$> 10^{-4}$
4–CO_2H	3.0×10^{-9}	$> 10^{-4}$
4–$CONH_2$	5.6×10^{-8}	$> 10^{-4}$
4–$CONHCH_3$	8.6×10^{-7}	$> 10^{-4}$
4–$CON(CH_3)_2$	1.1×10^{-6}	$> 10^{-4}$
2–CO_2H	$> 10^{-5}$	N.D.
3–CO_2H	5.3×10^{-7}	$> 10^{-4}$
4–CH_2CO_2H	3.2×10^{-8}	N.D.
4–OCH_2CO_2H	9.5×10^{-8}	$> 10^{-4}$

a: RADIOIMMUNOASSAY
b: NOT DETERMINED

Table 6

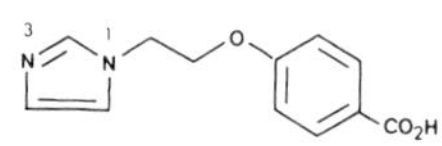

DAZOXIBEN

N–1 → CO_2H 8.1–8.8Å
N–3 → CO_2H 8.4–10.75Å

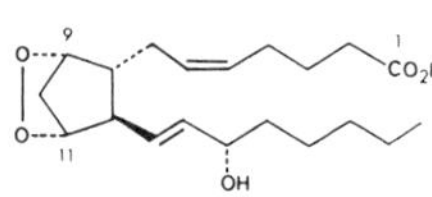

PGH_2

Fig. 2

the benzene ring and the carboxyl group is also detrimental. None of the compounds tested shows any activity against PGI_2 synthetase.

These results suggest that the distance between the imidazole ring and the carboxylic acid group may be important and that, for this series, the distance is optimal in dazoxiben. In a series of n-carboxyalkylimidazoles, Yoshimoto et al.[18,35] found optimal inhibition with side-chains containing from 7 to 10 carbon atoms. In a more extensive study of compounds incorporating aryl groups in the side chain, the Ono-Kissei group also concluded that the distance between the carboxyl and imidazole groups is important.[36] Based on an examination of Dreiding models an optimum distance of 8.5-9.0Å between N-1 of the imidazole and the carboxyl carbon was found. Clearly the distance depends on the conformation of the molecule, and molecules such as dazoxiben have an appreciable degree of flexibility. Preferred conformations for dazoxiben were determined using the CHEMLAB molecular modelling system and the corresponding N-1 to carboxyl distances were measured. A range of 8.1 to 8.8Å was found for conformers within 3kcal of the minimum (Fig. 2). Since it seems to be N-3 of the imidazole which is important for activity, it is probably more relevant to consider the overall N-3 to carboxyl distance. In this case a wider range of 8.4 to 10.75Å was found. Conformational calculations on inhibitor molecules with less flexibility would give a closer estimate of the optimum distance. Consideration of the suggested mechanism for TxA_2 formation and the inhibitor SAR suggests that the basic nitrogen and carboxyl groups of dazoxiben and related inhibitors may interact with the same sites on the enzyme as the C-9 oxygen and carboxyl respectively, of the substrate PGH_2.

In order to assess in vivo efficacy, dazoxiben was evaluated in anaesthetised rabbits.[16] In all in vivo studies, TxB_2 was generated in vitro in whole blood samples that were removed from the animal prior to, and following, administration of compound. The blood samples were allowed to clot at 37°C and the serum TxB_2 was determined by radioimmunoassay. Fig. 3 shows that a dose-dependent inhibition of thromboxane synthesis was obtained with dazoxiben. An intravenous dose of 1.0mg/kg produces almost complete inhibition and about 50% inhibition was still observed after a dose of 0.1mg/kg.

The time course of inhibition was studied after oral administration to conscious dogs using the same assay method for estimation of TxB_2 (Fig. 4). Administration of 3mg/kg of dazoxiben caused almost complete inhibition 1 hr. after dosing, falling to a 40% reduction relative to the pre-dose value after 6 hrs. Significant inhibition was also seen with a dose as low as 0.5mg/kg.[16]

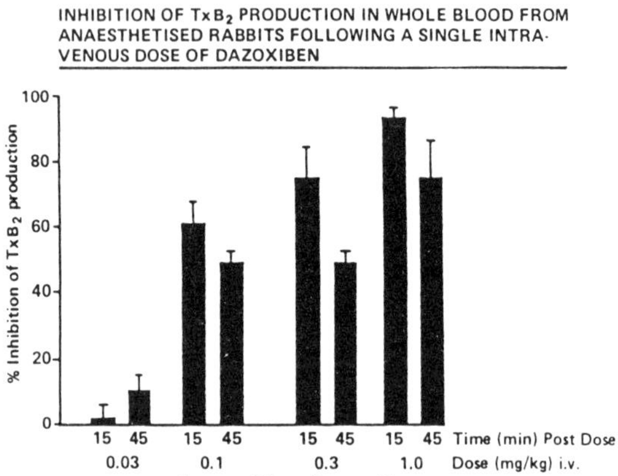

n = 5 Fig. 3 n = 3

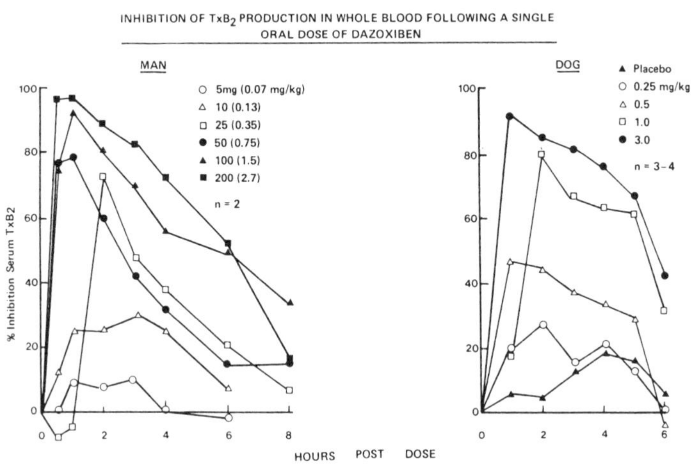

Fig. 4

An essentially similar picture was seen after oral administration to human volunteers.[37] Single doses of 100 and 200mg caused inhibition of TxB_2 production of more than 90% after 1 hr. and about 50% after 6 hrs. The blood samples from this study were also analysed for 6-keto $PGF_{1\alpha}$.[38] A dose-related increase in serum levels was found, providing evidence for diversion of the substrate PGH_2 to PGI_2 when TxA_2 synthesis is prevented. Recent studies have demonstrated a 3-4 fold increase in urinary excretion of 6-keto $PGF_{1\alpha}$ metabolites after oral administration of 200mg of dazoxiben to human volunteers.[39]

Clinical evaluation of dazoxiben in a number of indications is in progress and results will be reported elsewhere in due course.

As part of our follow up programme to dazoxiben, we investigated whether introduction of a carboxylic acid substituent into the 3-imidazolylmethylindole system would also favourably influence the activity/selectivity profile. A series of indole-1-alkanoic acid derivatives was prepared and enzyme inhibition data are summarised in Table 7. In this series, incorporation of a carboxylic acid substituent does not appear to increase activity against TxA_2 synthetase significantly. However, it is notable that none of the compounds shows activity against PGI_2 synthetase, which is in marked contrast to the situation with the corresponding analogues lacking the 1-carboxyalkyl substituents. Data for activity against other enzymes are less complete, but no compounds have been found to inhibit cyclo-oxygenase. Also, in the cases examined, introduction of the carboxylic acid substituent causes a substantial reduction in activity against 11β-hydroxylase.

INDOLE – I – ALKANOIC ACID DERIVATIVES
ENZYME INHIBITION IN VITRO

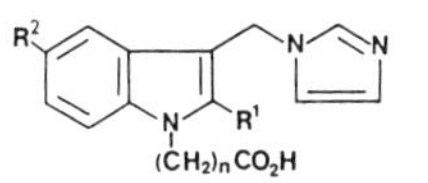

R^1	R^2	n	IC_{50}(M) TxA_2 SYNTHETASE [a]	IC_{50}(M) PGI_2 SYNTHETASE
H	H	1	3.0×10^{-8}	$> 10^{-4}$
H	H	2	1.9×10^{-8}	$> 10^{-4}$
CH_3	H	2	1.8×10^{-8}	$> 10^{-4}$
$CH(CH_3)_2$	H	2	8.2×10^{-9}	$> 10^{-4}$
H	CH_3	2	4.2×10^{-8}	$> 10^{-4}$
H	OCH_3	2	4.3×10^{-8}	$> 10^{-4}$
H	Cl	2	2.1×10^{-8}	$> 10^{-4}$
H	Br	2	1.2×10^{-8}	$> 10^{-4}$
H	$N(CH_3)_2$	2	8.2×10^{-8}	$> 10^{-4}$

a: RADIOIMMUNOASSAY

TABLE 7

UK-38,485 : ENZYME INHIBITION IN VITRO

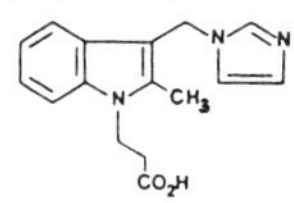

ENZYME	IC_{50}(M)
TxA_2 SYNTHETASE	1.8×10^{-8}
PGI_2 SYNTHETASE	$> 10^{-4}$
CYCLO-OXYGENASE	$> 10^{-4}$
11β – HYDROXYLASE	$> 10^{-4}$

Table 8

Evaluation of the compounds in anaesthetised rabbits showed that the 2-methylindole analogue UK-38,485 had activity at least as good as that of dazoxiben.[40] Full enzyme inhibition figures for this compound are summarised in Table 8.

The results of oral administration to conscious dogs are shown in Fig. 5. A high level of inhibition was found after a single dose of 1mg/kg. Even after a dose of 0.25mg/kg, the degree of inhibition after 6 hrs. was greater than that after 3mg/kg of dazoxiben, although the initial level of inhibition was lower. No activity was found after a dose of 0.1mg/kg. Further experiments have shown that 15 hr. after a dose of 1mg/kg, thromboxane synthesis is still inhibited by

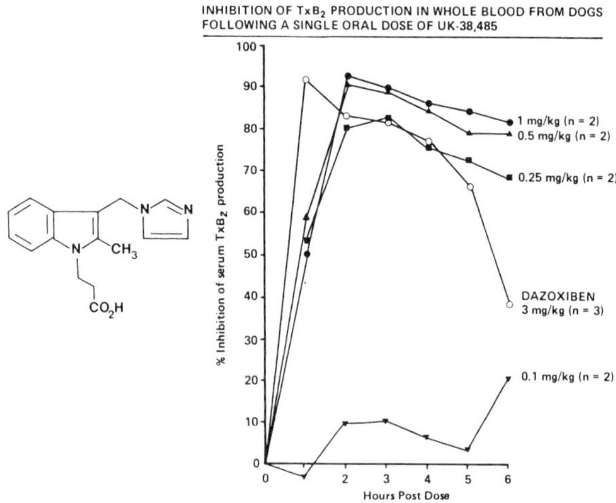

Fig. 5

50%. Initial clinical studies have confirmed the greater duration of action relative to dazoxiben.

In addition to our work with indoles related to UK-38,485, we were also interested in further modification of structures related to UK-38,322 and dazoxiben. In particular, we were interested in incorporating the heteroatom linkage into a heterocyclic ring system such as benzofuran. Cyclisation based on UK-38,322 would lead to a 5-imidazolylmethylbenzofuran, whereas cyclisation based on the dazoxiben structure would lead to a 2-imidazolylmethylbenzofuran (Fig. 6). The corresponding benzo[b]thiophenes and indoles were also of interest.

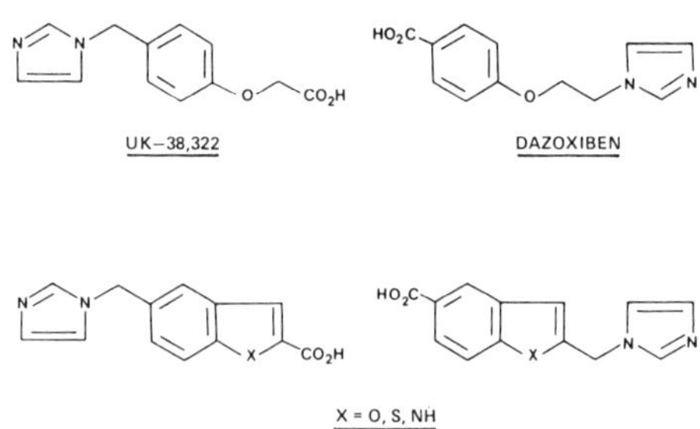

FIG. 6

The activity of 5-imidazolylmethyl-substituted benzoheterocycles against TxA_2 synthetase is shown in Table 9 together with results for the corresponding

INHIBITION OF TxA_2 SYNTHETASE IN VITRO

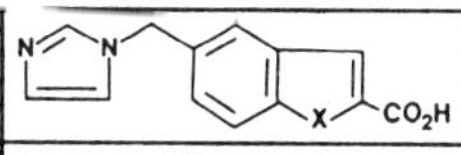

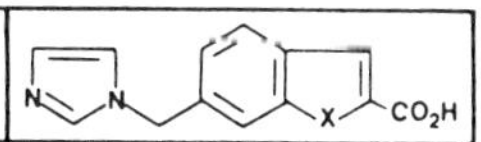

X	IC_{50}(M)	IC_{50}(M)
O	3.4×10^{-8}	2.0×10^{-8}
S	2.1×10^{-8}	1.8×10^{-8}
NH	5.0×10^{-9}	3.3×10^{-8}

Table 9

6-isomers. All have similar activity to UK-38,322 in vitro . Evaluation in anaesthetised rabbits showed that none of the compounds had greater potency or duration of action than dazoxiben, although one analogue, 6-imidazolylmethyl-benzo[b]thiophene-2-carboxylic acid, did show promising activity. A small number of substituted analogues of this compound was prepared to see if this activity could be improved upon (Table 10). Additional substitution caused a slight increase in activity in vitro but only the 3-methyl analogue UK-45,651 showed evidence of increased activity in the rabbit. However, further evaluation in conscious dogs showed that the compound offered no real

ACTIVITY AGAINST THROMBOXANE SYNTHETASE IN VITRO

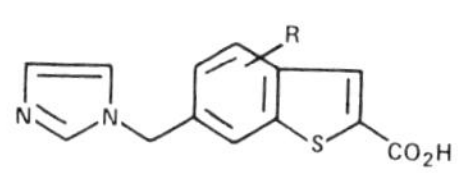

R	IC_{50} (M)
H	1.8×10^{-8}
4-OCH_3	5.0×10^{-9}
3-CH_3	5.2×10^{-9}
3-Cl	7.4×10^{-9}

Table 10

INHIBITION OF TxB_2 IN WHOLE BLOOD FROM DOGS FOLLOWING A SINGLE ORAL DOSE OF UK-45,651

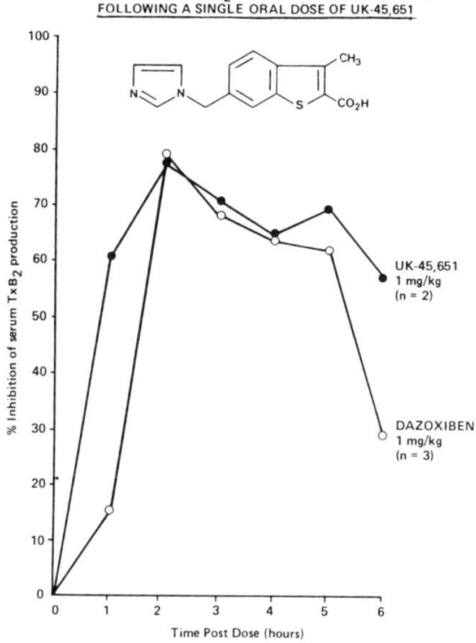

Fig. 7

improvement in potency over dazoxiben although there was a suggestion of a longer duration of action (Fig. 7).

Attention then turned to the isomeric 2-imidazolylmethyl substituted systems which can be considered as being derived from dazoxiben. The activity against TxA_2 synthetase in vitro is shown in Table 11 together with results for two isomeric 6-carboxylic acids. All the compounds have virtually identical potency. However, unlike the isomeric compounds already mentioned, several showed a high level of activity in vivo. The first compound of this type prepared, the 3-methylbenzo[b]thiophene analogue UK-47,852, was found to have

INHIBITION OF TxA_2 SYNTHETASE IN VITRO

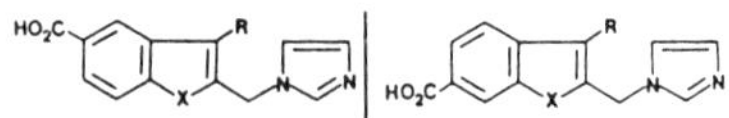

X	R	IC_{50}(M)	IC_{50}(M)
O	H	1.2×10^{-8}	–
O	CH_3	3.4×10^{-8}	–
O	Cl	3.1×10^{-8}	–
S	H	1.5×10^{-8}	–
S	CH_3	1.7×10^{-8}	1.8×10^{-8}
S	Cl	2.2×10^{-8}	4.0×10^{-8}
NH	CH_3	3.7×10^{-8}	–
NCH_3	CH_3	3.2×10^{-8}	–

TABLE 11

INHIBITION OF TxB_2 PRODUCTION IN WHOLE BLOOD FROM DOGS FOLLOWING A SINGLE ORAL DOSE OF UK-47,852

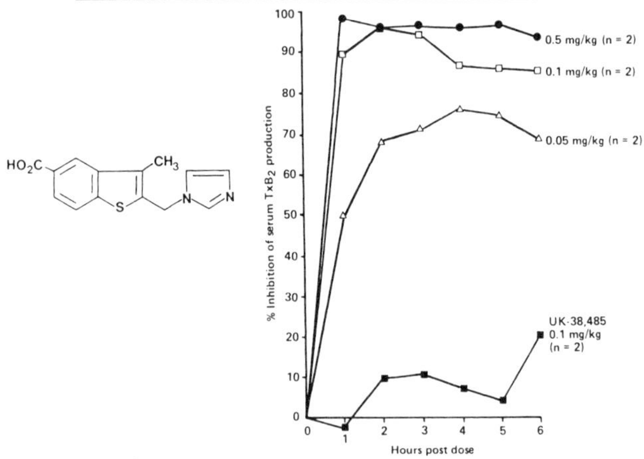

Fig. 8

excellent potency and duration of action in the anaesthetised rabbit. Thus, it produced complete inhibition of TxB_2 production 75 min. after an i.v. dose of 0.1mg/kg whereas 45 min. after the same dose of dazoxiben, inhibition was only 50%. The activity in conscious dogs following oral administration is shown in Fig. 8. A dose of 0.5mg/kg causes complete inhibition of TxB_2 formation for at least 6 hrs. Even a dose as low as 0.05mg/kg produces around 70% inhibition. The marked increase in potency over UK-38,485 is shown by comparison of the effect of administration of 0.1mg/kg of each compound.

Other benzo[b]thiophenes showed similar activity in vivo whereas the corresponding benzofurans were less potent, the level of activity more closely resembling that of UK-38,485. Fig. 9 shows a comparison of the activity of UK-47,852 with the same dose of the corresponding benzofuran and N-methylindole analogues in dogs. The benzofuran is much less active but good activity is shown by the indole. The corresponding indole lacking the N-methyl group did not show significant activity in the rabbit. Benzo[b]thiophene-6-carboxylic acid analogues showed similar activity to the corresponding 5-isomers in the conscious dog (Fig. 10).

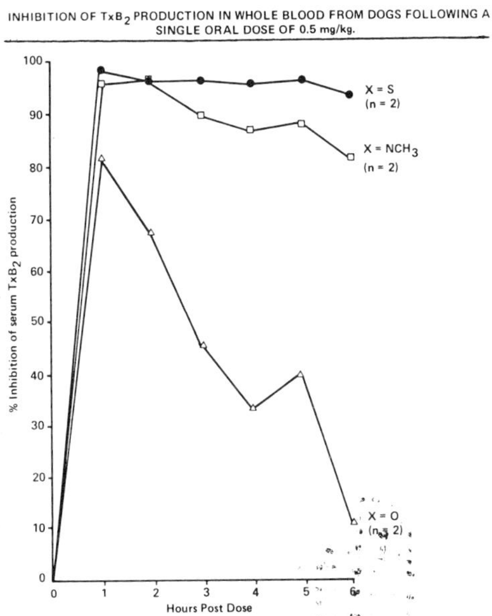

Fig. 9

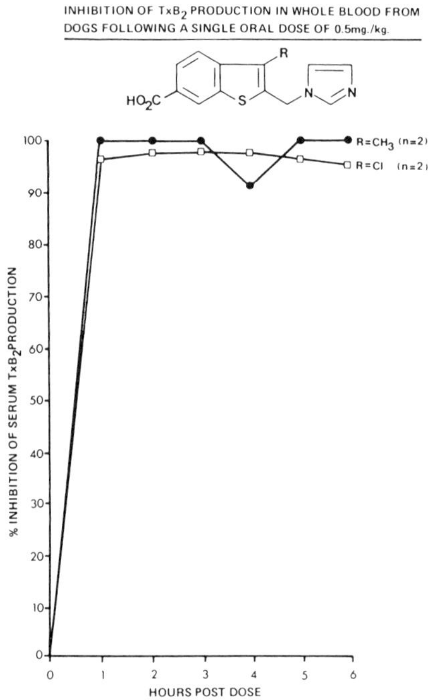

Fig. 10

As we and other groups[41,42] have shown that replacement of imidazole by a 3-pyridyl group generally retains activity against TxA_2 synthetase, we were interested in the preparation of 3-pyridyl analogues in the present series. The activity of a number of analogues against TxA_2 synthetase in vitro is summarised in Table 12. In all cases, activity matches that of the corresponding imidazoles. As in the imidazole series, the benzo[b]thiophenes proved superior in vivo to the benzofurans. The activity of the 3-methylbenzo[b]-thiophene UK-49,883 in the conscious dog is shown in Fig. 11. As with the imidazole analogue UK-47,852, a dose of 0.5mg/kg produces complete inhibition for at least 6 hrs and a high level of inhibition is shown after the lower dose of 0.05mg/kg.

Several of these imidazolylmethyl- and pyridylmethyl-substituted benzo[b]thiophenes have a duration of action at least comparable with that of UK-38,485 and further evaluation of compounds from the series is in progress.

INHIBITION OF TxA_2 SYNTHETASE IN VITRO

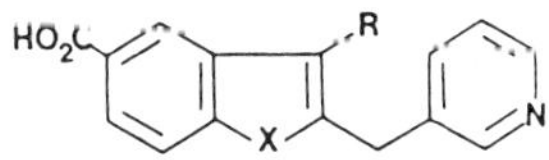

X	R	IC_{50}(M)
O	H	3.2×10^{-8}
O	CH_3	5.5×10^{-8}
S	H	1.6×10^{-8}
S	CH_3	2.6×10^{-8}

Table 12

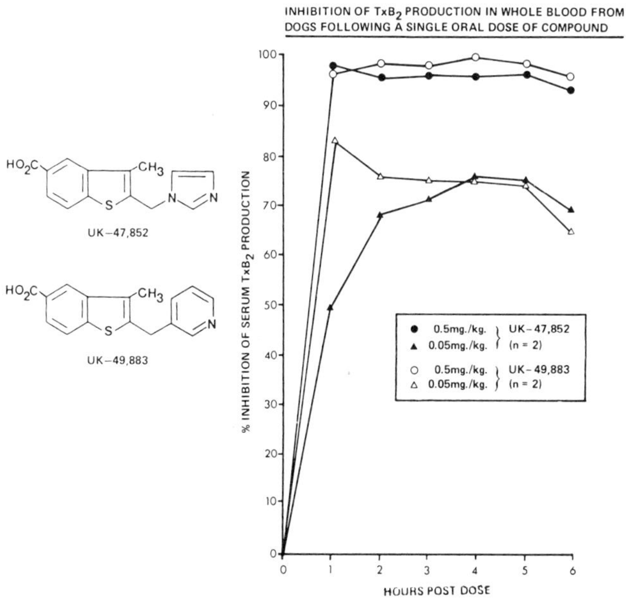

Fig. 11

In summary, we have discussed the evidence that TxA_2 synthetase is a cytochrome P-450 enzyme and described the design of potential inhibitors. Initially, selectivity problems were encountered because of inhibition of other cytochrome P-450 enzymes. Much improved selectivity and, in certain cases, increased potency against TxA_2 synthetase, was achieved by incorporation of a carboxylic acid group at a suitable distance from the imidazole ring. This led to the clinical candidates dazoxiben and UK-38,485. More recent work has identified a series of imidazolylmethyl- and pyridylmethyl-substituted benzo[b]thiophene-carboxylic acids, several of which have exceptional potency in vivo.

Acknowledgements

We wish to thank Messrs. G.N. Thomas, D.J. Tims, M.J. Pitcher, R. Gammon, A.J. Reason, Miss N. Rose and Miss P.H. Koza for the preparation of compounds and Drs. M.J. Parry, M.J. Randall, V.A. Alabaster, Messrs. E. Hawkeswood, R.I.R. Wilding and Mrs. G. Land for the biological testing. Thanks are also due to Dr. M.S. Tute for conformational calculations.

References

[1] M. Hamberg, B. Samuelsson, Proc. Nat. Acad. Sci. USA, 1974, 71, 3400.
[2] M. Hamberg, J. Svensson, B. Samuelsson, Proc. Nat. Acad. Sci. USA, 1975, 72, 2994.
[3] P. Needleman, S. Moncada, S. Bunting, J.R. Vane, M. Hamberg, B. Samuelsson, Nature, 1976, 261, 558.
[4] S. Moncada, R. Gryglewski, S. Bunting, J.R. Vane, Nature, 1976, 263, 663.
[5] R.J. Gryglewski, S. Bunting, S. Moncada, R.J. Flower, J.R. Vane, Prostaglandins, 1976, 12, 685.
[6] S. Moncada, R.J. Gryglewski, S. Bunting, J.R. Vane, Prostaglandins, 1976, 12, 715.
[7] S. Bunting, R. Gryglewski, S. Moncada, J.R. Vane, Prostaglandins, 1976, 12, 897.
[8] G.J. Dusting, S. Moncada, J.R. Vane, Prostaglandins, 1977, 13, 3.
[9] S. Moncada, E.A. Higgs, J.R. Vane, Lancet, 1977, 1, 18.
[10] F.P. Nijkamp, S. Moncada, H.L. White, J.R. Vane, Eur. J. Pharmacol., 1977, 44, 179.
[11] P. Needleman, A. Wyche, A. Raz, J. Clin. Invest., 1979, 63, 345.
[12] M. Verstraete, Br. J. Clin. Pharmacol., 1983, 15, 75.
[13] S. Moncada, S. Bunting, K. Mullane, P. Thorogood, J.R. Vane, Prostaglandins, 1977, 13, 611.
[14] P. Needleman, A. Raz, J.A. Ferrendelli, M. Minkes, Proc. Nat. Acad. Sci. USA, 1977, 74, 1716.
[15] P.E. Cross, R.P. Dickinson, M.J. Parry, M.J. Randall, Agents and Actions, 1981, 11, 274.
[16] M.J. Randall, M.J. Parry, E. Hawkeswood, P.E. Cross, R.P. Dickinson, Throm. Res., 1981, 23, 145.
[17] H.-H. Tai, B. Yuan, Biochem. Biophys. Res. Comms., 1978, 80, 236.
[18] T. Yoshimoto, S. Yamamoto, O. Hayaishi, Prostaglandins, 1978, 16, 529.
[19] A.L. Johnson, J.C. Kauer, D.C. Sharma, R.I. Dorfman, J. Med. Chem., 1969, 12, 1024.

[20]C.F. Wilkinson, K. Hetnarski, T.O. Yellin, Biochem. Pharmacol., 1972, 21, 3187.
[21]K.C. Leibman, E. Ortiz, Drug Metab. Dispos., 1973, 1, 775.
[22]C.F. Wilkinson, K. Hetnarski, L.J. Hicks, Pestic. Biochem. Physiol., 1974, 4, 299.
[23]C.F. Wilkinson, K. Hetnarski, G.P. Cantwell, F.J. Di Carlo, Biochem. Pharmacol., 1974, 23, 2377.
[24]T.D. Rogerson, C.F. Wilkinson, K. Hetnarski, Biochem. Pharmacol., 1977, 26, 1039.
[25]D.L. Cinti, M.B. Feinstein, Biochem. Biophys. Res. Comms., 1976, 73, 171.
[26]J.A. Turner, W. Herz, Experientia, 1977, 33, 1133.
[27]M.J. Parnham, P.C. Bragt, A. Bast, F.J. Zijlstra, Biochim. Biophys. Acta, 1981, 677, 165.
[28]V. Ullrich, M. Haurand, Adv. Prostag. Thrombox. Leukotr. Res., 1983, 11, 105.
[29]V. Ullrich, L. Castle, P. Weber, Biochem. Pharmacol., 1981, 30, 2033.
[30]H. Graf, H.H. Ruf, V. Ullrich, Angew. Chem. Int. Ed. Engl., 1983, 22, 487.
[31]D.L. De Witt, W.L. Smith, J. Biol. Chem., 1983, 258, 3285.
[32]R.J. Gryglewski, A. Zmuda, A. Dembinska-Kiec, E. Krecioch, Pharmacol. Res. Comms., 1977, 9, 109.
[33]R.J. Gryglewski, A. Zmuda, R. Korbut, E. Krecioch, K. Bieron, Nature, 1977, 267, 627.
[34]T.E. Temple, G.W. Liddle, Ann. Rev. Pharmacol., 1970, 10, 199.
[35]N. Kayama, K. Sakaguchi, S. Kaneko, T. Kubota, T. Fukuzawa, S. Kawamura, T. Yoshimoto, S. Yamamoto, Prostaglandins, 1981, 21, 543.
[36]K. Iizuka, K. Akahane, D. Momose, M. Nakazawa, T. Tanouchi, M. Kawamura, I. Ohyama, I. Kajiwara, Y. Iguchi, T. Okada, K. Taniguchi, T. Miyamoto, M. Hayashi, J. Med. Chem., 1981, 24, 1139.
[37]H.M. Tyler, C.A.P.D. Saxton, M.J. Parry, Lancet, 1981, 1, 629.
[38]M.J. Parry, M.J. Randall, H.M. Tyler, E. Myhra, J. Dale, E. Thaulow, Lancet, 1982, 2, 164.
[39]G.A. Fitzgerald, A.R. Brash, A.K. Pedersen, Thromb. Haemostasis, 1983, 50, 280.
[40]M.J. Parry, M.J. Randall, E. Hawkeswood, P.E. Cross, R.P. Dickinson, Brit. J. Pharmacol., 1982, 77, 547P.
[41]H.H. Tai, C.L. Tai, N. Lee, Arch. Biochem. Biophys., 1980, 203, 758.
[42]T. Tanouchi, M. Kawamura, I. Ohyama, I. Kajiwara, Y. Iguchi, T. Okada, T. Miyamoto, K. Taniguchi, M. Hayashi, K. Iizuka, M. Nakazawa, J. Med. Chem., 1981, 24, 1149.

Agonists, Analogs, and Modulators of Arachidonic Acid Cascade as Therapeutic Agents

By U. Axen

THE UPJOHN COMPANY, KALAMAZOO, MICHIGAN 49001, U.S.A.

Almost twenty-five years have passed since the first prostaglandins (PGs), namely prostaglandin E_1 (PGE_1) and prostaglandin $F_1\alpha$ ($PGF_1\alpha$), have been isolated from sheep prostate glands and obtained in crystalline form[1]. Since then the field has matured scientifically and today we are using the arachidonic acid cascade in many areas as the basis for research as a well-defined biochemical/physiological concept. Right from the beginning pharmaceutical chemists, biochemists, pharmacologists and physicians have been enticed by the promise of therapeutic applications resulting from this fascinating development. Scientifically, prostaglandin and arachidonic acid cascade research have outpaced the most optimistic expectations; almost 70 metabolites (prostaglandins, thromboxanes, leukotrienes, hydroxylated fatty acids) of arachidonic acid have been identified; to that we have to add metabolites of the corresponding eicosa-trienoic and - pentaenoic acids. The pathways of their biogenesis and biodegradation are well defined, we know much about their physiological, pathophysiological and pharmacological actions. Practical chemical methods are in hand to produce essentially all of these compounds in desired amounts and to synthesize essentially any analog one reasonably can imagine. This knowledge and know-how is available in a library of,by now,over 50,000 publications concerning prostaglandins, other eicosanoids and their modulators.

As much as the science of eicosanoids has outpaced early prognoses, almost as much have therapeutic applications lagged behind early promise and widely publicized expectations. Why? Were our expectations unjustified? I don't believe so. Rather, I believe we have overestimated our present ability to translate laboratory findings into therapeutic modalities. Certainly, early optimism that prostaglandins or eicosanoid modulators would prove to be magic bullets whose therapeutic applications would become self-evident once they reached the clinic, has not been fulfilled; at least not yet. More likely, their full value as therapeutic modalities will have to be elucidated through meticulous and tedious hard work, often requiring development of new methods and tools as their evaluation progresses. We still are lacking many of the necessary tools, i.e. relevant pharmacological models, diagnostic precision, and well-defined predictive clinical endpoints, especially in those areas where theoretically, at least, eicosanoids could deliver the most dramatic and perhaps the most needed breakthroughs. Obviously, breakthroughs are desired most in areas where effective drug treatment presently is not available; but this very fact, i.e. the absence of proven effective treatment, makes development of efficacious therapeutic agents difficult and labyrinthian: positive controls are missing for development and verification of laboratory models, proven and generally accepted clinical methods for evaluation of efficacy are lacking, definitions of homogeneous patient populations are often inadequate, and lack of alternative treatment often poses ethical questions regarding the use of placebo in scientifically designed controlled studies in severe diseases where preliminary open trials seem to have shown positive results.

Not surprisingly, such difficulties were largely absent in the area where prostaglandins first reached product status. The first prostaglandin approved for human use was $PGF_2\alpha$, which is marketed under the trademark PROSTIN F2 ALPHA® for termination of pregnancy. Other prostaglandins followed for related indications: PGE_2 (PROSTIN E2®) and (15S)-15-methyl-$PGF_2\alpha$-THAM (PROSTIN/15 M®), the first marketed synthetic analog of a natural prostaglandin[2]. Recently, Schering and Ono have introduced synthetic analogs, N-methanesulfonyl-16-phenoxy-ω-tetranor-PGE_2 amide and 16,16-dimethyl-*trans*-Δ^2-PGE_1 methyl ester, respectively, for obstetrical/gynecological uses. A new generation of prostaglandin analogs is under clinical evaluation for cervical dilation, menstrual regulation and related indications; Upjohn's entry is a chemically stable PGE_2 analog, 9-deoxo-9-methylene-16,16-dimethyl-PGE_2, which combines the more favorable therapeutic ratio of E_2-type prostaglandins with the chemical stability of $PGF_2\alpha$ analogs[3].

For acute clinical uses of prostaglandins in obstetrics and gynecology, the underlying physiology is well understood, diagnosis is generally straightforward, endpoints for clinical efficacy are generally unequivocal and laboratory models and assays are predictive. Accordingly, prostaglandin therapy in this area is rapidly maturing. In another area where clinical evaluation of prostaglandins is well advanced, although product status has not yet been achieved, similar criteria apply, although not to the same degree. The ability of natural prostaglandins (PGE_1, PGA_1) to inhibit gastric acid secretion stimulated by food, histamine, pentagastrin, and other stimuli was first discovered in rats[4] and dogs[5]. This effect was confirmed in man[6,7]; however, the naturally occurring prostaglandins are only active when given by continuous intravenous infusion. Fortunately, it turned out that prostaglandin analogs methylated at the 15 or 16 position showed inhibition of acid secretion after oral administration[8-11]. (15R)-15-Methyl-PGE_2 (Arbaprostil, ARBACET®)[12], is under extensive clinical development by Upjohn. In man 100 mcg of ARBACET, given orally, significantly reduced meal-stimulated acid secretion for over 3 hours[13]; reduction of acid secretion occurred in a dose response fashion[14]. Healing of gastric ulcers with (15R)-15-methyl-PGF_2 had been reported earlier[15]. In a recent multi-center double-blind clinical trial on 173 patients with active, endoscopically proven duodenal ulcers, ARBACET, 100 mcg orally 4 times a day, was compared with placebo[16]. After 14 days, 37% of the patients receiving drug were completely healed compared to 12% in the placebo group. After 28 days the ratio was 67% to 39%. Recently, three cases have been reported where ARBACET (50 mcg every 6 hours) had been given to patients with severe upper gastrointestinal hemorrhage, refractory to cimetidine, high dose antacid and vasopressin[17-19]. Beneficial effects were noted after 14 hours and significant or complete healing was observed after 10 to 14 days of therapy.

The ulcer healing effects of prostaglandins are thought to be the result of their ability to inhibit gastric secretion. But many prostaglandins, even many of those which do not inhibit acid secretion, have the additional ability to protect the stomach and the intestine from becoming inflamed or necrotic when exposed to noxious agents, e.g. non-steroidal anti-inflammatory drugs, alcohol, hydrochloric acid and others[20,21]. This phenomenon, pioneered and developed by A. Robert, has been coined "cytoprotection". It occurs at much lower doses of prostaglandins than those needed for inhibition of acid secretion and has been demonstrated in man with PGE_2[22-26] and ARBACET[27]. The implications for therapeutic and prophylactic treatment are obvious.

An area where translation of laboratory findings into clinical applications did not proceed as straightforwardly is the cardiovascular area. That prostaglandins lower blood pressure was known from the beginning; in fact, that property contributed to their discovery. In the mid-sixties it was found that PGE_1 inhibits aggregation of platelets[28-30]; in fact, PGE_1 was the most potent known inhibitor of platelet aggregation until the advent of prostacyclin. In spite of these early discoveries only two prostaglandins have reached the stage of marketed drugs, namely PGE_1 and prostacyclin. PGE_1 (PROSTIN VR Pediatric™ Sterile Solution) is marketed for maintaining patency of the ductus arteriosus until corrective surgery can be performed in neonates with congenital heart defects, and recently prostacyclin has been introduced by Upjohn (CYCLO-PROSTIN® Sterile Powder) and Wellcome for use in procedures where extracorporeal circulation is necessary: cardiopulmonary by-pass, charcoal hemoperfusion and renal dialysis. In both cases approved indications closely correspond to findings in laboratory models and clinical endpoints which provided the basis for regulatory approval, were readily measurable in acute clinical trials. Coceani and Olley published in 1973 that PGE_1 and PGE_2 dilate the ductus arteriosus of lamb[31]. This suggested that PGE_1 and PGE_2 may be useful in neonates with congenital heart defects where patency of the ductus is vital until corrective surgery can be performed. Clinical efficacy could be measured directly by monitoring blood oxygenation; therefore, optimal doses could be established relatively readily[32-36]. Similarly, the key endpoint used for prostacyclin in extracorporeal circulation was the same as the one used in the laboratory in in vitro studies, namely the inhibition of platelet aggregation. Optimal doses could be established rather readily in acute trials by monitoring effects on platelet aggregation and blood pressure[37].

The story becomes more complex for the one additional cardiovascular indication for which a prostaglandin is marketed. PGE_1 is marketed in Japan for treatment of arteriosclerosis obliterans and thromboangiitis obliterans (Buerger's disease) by intraarterial administration. In a multi-clinic double-blind study high dose PGE_1 (0.15 nanograms/kg/min for 2-6 weeks by continuous intraarterial infusion) was compared with low dose PGE_1 (0.05 nanograms/kg/min) and inositol niacinate in 65 patients[38]. Improvement of ischemic ulcers was judged to be superior in the high dose group when compared with the low dose group and the inositol niacinate group;

however, statistical significance was only reached when the high dose PGE_1 group was compared with the combined low dose PGE_1 and inositol niacinate groups.

In a recent prospective, randomized, double-blind multi-center study we were not able to demonstrate efficacy of intravenously administered PGE_1 in the treatment of ischemic leg ulcers due to peripheral vascular occlusive disease (PVOD)[39]. One hundred and twenty patients were randomized into treatment (57) and placebo (63) groups. Both groups were matched according to age, race, sex, complications such as diabetes or hypertension and number of ulcers per patient. Approximately 50% improvement, i.e. decreased ulcer size or healing of ulcers, was observed in both the PGE_1 treatment group and the placebo group. The same was true when subpopulations were analyzed separately, namely diabetic patients, non-diabetic patients, patients with distal lesions and patients with proximal lesions. Ulcer healing had been chosen as the key endpoint because, short of gangrene, ischemic ulcers are the most severe manifestation of PVOD and healing of ulcers can be measured more objectively than alleviation of rest pain or claudication.

The unexpected negative result of this study raises several questions. First of all, was the intravenous route the correct route? It is known that PGE_1 is rapidly metabolized in the lung; the intraarterial route, therefore, would have been more logical. However, several reports in the literature, describing studies with small number of patients, suggested that PGE_1 given intravenously may be beneficial in the treatment of PVOD[40]. In view of these reports it was considered difficult to convince investigators to conduct a placebo controlled study where placebo was to be given for 72 hours up to three times by intraarterial infusion to 50% of the patients. Secondly, was the disease advanced too far for drug treatment when manifested by ischemic ulcers? Are alleviation of rest pain or claudication more realistic endpoints? Certainly a reasonable thesis as far as the disease process is concerned. However, these endpoints are very subjective and the palcebo rate can easily reach proportions where it becomes impossible to demonstrate superiority of drug treatment with a manageable number of patients. The third question relates to dose and dose regimen. Normally, if efficacy cannot be demonstrated at the highest tolerated dose given as long as possible in acceptable medical practice, it is unlikely that a lower dose will prove to be efficacious. However, this is not always the case; if the target system becomes refractory with time in a dose related fashion, a lower dose and/or shorter duration of administration may be more effective; or, intermittent high dose administration may be more effective than continuous low dose infusion. For instance, Sinzinger et al.[41] and Dembinska-Kiec et al.[42] reported that platelets become less sensitive to prostacyclin during continuous long-term infusion and a rebound phenomenon may even occur after discontinuation of long-term infusion. Detailed dose finding studies, obviously, answer such questions if unambiguous results can be obtained in short-term acute trials; however, they are ruled out as a practical matter if large numbers of patients and months or years of treatment and observation are required to establish efficacy. Fourth and last, the purpose of preliminary, open, uncontrolled, small patient studies should be addressed in this context. These preliminary exploratory studies are necessary and have their value. They stimulate the evaluation of existing and new drugs for new, unexplored indications; but they need to be taken as suggestions only for future clinical research. Definitive conclusions can rarely be drawn in small patient studies in areas where placebo effects may run 50% or higher. If overinterpreted and mistaken as proof of efficacy, preliminary accounts may make extensive definitive trials difficult or impossible; because of ethical considerations clinical investigators may find it unacceptable to compare placebo with a new drug already reported as efficacious.

These considerations apply in various degrees to most of the cardiovascular indications which have been discussed for prostaglandins. The list is too long to mention here; the more significant or promising indications include Raynaud's phenomenon, angiographic enhancement, skin graft healing, acute respiratory distress, pulmonary hypertension, unstable and stable angina, congestive heart failure, arterial and veinal thrombosis, ischemic stroke and atherosclerosis. PGE_1, prostacyclin and several prostaglandin analogs are being explored clinically for many of these indications. However, in many instances challenges are even greater than with PVOD: many of these diseases are less well defined and it is more difficult to diagnose homogeneous patient populations for clinical trials. Obviously, if a clinical trial population includes an unresponsive subgroup the total number of patients needs to be increased to reach statistical significance for an efficacy decision.

In addition, many of these indications require chronic treatment and/or long-term follow-up, requiring large investments before clinical efficacy can be established. Expeditious evaluation and development of prostaglandins for cardiovascular indications require a four-pronged attack: pharmaceutical development of the drug entities; development of laboratory models relevant for target diseases which reduce the need for large numbers of exploratory clinical studies and which encourage investments in long-term clinical trials because of their predictive value; development of better diagnoses which assure more homogeneous patient populations; and development of quantifiable endpoints which can be measured in acute studies and are accepted as predictive for long-term beneficial effects such as prevention or retardation of atherosclerosis or ischemic diseases of the heart or brain.

Obviously, this is also true for arachidonic acid cascade modulators to the extent they represent novel types of therapeutic agents. The utility of non-steroidal antiinflammatory drugs and, therefore, methods for their clinical evaluation, were well established before their mechanism of action, i.e. inhibition of the elaboration of arachidonic acid into prostaglandins, became known[43]. But two other groups of modulators which are presently of high interest, namely inhibitors of thromboxane and leukotriene synthetases, undoubtedly will face similar challenges as the prostaglandins. Thromboxane synthetase inhibitors have the theoretical advantage over cyclooxygenase inhibitors that they prevent the formation of thromboxane A_2 only, while not interfering with the formation of prostacyclin. For certain applications they also have a potential advantage over prostacyclin-type agonists: they may prove to have similar effects as prostacyclin on the platelet system without the profound vasodilator action seen with prostacyclin and its biologs. Johnson and his colleagues recently have synthesized a selective thromboxane synthetase inhibitor, the sodium salt of 5-(3'-Pyridinylmethyl) – benzofuran-2-carboxylic acid, U-63557A, which has shown interesting pharmacological and pharmacodynamic properties in laboratory models[44]. Gorman, Aiken and Spilman found that oral administration to Rhesus monkeys, 10 mg/kg twice a day, reduced ex vivo arachidonic acid or collagen stimulated thromboxane synthesis by 75% of control. Like prostacyclin, U-63557A prevented platelet aggregation in the partially obstructed circumflex coronary artery in the dog; this effect was reversed by indomethacin, indicating that the prevention of platelet aggregation was at least partially due to changing the endogenous ratio of prostacyclin to thromboxane. The same observations were made when U-63557A was given orally. The challenge now is to find or develop laboratory models which provide guidance for translating these exciting findings into therapeutic applications.

While the cyclooxygenase branch of the arachidonic acid cascade has been the subject of extensive research for over ten years, the lipoxygenase branch only recently has moved into the limelight with the discovery that slow reacting substance of anaphylaxis (SRS-A) is a mixture of certain arachidonic acid metabolites, now called leukotrienes[45-49]. The recognition that SRS-A is a part of the arachidonic acid cascade suggested prostaglandin-type structures as a possible source of selective inhibitors. Bach, Smith and colleagues found that 6,9-deepoxy-6,9-(phenylimino)-$\Delta^{6,8}$-PGI_1, U-60257, inhibits leukotriene C and D synthesis in ionophore-challenged peritoneal mononuclear cells from rat; the same was found in sensitized and then antigen-challenged human lung fragments[50,51]. Neither release of histamine nor formation of thromboxane (cyclooxygenase pathway) were affected at concentrations of U-60257 which inhibited formation of leukotrienes C and D. The precise site or sites of action of U-60257 are still being elucidated; however, Sun and McGuire recently found that U-60257 inhibits the 5-lipoxygenase of human polymorphonuclear leukocytes at micromolar concentrations, demonstrating that this first step of the leukotriene pathway is at least one of the sites being inhibited by U-60257[52]. In addition, U-60257 also seems to be a competitive leukotriene antagonist as could be demonstrated on the superfused guinea pig ileum[50].

SRS-A is believed to play an important mediator role in asthmatic reactions[53], suggesting the application of leukotriene inhibitors for alleviation of symptoms of asthma and related diseases. In at least one pharmacological model of allergy, the Ascaris-sensitive Rhesus monkey, Johnson et al. found that U-60257 effectively inhibited the bronchopulmonary effects normally elicited by Ascaris antigen challenge[54].

But leukotriene inhibitors may find broader applications than the treatment of asthmatic and other allergic symptoms. In the short period since their characterization reports have appeared which implicate leukotrienes in acute respiratory distress syndrome[55], bronchitis[56], cystic fibrosis[57], arthritis[58-60], inflammatory bowel disease[61], and psoriasis[62]. Obviously, we are just at the beginning of a whole new era of physiological, pathophysiological and pharmacological discoveries.

Other potential therapeutic applications for agonists and modulators of the arachidonic acid cascade are emerging. To mention only two: Lund et al. discovered that continuous infusion of PGE_1 in dog caused thickening of the long bone of the limb, a finding which deserves further exploration[63], and, in 1981, a symposium was organized which tried to bring together the numerous and sometimes contradictory reports about possible roles of prostaglandins and modulators in cancer therapy[64].

Medicinal chemists will have the continued challenge and opportunity to design arachidonic acid cascade analogs and modulators with improved therapeutic ratios and better pharmaceutical and pharmacodynamic properties; biochemists undoubtedly will expand and refine our knowledge about the pathways of arachidonic acid metabolism; but the most formidable challenge rests with pharmacologists and clinicians to define therapeutic roles of these agonists and modulators, and in the process to invent, design and develop the necessary models and methods.

References

[1] S. Bergström and J. Sjövall, Acta Chem. Scand., 1960, 14, 1693, 1701, 1706.

[2] For a recent review see "The Importance of Prostaglandins in Obstetrics and Gynecology", M. Bygdeman, V. Lundström and U. Ulmsten, Editors, Acta Obstet. Gynecol. Scand., 1983, Supplement 113.

[3] F. A. Kimball, G. L. Bundy, A. Robert and J. R. Weeks, Prostaglandins, 1979, 17, 657.

[4] A. Robert, J. E. Nezamis and J. P. Phillips, Gastroenterology, 1968, 55, 481.

[5] J. E. Nezamis, A. Robert and D. F. Stowe, J. Physiol. (Lond.), 1971, 218, 369.

[6] D. E. Wilson, C. Phillips and R. A. Levine, Gastroenterology, 1971, 61, 201.

[7] M. Classen, H. Koch, J. Bickhardt, G. Topf and L. Deming, Digestion, 1971, 4, 333.

[8]A. Robert, B. Nylander and S. Andersson, Life Sci., 1974, 14, 533.

[9]A. F. Ippoliti, J. I. Isenberg, V. Maxwell and J. H. Walsh, Gastroenterology, 1976, 70, 488.

[10]F. W. K. Chen, H. S. Teck and S. M. M. Karim, Prostaglandins, 1977, 13, 115.

[11]S. J. Konturek, N. Kwiecien, J. Swiereczek, J. Oleksy, E. Sito and A. Robert, Gastroenterology, 1976, 70, 683.

[12]E. W. Yankee, U. Axen and G. L. Bundy, J. Am. Chem. Soc., 1974, 96, 5865.

[13]W. Peterson, M. Feldman, I. Taylor and M. Bremer, Dig. Dis. Sci., 1979, 24, 381.

[14]A. Robert, G. Kane and S. B. Reele, Gut, 1981, 22, 728.

[15]W. P. Fung and S. M. M. Karim, Med. J. Aust., 1976, 2, 127.

[16]G. Vantrappen, J. Janssens, T. Popiela, J. Kulig, G. N. J. Tytgat, K. Huibregtse, R. Lambert, J. P. Pauchard and A. Robert, Gastroenterology, 1982, 83, 357.

[17]J. B. Weiss, G. W. Peskin and J. I. Isenberg, Gastroenterology, 1982, 82, 558.

[18]J. S. Groeger, S. J. Dazza, G. C. Carlon, A. D. Turnbull, M. K. Pierri and W. S. Howland, Critical Care Medicine, 1982, 10, 486.

[19]I. Grannikopoulos, P. Ringel, D. Zai, R. Berko, P. E. Lim, R. McCandlis, D. Griffith and P. Bright-Asare, Gastrointestinal Endoscopy, 1982, 28, 127.

[20]A. Robert, J. E. Nezamis, C. Lancaster and A. J. Hanchar, Gastroenterology, 1979, 77, 433.

[21]A. Robert, Gastroenterology, 1979, 77, 761.

[22]M. M. Cohen, Lancet, 1978, ii, 1253.

[23]M. M. Cohen, G. Cheung and D. M. Lyster, Gut, 1980, 21, 602.

[24]M. M. Cohen, Prostaglandins, 1981, 21, 155.

[25]C. Johansson, B. Kollberg, R. Nordeman and S. Bergström, Lancet, 1979, i, 317.

[26]J. N. Hunt and D. R. Franz, Dig. Dis. Sci., 1981, 26, 301.

[27]D. A. Gilbert, A. D. Feld, F. E. Silverstein, C. Weinberg and D.R. Saunders, Gastroenterology, 1981, 80, 1155.

[28] J. Kloeze, in "Nobel Symp. 2, Prostaglandins", S. Bergström and B. Samuelsson, Editors, Almqvist and Wiksell, Stockholm, 1967, 241.

[29] N. Chandrasekhar, Blood, 1967, 30, 554.

[30] P. R. Emmons, J. R. Hampton, M. J. G. Harrison, A. J. Honour and J. R. A. Mitchell, Br. Med. J., 1967, 2, 468.

[31] F. Coceani and P. M. Olley, Can. J. Physiol. Pharmacol., 1973, 51, 220.

[32] M. D. Freed, M. A. Heymann, A. B. Lewis, S. L. Roehl and R. C. Kensey, Circ., 1981, 64, 899.

[33] M. A. Heymann, Am. Heart J., 1981, 101, 837.

[34] P. M. Olley and F. Coceani, Ann. Rev. Med., 1981, 32, 375.

[35] L. A. Linday and M. A. Engle, Ped. Ann., 1981, 10, 29.

[36] A. R. Levin, Ped. Ann., 1981, 10, 16.

[37] For a recent review see O. I. Linet, Postgraduate Medicine, 1982, 72, 105.

[38] S. Sakaguchi, A. Kusaba, Y. Mishima, K. Kamiya, A. Nishimura, K. Furukawa, S. Shionoya, M. Kawashima, T. Katsumura and A. Sakuma, Vasa, 1978, 7, 263.

[39] J. J. Schuler, D. P. Flanigan, J. W. Holcroft, J. J. Ursprung, J. S. Mohrland and J. Pyke (manuscript in preparation).

[40] For recent summary articles and references see "Prostaglandins in Clinical Medicine, Cardiovascular and Thrombotic Disorders", K. K. Wu and E. C. Rossi, Editors, Year Book Medical Publishers, Inc., Chicago, London, 1982.

[41] H. Sinzinger, K. Silberbauer, A. K. Horsch and A. Gall, Prostaglandins, 1981, 21, 49.

[42] A. Dembinska-Kiec, A. Zmuda, L. Grodzinska, K. Biedron, M. Basista, A. Kedzior, E. Kostka-Trabka, E. Telesz and T. Zelazny, Prostaglandins, 1981, 21, 827.

[43] J. R. Vane, Nat. New Biol., 1971, 231, 232.

[44] R. R. Gorman, R. A. Johnson, J. W. Aiken and C. H. Spilman, Prostaglandins, (in press).

[45]R. C. Murphy, S. Hammarström and B. Samuelsson, Proc. Natl. Acad. Sci. USA, 1979, 76, 4275.

[46]H. R. Morris, G. W. Taylor, P. J. Piper, M. N. Samhoun and J. R. Tippins, Prostaglandins, 1980, 19, 185.

[47]L. Örning, D. Widegran, S. Hammarström and B. Samuelsson, Proc. Natl. Acad. Sci. USA, 1980, 77, 2014.

[48]M. K. Bach, J. R. Brashler, S. Hammarström and B. Samuelsson, J. Immunol. 1980, 125, 115.

[49]M. K. Bach, J. R. Brashler, S. Hammarstrom and B. Samuelsson, Biochem. Biophys. Res. Commun., 1980, 93, 1121.

[50]M. K. Bach, J. R. Brashler, H. W. Smith, F. A. Fitzpatrick, F. F. Sun and J. C. McGuire, Prostaglandins, 1982, 23, 759.

[51]H. W. Smith, M. K. Bach, A. W. Harrison, H. G. Johnson, N. J. Major and M. A. Wasserman, Prostaglandins, 1982, 24, 543.

[52]F. F. Sun and J. C. McGuire, Prostaglandins, 1983 (in press).

[53]W. E. Brocklehurst, Adv. Drug. Res., 1970, 5, 109.

[54]H. G. Johnson, M. L. McNee, M. K. Bach and H. W. Smith, Int. Archs Allergy appl. Immun., 1983, 70, 169.

[55]M. L. Ogletree, J. R. Snapper and K. L. Brigham, Am. Physiol. Soc., Fall 1982, abstracts.

[56]L. W. Turnbull, L. S. Turnbull, J. Crofton and A. B. Kay, Lancet, 1978, 2, 184.

[57]O. Cromwell, M. J. Walport, H. R. Morris, G. W. Taylor, M. E. Hodson, J. Batten and A. B. Kay, Lancet, 1981, 2, 164.

[58]R. A. Lewis, J. M. Drazen, J. C. Figueiredo, E. J. Corey and K. F. Austen, Int. J. Immunopharmacol., 1982, 4, 85.

[59]L. B. Klickstein, C. Shapleigh and E. J. Goetzl, J. Clin. Invest., 1980, 66, 1166.

[60]E. M. Davidson, S. A. Rae and M. J. H. Smith, J. Pharm. Pharmacol., 1982, 34, 410.

[61]W. F. Stenson and E. Lobos, J. Clin. Invest., 1982, 69, 494.

[62]S. D. Brain, R. D. R. Camp, P. M. Dowd, A. K. Black, P. M. Woolard, A. I. Mallet and M. W. Greaves, Lancet, 1982, 2, 762.

[63] J. E. Lund, W. P. Brown and L. Tregerman, in "Prostaglandins in Clinical Medicine, Cardiovascular and Thrombotic Disorders", K. K. Wu and E. C. Rossi, Editors, Year Book Medical Publishers, Inc., 1982, 93.

[64] See "Prostaglandins and Cancer: First International Conference", T. J. Powles, R. S. Bockman, K. V. Honn, P. Ramwell, Editors, Prostaglandins and Related Lipids, Vol. 2, A. R. Liss, Inc., New York, 1982.

Synthesis of Some 13-Azaprostanoids and their Actions at Thromboxane Receptors

By E. W. Collington*, H. Finch, P. Hallett, P. Hunt, T. Parkhouse, D. Reynolds, L. M. Smith, and C. J. Wallis

GLAXO GROUP RESEARCH LTD., WARE, HERTS. SG12 0DJ, U.K.

Our interest in prostaglandins at Glaxo started a number of years ago as a consequence of some collaborative work with the University of Salford. The collaboration was concerned with the development of a new synthetic route to the natural prostanoids by making use of the stereochemical control offered by the rigid tricyclo[3.2.0.0.^{2,7}]heptanone system (1) in reactions with nucleophiles. Regiospecific attack at C1 of the tricyclic ketone (1) with simple nucleophiles (R^- = cyanide, ethoxide) results in cleavage to give the norbornanones (2) in good yield[1]. The successful extension of this homoconjugate addition reaction to include organometallic reagents, in particular the heterocuprate reagent (3) derived from the natural prostanoid β-chain, enabled the preparation of the norbornanone (2a) which was subsequently elaborated in 4 steps into $PGF_{2\alpha}$[2] (Scheme 1).

The possibility that Michael type attack at C1 on the tricyclic ketone (1) might well be extended to other nucleophiles, in particular amines and sulphonamides, was to us attractive, since this could offer a relatively short, stereoselective synthesis of a variety of novel 13-aza-prostanoids of potential biological interest.

Scheme 1

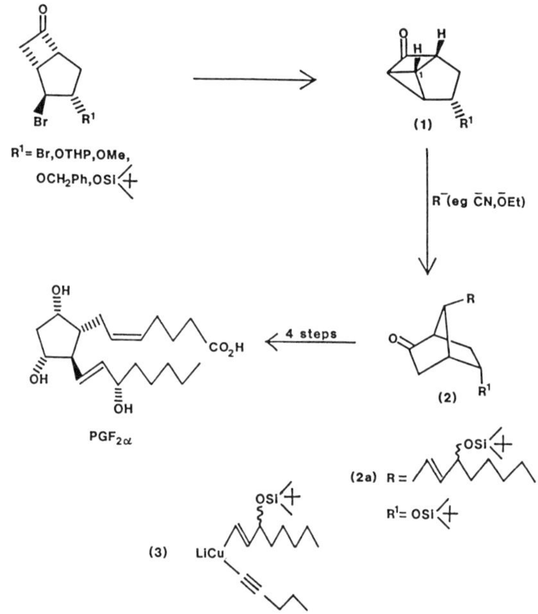

At the outset of this work a major consideration was to try and simplify some of the problems of stereochemistry and known sensitivity of the functional groups prevalent in prostaglandin synthesis. With this in mind we chose initially to make the benzyloxytricycloheptanone[3] (5) and examine its cleavage using morpholine (Scheme 2).

The readily available benzyloxybromide (4a), whose preparation has been previously described[1], was treated with potassium tert-butoxide at low temperature to give a >90% yield

Scheme 2

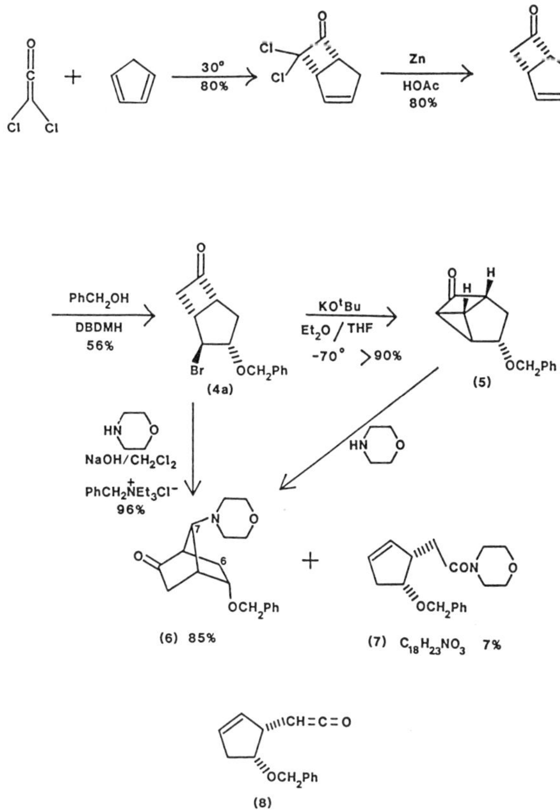

DBDMH = 1,3-Dibromo-5,5-dimethyl hydantoin

of the required tricyclic ketone (5) as a white crystalline solid, which could be stored indefinitely without decomposition at -20^0. Cleavage of the ketone (5) with an excess of morpholine at room temperature gave the bicyclo[2.2.1]heptan-2-one (6) in 85% yield. The configuration of the morpholino group in ketone (6) was established from its n.m.r. spectrum (C_6D_6) by the appearance of W coupling between the 6-endo proton and the 7-proton (J 1.2-1.5 Hz). In addition to ketone (6) there was also obtained from this reaction 7% of a compound of molecular formula $C_{18}H_{23}NO_3$, which we subsequently characterised as the cyclopentene-amide (7). We believe the product (7) is most likely formed via the ketene intermediate (8) derived by thermal ring cleavage of ketone (5). Although we made several unsuccessful attempts to isolate a ketene, we did observe that when a solution of the ketone (5) in carbon tetrachloride stocd for a few hours at room temperature it displayed a characteristic 2150cm^{-1} band in its i.r. spectrum.

The possibility that preformation and isolation of the tricyclic ketone (5) might not be essential, was substantiated when treatment of bromo-ether (4a) with an excess of morpholine, in acetone or in the presence of a sodium hydroxide-dichloromethane mixture under phase transfer catalysis, afforded the ketone (6) in 96% yield. Under these conditions formation of the amide (7) was not evident.

Baeyer-Villiger oxidation of the norbornanone (6) with buffered hydrogen peroxide was slow but gave rise as expected to two bicyclo[3.2.1]octanones (9a) and (10), from which the desired lactone (9a) was isolated by column chromatography in 26% yield. The subsequent use of commercial 40% peracetic acid in acetic acid-water as oxidant was found to be far superior, and as an added bonus the 'unwanted' lactone (10) was, in this medium, preferentially hydrolysed to its hydroxy-acid (12) making separation and isolation of the desired isomer (9a) relatively easy. Confirmation that lactone (9a) was isomerically pure was established by its cleavage with methanol-hydrogen chloride to the ester (11) and comparison of its g.l.c. retention time with that for the isomer (13) obtained by esterification of the hydroxy-acid (12). The conversion of aminonorbornanones of type (6) into their corresponding lactones turned out to be of some interest and is discussed separately later.

Scheme 3

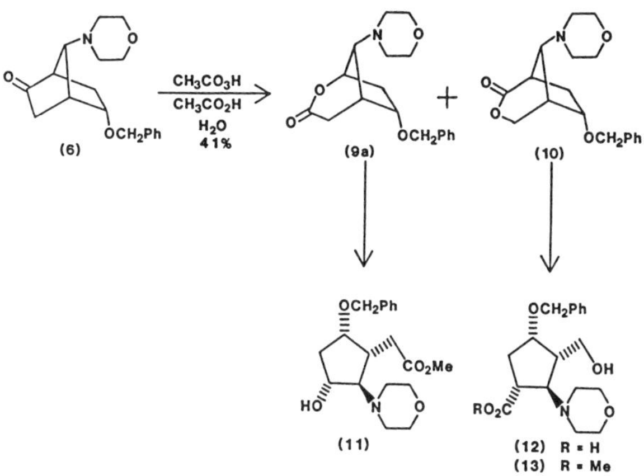

Partial reduction of the lactone (9a) using diisobutyl-aluminium hydride (dibah) produced the lactol (14) in quantitative yield which from n.m.r. and i.r. analysis was shown to exist predominantly as the open-chain tautomer (15). The aldehyde (15), without purification, was condensed with the ylide (16) (preformed from 4-carboxybutyltriphenyl phosphonium bromide using potassium tert-butoxide in tetrahydrofuran) and the crude product was esterified with methanol-sulphuric acid to give the prostanoid (18a). Purification of the prostanoid (18a) was best achieved by treating the crude mixture with acetic anhydride in pyridine, chromatographically isolating the pure acetate (17) formed, and then regenerating the hydroxy-prostanoid (18a) by treatment with potassium carbonate. G.l.c. data indicated that the prostanoid (18a) had a $C_{5,6}$ cis:trans isomer ratio of 9:1. These compounds (17) and (18a) could be further modified; viz. oxidation of alcohol (18a) using "activated" DMSO gave rise to the keto-ester (19), whereas saponification of ester (17) followed by oxidation with Jones reagent afforded the corresponding keto-acid (20). The relevance of these transformations becomes apparent from consideration of the biology which is summarised later.

Scheme 4

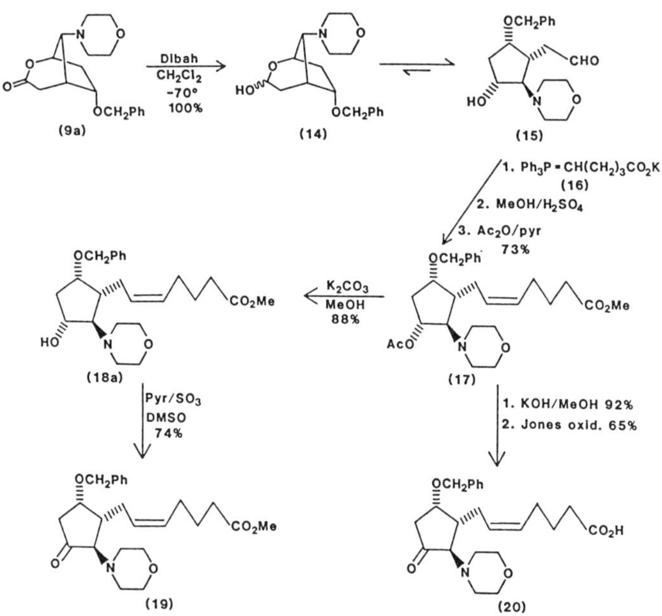

The overall synthetic route to 13-azaprostanoid (18a) from bicycloheptenone (21) is summarised in Scheme 5.

Scheme 5

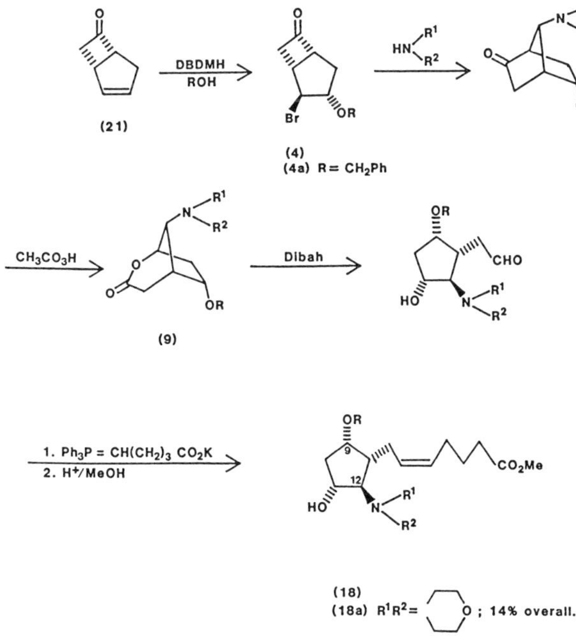

The synthesis proved particularly notable in several respects. It controls the stereochemistry at every centre on the cyclopentane ring, it proceeds in very good yield, and it utilises a versatile intermediate bromide (4). Thus, other bromides (4) which have been successfully employed include R=Me, COMe, CH_2OMe, THP and $SiMe_2{}^tBu$. Subsequent <u>in situ</u> cyclisation and cleavage have been examined with a diversity of both cyclic and acyclic secondary amines which have led ultimately to a wide variety of C9 and C12 (prostanoid numbering) modified azaprostanoids. A representative illustration is shown in Scheme 6. Noteworthy in this sequence is that Baeyer-Villiger oxidation of norbornanone (22) presumably leads initially to the hydroxy δ-lactone which undergoes spontaneous rearrangement to the γ-lactone (23) under the reaction conditions, thus allowing ready access into the 9-hydroxy series of 12-aminoprostanoids.

Scheme 6

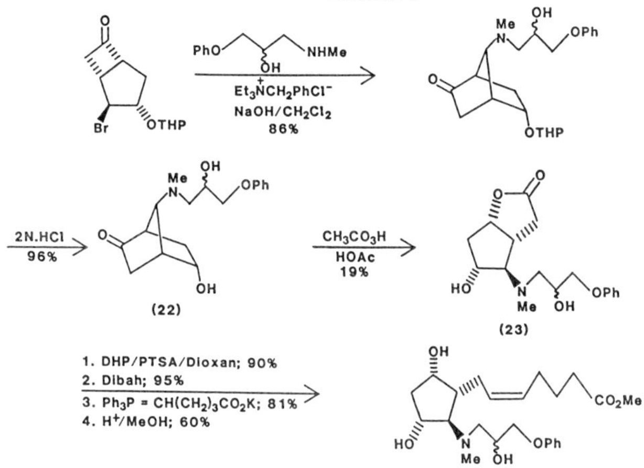

Encouraged by the successful exemplification of a number of appropriately functionalised aminoprostanoids and their accompanying biological activity, we extended our interest in the area by examining another group of nitrogen containing nucleophiles, namely the sulphonamides. Exposure of the bromo-ether (4b) to the readily available sulphonamides (24a) and (24b) in the presence of potassium tert-butoxide took the expected course to afford the corresponding norbornanones (25a) and (25b) in respectable yields. These norbornanones were converted by the standard reactions previously described into the azaprostanoids (26), the hydroxyl groups of which were oxidised to the ketones (27) using Jones reagent.

Scheme 7

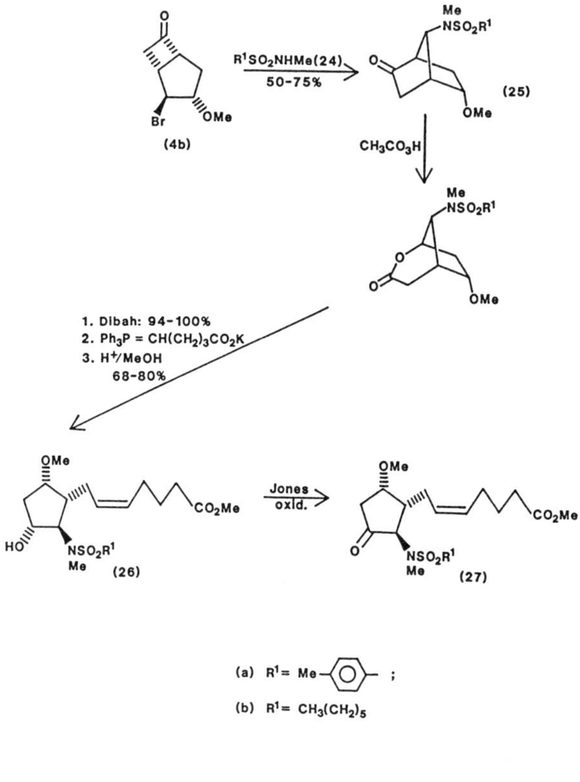
R¹SO₂NHMe(24)
50-75%
(4b)
(25)
CH₃CO₃H
1. Dibah: 94-100%
2. Ph₃P = CH(CH₂)₃CO₂K
3. H⁺/MeOH
68-80%
(26)
Jones oxid.
(27)
(a) R¹= Me-⟨O⟩- ;
(b) R¹= CH₃(CH₂)₅

As expected rearrangement of a sulphonamido δ-lactone proceeded smoothly into the corresponding γ-lactone which, as illustrated in Scheme 8, may serve as a key intermediate in the synthesis of both E- and A-ring type prostanoid analogues.

Scheme 8

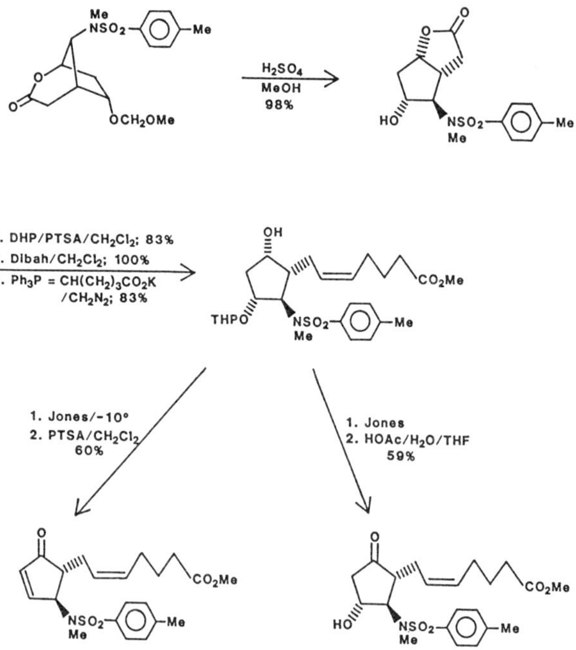

Although the above general sequence was clean and had the capability of providing a range of modified azaprostanoids in good overall yield, the introduction of each new aza moiety necessitated a relatively lengthy total synthesis. It therefore not only became desirable but clearly more economical to establish late stage synthons which would allow the selective introduction of a number of aza substituted β-side chains. Two such target structures given our attention were the amines (28) and (29).

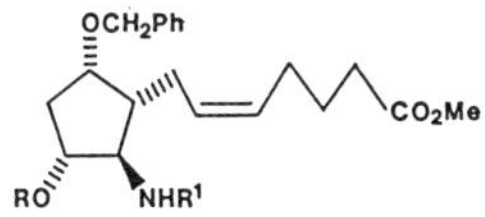

R = H or protecting group.

(28) $R^1 = Me$
(29) $R^1 = H$

The acetoxy methylamino synthon (28a) was obtained as the major component of a two part mixture (28a:31, 92:8) via treatment of the readily available protected amine[4] (30) with trichloroethyl chloroformate, followed by reduction of the resulting carbamate with zinc. The accompanying dichloro-urethane (31) presumably arises via attendant protonation of the intermediate zinc alkyl. Unfortunately, modifying the reaction conditions failed to have any effect on the production of this impurity. However, this did not preclude the use of the crude mixture since only the amine (28a) could be elaborated further (vide infra).

Scheme 9

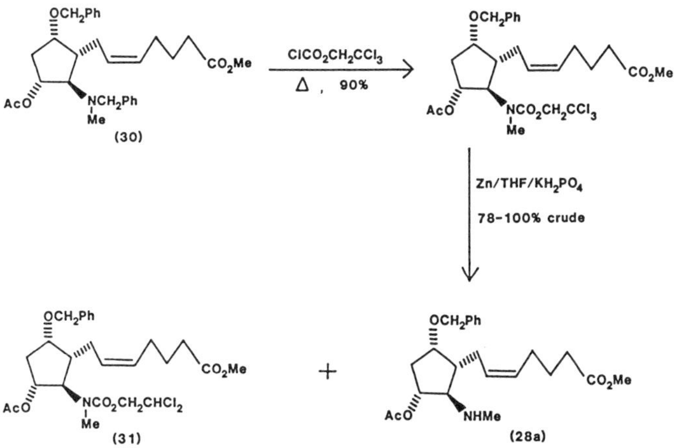

It was also envisaged that we would generate the amino group in structure (29) as the final step, and so our prime consideration became the choice of an appropriately disguised amino function which was not only capable of surviving exposure to all the reagents used in the general synthetic route, but was also capable of being readily deblocked at the appropriate time. The azido group appeared to fulfil these criteria (Scheme 10).

Scheme 10

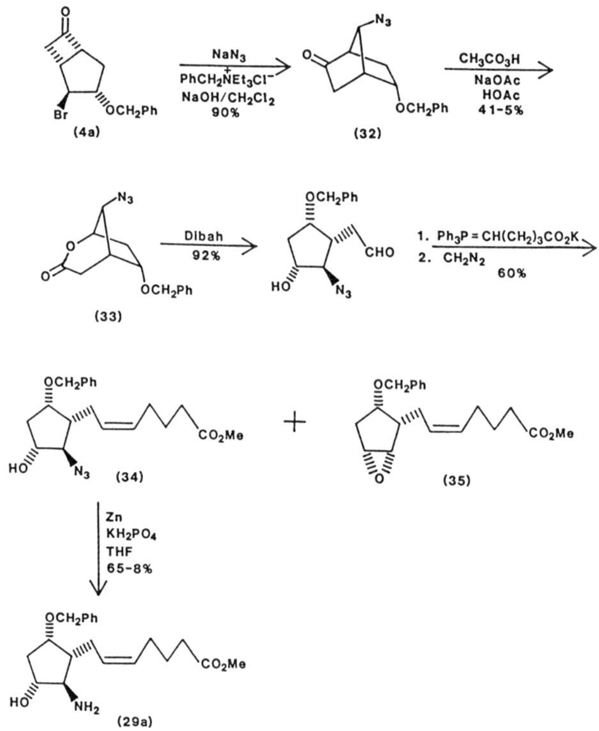

Conversion of the bromo-ether (4a) into the required norbornanone (32) was successfully carried out with sodium azide under strongly basic conditions in the presence of a phase transfer catalyst. The lactone (33), obtained from ketone (32) by treatment with buffered peracetic acid, was then converted into the azido prostanoid (34) using standard methodology. Interestingly, as well as the expected product from the Wittig reaction we also isolated small amounts (≤10%) of the epoxide (35), presumably formed by internal displacement of the azido group under the basic reaction conditions. Reduction of the azide (34) led to the required amino alcohol (29a).

The versatility of these amino prostanoids (28a, 29a) in synthesising a number of analogue types is demonstrated in Scheme 11. Thus, they may be used to prepare 15-hydroxy derivatives (36) by reaction with epoxides, sulphonamides (37) by reaction with sulphonyl chlorides, amides (38) by reaction with acids or acid chlorides, and, in the case of the primary amine, cyclic amino compounds (39) by reaction with α,ω-dibromo alkanes. Furthermore, these Δ^5-cis compounds (X=cis-CH=CH) can easily be reduced using hydrogen over palladium, or isomerised to the predominantly Δ^5-trans compounds either by using thiyl radicals generated thermally utilising thiophenol and the radical initiator azobisisobutyronitrile[5], or by using toluene-p-sulphinic acid[6] in dioxan.

Finally, another class of readily prepared synthon [representative structure (40)] proved extremely useful in providing an entry into analogues with variations at C9. Elaboration was carried out by established alkylation procedures (RBr/NaH/DMF) followed by acid catalysed removal of the protecting group (Scheme 12).

Scheme 11

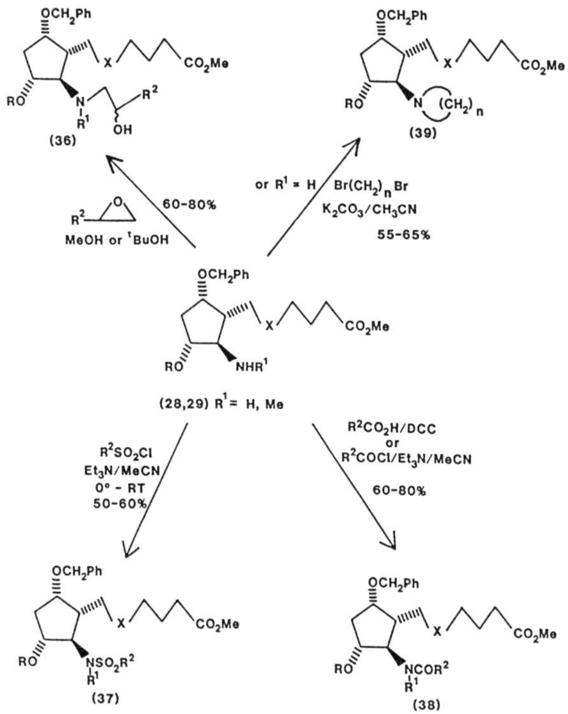

Scheme 12

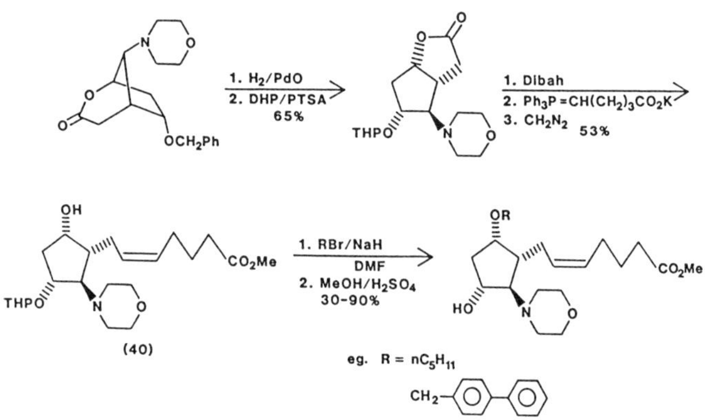

Rearrangement of 7-anti-aminonorbornanones

A particularly important step in our synthesis of 13-azaprostanoids is the Baeyer-Villiger oxidation since two ring expanded products are formed and only one is of use to us. The relative isomer ratios have been shown to be influenced by a number of factors but no general or specific pattern has emerged which would enable us to predict the outcome with any degree of certainty. However, the course of the conversion (for the aminonorbornanones) and the ready isolation of the desired isomer (9) in high purity is worthy of comment.

Under the standard conditions of the Baeyer-Villiger reaction the amino ketone (41) is fairly rapidly (<2h at 20^0) and cleanly converted initially into its N-oxide (42). This was confirmed by regeneration of the starting material on addition of an excess of sodium metabisulphite. The N-oxide (42) subsequently undergoes slow (12-24h) oxidative rearrangement to a mixture of lactone N-oxides (43) and (44), in which the isomer (44), in the aqueous acidic environment, suffers hydrolysis followed by elimination of a hydroxylamine to afford the α,β-unsaturated acid (47). On work-up through sodium metabisulphite solution the remaining N-oxide lactones are reduced to the amino lactones (9) and (45), and further stirring brings about selective and virtually complete hydrolysis of the lactone (45) to the amino acid (46), thus making isolation of the required lactone isomer (9) a simple matter.

The more rapid hydrolysis of the 3-oxabicyclooctanone system relative to the 2-oxabicyclooctanone system under these conditions has been rationalised on the basis of there being less severe congestion within the tetrahedral intermediate. If hydrolysis is assumed to occur by an $A_{AC}2$ mechanism[7] then

Scheme 13

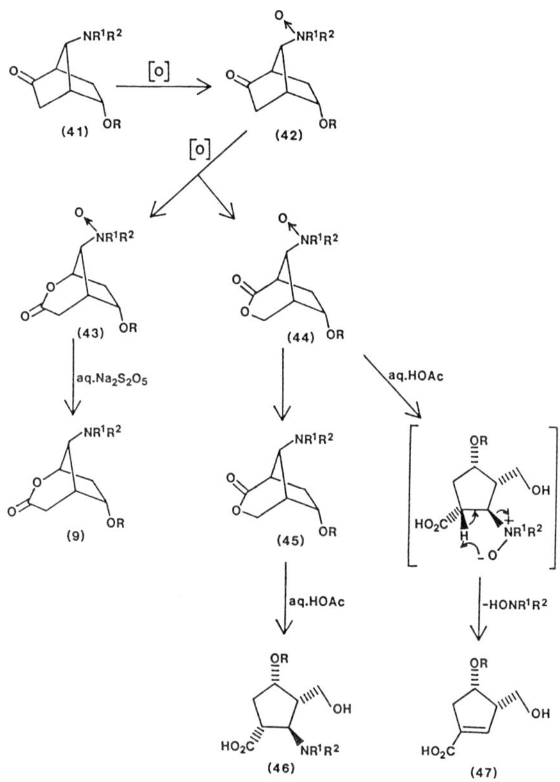

tetrahedral intermediates must be invoked which in their most stable conformations will be in the chair form. In intermediate B an unfavourable diaxial interaction occurs between a hydroxyl group and the ethylene bridge with its associated substituent. Steric crowding in intermediate A is less severe, and assuming this is also true for the corresponding transition state, accounts for the more rapid hydrolysis of lactones (44) and (45).

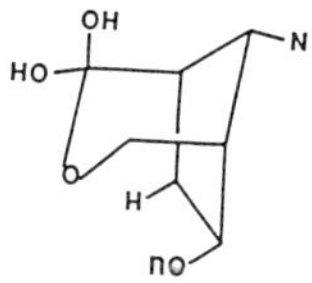

A

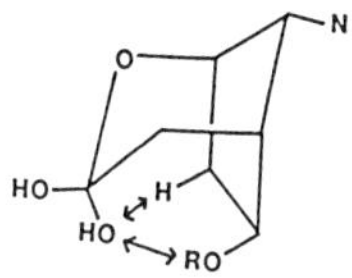

B

<u>Biological Results and Discussion</u>

Developments over the last few years have shown that two arachidonate metabolites, prostacyclin (PGI_2) and thromboxane A_2 (TxA_2), play a crucial role in the regulation of platelet-vessel wall interactions. TxA_2, which is the major product of platelet endoperoxide metabolism, is a very potent vasoconstrictor and platelet aggregating agent and is largely responsible for the irreversible, secondary phase of platelet aggregation. Consequently, antagonism of TxA_2-induced platelet aggregation has been proposed as a possible means of developing treatments for thrombotic disorders[8], and compounds identified with this profile of action could therefore be of potential therapeutic value.

The preliminary biological screening of the 13-azaprostanoids comprised of examining their effect on collagen-induced aggregation (which is largely TxA_2 mediated[9]) of human platelet rich plasma and their behaviour on a range of isolated smooth muscle preparations selectively sensitive to PGE_2, $PGF_{2\alpha}$ or TxA_2, <u>e.g.</u> guinea-pig and chick ileum, dog iris sphincter muscle, guinea-pig lung strip, dog saphenous vein and rat aorta, all of which contract in response to prostanoids[10].

Depending upon which ring type and β-chain was present in the molecule the 13-azaprostanoids displayed a variety of activities. A number of compounds were of particular interest since they revealed a profile of action consistent with selective activity at thromboxane receptors, e.g. activity on guinea-pig lung strip (or rat aortic strip) and platelets.

Guinea-pig lung strip is highly sensitive to the agonist actions of TxA_2. However, instability and the necessity for biosynthetic preparation make TxA_2 unsuitable for routine biological study. As an alternative, in early studies, $PGF_{2\alpha}$ was used to contract this tissue as we have shown that its action, whilst being relatively weak, is nevertheless mediated via TxA_2 receptors[10,11]. More recently however, U46619, which we have characterised[12] as a stable, potent and selective TxA_2 mimetic, has been the agent of choice.

Tables 1 and 2 show the effect of a representative sample of sulphonamides on the guinea-pig isolated lung strip. Sulphonamides (26a, 48-50) showed increasing degrees of agonist activity with the trans-derivative (50) being 50 times more potent than $PGF_{2\alpha}$ as a contractile agent. Changing the alkoxy group at C9 to phenylmethoxy generated a profile of mixed agonist-antagonist activity (see 51 and 52), which could be converted into pure antagonist activity by oxidation at C11 (see 53 and 54). Similarly, in the n-hexyl series the presence of the 11-keto function was also found to confer an antagonist pharmacological profile (see 27b and 57). None of these sulphonamide analogues induced platelet aggregation or had any effect on collagen-induced platelet aggregation.

The amino analogues reported in Table 3 were of greater interest. In addition to observing potent inhibitory activity on the lung strip (see 61-63,19) these same compounds were effective inhibitors of collagen-induced platelet aggregation.

TABLE 1 - Activities of [N-Methyl-N[(4-methylphenyl)sulphonyl]amino]cyclopentyl Esters on Guinea-pig Isolated Lung Strip

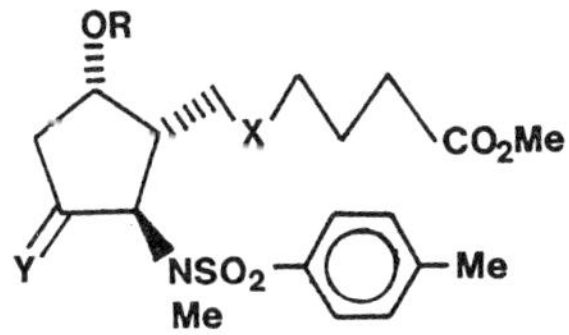

				G.P. ISOLATED LUNG STRIP[a]	
No.	R	X	Y	Equipotent Concentration ($PGF_{2\alpha}$=1)[b]	Inhibition of $PGF_{2\alpha}$-induced Contractions Concentration Ratio at 10μg/ml[c]
48	H	cis CH=CH	....OH,H	8	-
26a	Me	cis CH=CH	....OH,H	1	-
49	Me	$(CH_2)_2$	....OH,H	0.04	-
50	Me	trans CH=CH	....OH,H	0.02	-
51	CH_2Ph	cis CH=CH	....OH,H	23	49
52	CH_2Ph	trans CH=CH	....OH,H	15	100
53	CH_2Ph	cis CH=CH	O	-	11
54	CH_2Ph	trans CH=CH	O	-	30

a) Suspended in an organ bath containing modified Krebs solution with added antagonists[14].

b) Concentration-effect curves were obtained for contractile responses to $PGF_{2\alpha}$ at 30min intervals until sensitivity was constant. A further concentration-effect curve was then obtained 30min later for each azaprostanoid. Since on lung it was not possible to obtain maximal contractile responses to $PGF_{2\alpha}$ due to low potency, equipotent concentrations were obtained by dividing the EC_x for the test prostanoid by the EC_x for $PGF_{2\alpha}$, where EC_x is an arbitrary point on the straight portion of the concentration-effect curve.

c) Antagonist activity was determined by repeating $PGF_{2\alpha}$ concentration-effect curves at 30min intervals until constant, then repeating the curve in the presence of each azaprostanoid, with a contact time of 30min. Degree of antagonism expressed as concentration ratio, determined by dividing EC_x for $PGF_{2\alpha}$ in the presence of azaprostanoid by that in its absence.

TABLE 2 - Activities of [N-(Hexylsulphonyl)-N-methylamino]cyclopentyl Esters on Guinea-pig Isolated Lung Strip

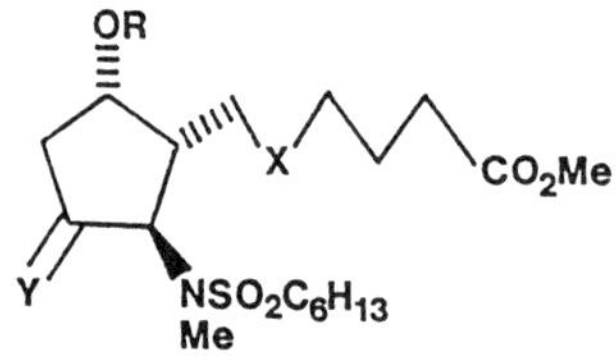

No.	R	X	Y	G.P. ISOLATED LUNG STRIP: Equipotent Concentration ($PGF_{2\alpha}$=1)	G.P. ISOLATED LUNG STRIP: Inhibition of $PGF_{2\alpha}$-induced Contractions Concentration Ratio at 10µg/ml
55	Me	$(CH_2)_2$	....OH,H	0.55	-
56	Me	trans CH=CH	....OH,H	0.25	-
27b	Me	cis CH=CH	O	-	7.4
57	CH_2Ph	cis CH=CH	O	-	4

Compound (19), AH 19437, was chosen for more detailed evaluation and full biological data have been published elsewhere[13]. In brief, the results confirm that AH 19437 is a specific thromboxane receptor blocking drug. At 3×10^{-7} - 7.2×10^{-5}mol/l it caused concentration-related parallel shifts to the right of concentration-effect curves for U46619 on guinea-pig lung strip with a pA_2 value of 6.5. Inhibitory concentration-effect curves for AH 19437 (10^{-7} - 10^{-4}mol/l) obtained against aggregation of human platelets induced by TxA_2 (generated from 1.2×10^{-7}mol/l PGH_2) and U46619 (3×10^{-6}mol/l) gave EC_{50} values (molar concentration of compound causing 50% inhibition of aggregation) of 4.65×10^{-6} and 8.5×10^{-6}mol/l respectively. The compound at 10^{-4}M had no effect on ADP-, adrenaline- or 5-HT-induced aggregation, and was devoid of significant inhibitory activity against fatty acid cyclo-oxygenase, thromboxane synthetase, cAMP phosphodiesterase or prostacyclin synthetase.

TABLE 3 - Antagonist Activities of Selected Aminocyclopentyl Esters on Guinea-Pig Isolated Lung Strip and Aggregation of Human Platelets

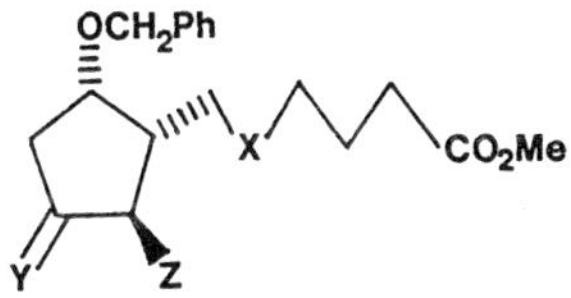

No.	X	Y	Z	G.P. ISOLATED LUNG STRIP: Inhibition of $PGF_{2\alpha}$-induced Contractions Concentration Ratio at 10μg/ml (or 1μg/ml*)	Inhibition of Collagen-Induced Aggregation of Human Platelets[d] EC_{50} μg/ml
58	cis CH=CH	..OH,H	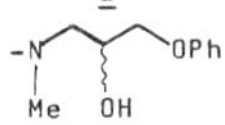	8	75
59	$(CH_2)_2$	..OH,H	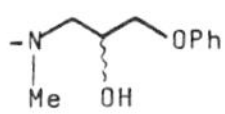	15	47
60	cis CH=CH	..OH,H	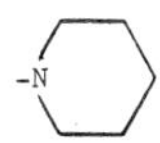	3*	-
61	cis CH=CH	O	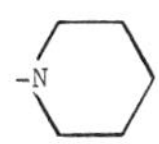	100	75
62	trans CH=CH	O	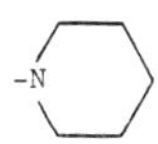	>100	12
63	$(CH_2)_2$	O	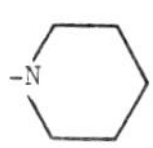	85	11
19	cis CH=CH	O	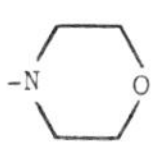	5.7*[e]	11[e]

d) Platelet aggregation[15] induced by collagen (1μg/ml) was studied using a Payton aggregometer. Test compounds pre-incubated for 5min at 37°C.

e) See text.

The structural feature common to the azaprostanoids in Table 3 is the benzyloxy group which may promote antagonist activity by simply functioning as a bulky lipophilic group. To explore this possibility we examined a number of morpholino analogues with appropriate modifications at C9, a selection of which is shown in Table 4 (64-70). Although the results indicate there is no firm correlation simply between size, lipophilicity and activity, there does appear to be a trend for increased blocking potency at smooth muscle thromboxane receptors with increased lipophilicity and perhaps a decrease in the relative ability of the p-substituent to donate electrons to the aromatic ring. The emergence of compounds (66-68) as interesting leads prompted us to prepare the corresponding acids (71-73). From the results shown in Table 5 compounds (71) and (72) reflect the importance of introducing this group since they possess potent blocking activity at thromboxane receptors in both smooth muscle and platelets. In contrast, the p-cyclohexyl acid (73) is relatively much weaker at the platelet receptor, and why this is so is unclear. That the biological profile is dependent on the appropriate combination of substituents is further illustrated by the C12 modified analogues (74) and (75).

TABLE 4 - Antagonist Activities of Some C9 Ether Analogues of Compound (19) on Guinea-Pig Isolated Lung Strip and Aggregation of Human Platelets

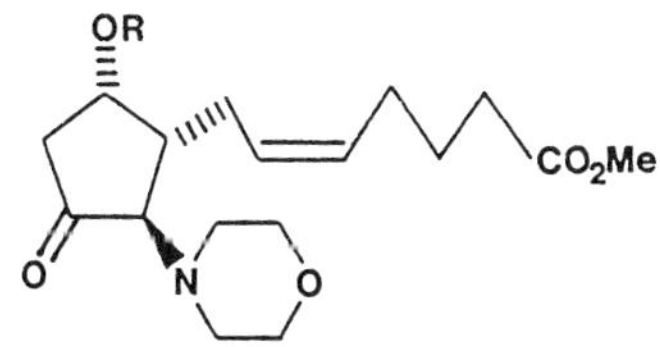

No.	R	Inhibition of $PGF_{2\alpha}$ (or U46619)-induced contractions of G.P. lung strip Relative potency (AH 19437 = 1)	Inhibition of collagen-induced aggregation of human platelets Relative potency (AH 19437 = 1)
19	$-CH_2Ph$	1	1
64	$-(CH_2)_2Ph$	1.3	0.4
65	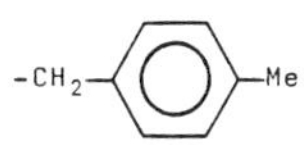	3	1.6
66	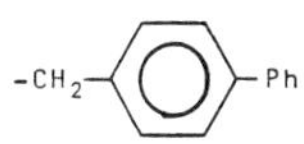	15	1
67	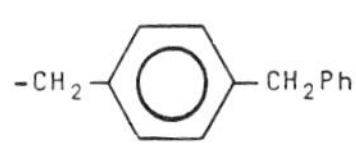	10	2.6
68	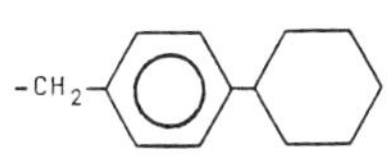	12	1
69		5	0.6
70	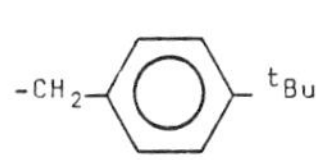	6	1.6

TABLE 5 - Antagonist Activities of Selected Aminocyclopentyl Acids on Guinea-Pig Isolated Lung Strip and Aggregation of Human Platelets

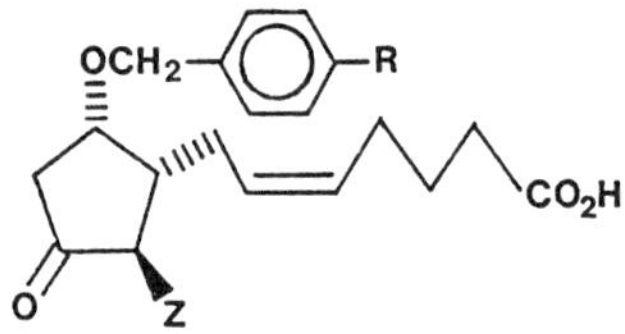

No.	Z	R	Guinea-Pig Lung Strip pA_2 (vs U46619)[f]	Inhibition of Collagen-Induced Aggregation of Human Platelets (PRP). Relative Potency (AH 19437=1)	Human Whole Blood[16] pA_2 (vs U46619)[f]
71	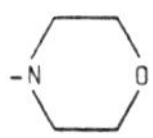	Ph	8.1[h]	>30	7.78[g]
72	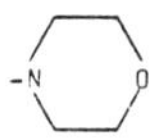	$-CH_2Ph$	7.6[g]	24	7.39[g]
73	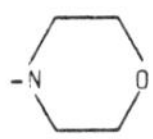	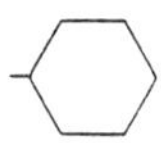	7.9[g]	2.7	
74	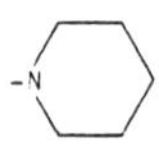	$-CH_2Ph$	7.0[i]	18	
75		$-CH_2Ph$	8.5[i]	2.6	

f) pA_2 values were determined by the method of Arunlakshana and Schild[17]. The pA_2 values represent the negative logarithm of the molar concentration of drug which produced a two-fold shift to the right of the U46619 concentration-effect curve.

g) Slopes of the Schild plots were not significantly different from unity and indicate competitive antagonism.

h) Slope of the Schild plot, 1.6.

i) pA_2 calculated using single antagonist concentration only [10^{-7}mol/l (74), 10^{-6}mol/l (75)] assuming unit slope for Schild regression.

References

[1]S.M. Roberts, J.C.S. Chem. Comm., 1974, 948.

[2]T.V. Lee, S.M. Roberts, M.J. Dimsdale, R.F. Newton, D.K. Rainey and C.F. Webb, J.C.S. Perkin I, 1978, 1176.

[3]T.V. Lee, S.M. Roberts and R.F. Newton, J.C.S. Perkin I, 1978, 1179.

[4]Prepared by acetylation of the corresponding 11-hydroxy compound which in turn was prepared in 13% overall yield from bromide (4a) and N-methylbenzylamine by the standard route.

[5]D.S. Sgoutas and F.A. Kummerow, Lipids, 1969, 4, 283.

[6]T.W. Gibson and P. Strassburger, J. Org. Chem., 1976, 41, 791

[7]Ingold, "Structure and Mechanism in Organic Chemistry", Cornell University Press, NY, (2nd Ed., 1969), p. 1129.

[8]S.T. Kam, P.S. Portoghese, W.E. Dunham and J.M. Gerrard, Prostaglandins and Medicine, 1979, 3, 279.

[9]L.C. Best, T.K. Holland, P.B.B. Jones and R.G.G. Russell, Thrombosis and Haemostasis, 1980, 43, 38.

[10]I. Kennedy, R.A. Coleman, P.P.A. Humphrey, G.P. Levy and P. Lumley, Prostaglandins, 1982, 24, 667.

[11]I. Kennedy, R.A. Coleman, P.P.A. Humphrey and P. Lumley, Advances in Prostaglandin, Thromboxane and Leucotriene Research, 1983, 11, 327.

[12]R.A. Coleman, P.P.A. Humphrey, I. Kennedy, G.P. Levy and P. Lumley, Br. J. Pharmac., 1980, 68, 127P.

[13]R.A. Coleman, E.W. Collington, H.P. Geisow, E.J. Hornby, P.P.A. Humphrey, I. Kennedy, G.P. Levy, P. Lumley, P.J. McCabe and C.J. Wallis, Br. J. Pharmac., 1981, 72, 524P; H.P. Geisow, E.J. Hornby and P.J. McCabe, Br. J. Pharmac., 1981, 73, 219P; R.A. Coleman, P.P.A. Humphrey, I. Kennedy, G.P. Levy and P. Lumley, Br. J. Pharmac., 1981, 73, 258P; P. Lumley and P.P.A. Humphrey, Abstracts of the VIth

International Congress on Thrombosis of the Mediterranean League against Thromboembolic Diseases, Monte Carlo, October 1980.

[14] R.A. Coleman, P.P.A. Humphrey, I. Kennedy, G.P. Levy and P. Lumley, Br. J. Pharmac., 1980, 69, 265P.

[15] G.V.R. Born, Nature, 1962, 194, 927.

[16] P. Lumley and P.P.A. Humphrey, J. Pharmacol. Methods, 1981, 6, 153.

[17] O. Arunlakshana and H.O. Schild, Br. J. Pharmac., 1959, 44, 48.